Hydrodynamics of Ocean Wave-Energy Utilization

International Union of Theoretical and Applied Mechanics

Hydrodynamics of Ocean Wave-Energy Utilization

IUTAM Symposium
Lisbon/Portugal 1985

Editors
D. V. Evans and **A. F. de O. Falcão**

With 248 Figures

Springer-Verlag
Berlin Heidelberg New York Tokyo

Dr. David V. Evans
Department of Mathematics
University of Bristol
University Walk
Bristol BS8 1TW
UK

Prof. António F. de O. Falcão
Mechanical Engineering Dept.
Instituto Superior Técnico
Technical University of Lisbon
1096 Lisboa Codex
Portugal

ISBN-13:978-3-642-82668-9 e-ISBN-13:978-3-642-82666-5
DOI: 10.1007/978-3-642-82666-5

Library of Congress Cataloging in Publication Data
IUTAM Symposium (1985 : Lisbon, Portugal)
Hydrodynamics of ocean wave-energy utilization.
Bibliography: p. Includes index.
1. Ocean wave power – Congresses.
2. Hydrodynamics – Congresses.
I. Evans, D. V. (David V.),
II. Falcão, A. F. de O. (António F. de O.),
III. Title.
TK1081.I88 1985 621.31'2134 85-27791
ISBN-13:978-3-642-82668-9 (U.S.)

2161/3020-5 4 3 2 1 0

General Information

International Scientific Committee

K. Budal, Norway

L. Claeson, Sweden

D.V. Evans, U.K., Co-Chairman

A.F. de O. Falcão, Portugal, Co-Chairman

J.-P. Germain, France

J. Lighthill (ex-officio), U.K.

M.E. McCormick, U.S.A.

C.C. Mei, U.S.A.

S. Takesawa, Japan

Scientific Organization

International Union of Theoretical and Applied Mechanics (IUTAM)

President: Sir James Lighthill, U.K.

Secretary-General: Professor Werner Schiehlen, F.R.G.

Supporting Organizations

Universidade Técnica de Lisboa

Centro de Termodinâmica Aplicada e Mecânica dos Fluidos da UTL

Instituto Nacional de Investigação Científica

Junta Nacional de Investigação Científica e Tecnológica

Laboratório Nacional de Engenharia e Tecnologia Industrial

Fundação Calouste Gulbenkian

Local Organizing Committee

A.F. de O. Falcão

Maria Teresa Pontes

A.J.N.A. Sarmento

Preface

The papers which follow were presented at an International Symposium held in Lisbon from 8-11 July 1985 on the Hydrodynamics of Ocean Wave-Energy Utilization and sponsored by the International Union of Theoretical and Applied Mechanics.

The subject of the Symposium embraced wave statistics, numerical methods, theoretical, experimental and field studies of wave-energy devices.

The idea of extracting useful energy from ocean waves continues to attract the curiosity of scientists and engineers in many parts of the world as the following papers indicate. Increasingly the trend is towards smaller devices suitable for use near remote island communities where wave power, as an alternative to costly diesel fuel for electric generators, is already very competitive in economic terms.

The decision to build two different prototype wave-power devices into the cliffs off Bergen in Norway has provided a welcome impetus to the field, stimulating a large amount of theoretical work on oscillating water column-type devices. In particular phase control methods - in which force and velocity of a rigid body, or pressure and volume flux across a turbine are matched in phase to achieve maximum power output - rightfully occupy a central place in the papers that follow.

In addition to the established workers in the field, a new generation of wave-energy enthusiasts is emerging, learning from the mistakes of others and contributing exciting ideas of both a conceptual and practical nature.

There is no doubt that wave power - already used to power navigation buoys, and for the desalination of sea-water - is here to stay. We believe that the high level of science and commitment evident in the following papers will ensure an increasingly important role for wave-energy as a renewable energy resource for the world. As editors, we are grateful to all the authors for their expert contributions.

We wish to thank the International Union of Theoretical and

Applied Mechanics and its President Sir James Lighthill for taking the initiative in arranging the Symposium and for the assistance offered to us during the planning.

We want to express our thanks to the members of the International Scientific Committee and of the Local Organizing Committee for their contribution to making the Symposium the success it was.

Finally we must not fail to thank the Calouste Gulbenkian Foundation for their generous hospitality and all the services which they so graciously provided. We are also very grateful to the other organizations that offered their support.

D.V. Evans
A.F. de O. Falcão
Editors

Participants

(Authors are identified by an asterisk)

AMBLI, Nils, Kvaener Brug A/S, Box 3610 Gb, 0135 Oslo 1, Norway

ANGEL, R.K., Engineering Dept., University of Reading, Reading, Berkshire RG6 2AX, United Kingdom

*BELLAMY, Norman W., Dept. of Electrical and Electronic Engineering, Coventry Polytechnic, Priory St, Coventry, United Kingdom

*BERTOTTI,L., ASCO, Via Beato Pellegrino 14/20, Padua, Italy

BONKE, Knut, Kvarner Brug A/S, Box 3610, 0135 Oslo 1, Norway

BRANCO, J.N.R., Dept. Eng. Mecânica, Instituto Superior Técnico, Av. Rovisco Pais, 1096 Lisboa, Portugal

*BUDAL, Kjell, Department of Physics and Mathematics, University of Trondheim, N-7034 Trondheim, Norway

BURDETTE, Ernest L., Triton Systems, Inc. 109 East Scenic Drive, Pass Christian, MS 39571 U.S.A.

CAMPOS, J.A.C.F., Dept. Engenharia Mecânica, Instituto Superior Técnico, Av. Rovisco Pais,1096 Lisboa Codex, Portugal

CARMICHAEL, A.D., Department of Ocean Engineering, MIT, Cambridge, MA 02139, U.S.A.

*CAVALERI, Luigi, IDGM-CNR 1364 San Paolo, 30125 Venezia, Italy

*CHEN, Guo-Kun, Shipbuilding Department, South China Institute of Technology, Guangzhou, Guangdong, People's Republic of China

CLAESON, K. Lennart, Technocean AB, Box 9006, S-400 70, Gothenburg, Sweden

CLEMENT, A., ENSM, 1, Rue de la Noe, 44072 Nantes Cedex, France

COSTA, F. Vasco, Consulmar, R. Joaquim Antônio de Aguiar, 27-9.D, 1000 Lisboa, Portugal

DAVIES, P.G., ETSU, Harwell, Didcot, Berkshire, United Kingdom

DAVIS, John P., Dept. of Civil Engineering, University of Bristol, Queen's Building, University Walk, Bristol BS8 1TR, United Kingdom

*ERTEKIN, R. Cengiz, Shell Development Company, PORD, 3727 Bellaire

Blvd., P.O.Box 481, Houston, Texas 77001, USA

*EVANS, D.V., School of Mathematics, University of Bristol, Bristol BS8 1TW, United Kingdom

*FALCÃO, A.F.O., Instituto Superior Técnico, Dept. Eng. Mecânica, Av. Rovisco Pais, 1096 Lisboa Codex, Portugal

*FALNES, Johannes, Dept. of Experimental Physics, University of Trondheim, N-7034 Trondheim-NTH, Norway

*FARLEY, Francis J.M., Dept. of Engineering, University of Reading, Whiteknights Reading RG6 2AY, United Kingdom

*FERDINANDE, V., Dienst voor Scheepsbouwkunde, RUG, Grotesteenweg, Noord 2, B-9710 Gent (Zwijnaarde), Belgium

*FORSBERG, Jan, Technocean i Goteborg AB, Box 9006, 400 70 Goteborg, Sweden

FREDRIKSON, Gunnar, Interproject Service A/B, S-64033 Bettna, Sweden

*FRENCH, Michael Joseph, Department of Engineering, University of Lancaster, Lancaster, United Kingdom

FUKUDA, Jun-ichi, Imagawa 2-1-19, Chuo-ku, Fukuoka 810, Japan

GARRIDO, M.S., Instituto Superior Técnico, Dept. de Engenharia Electrotécnica e de Computadores, Av. Rovisco Pais, 1096 Lisboa Codex, Portugal

GATO, L.M.C., Instituto Superior Técnico, Dept. Engenharia Mecânica, Av. Rovisco Pais, 1096 Lisboa Codex, Portugal

GERMAIN, Jean Pierre, Institut de Mecanique de Grenoble, Domaine Universitaire, 38402 Saint-Martin-d'Heres, France

*GREENHOW, Martin, Dept. of Mathematics, University of Manchester, Manchester, United Kingdom

*HAGERMAN Jr., George M., Gibbs & Cox, Inc., 1235 Jefferson Davis Hwy., Arlington, VA 22202, USA

HEARN, Grant E., Department of Naval Architecture, The University, Newcastle-upon-Tyne, United Kingdom

HERLAND, Jan, Kvaerner Brug A/S, Box 3610 Gb, 0135 Oslo 1, Norway

*HICKS, Douglas C., University of Delaware, College of Marine Studies, Newark, Delaware 19716, USA

*HOSKIN, Ruth, Department of Mathematics, University of Reading, P.O.Box 220, Reading, Berks., RG6 2AX, United Kingdom

*HOTTA, Hitoshi, Japan Marine Science and Technology, Center, 2-15, Natsushima-cho, Yokosuka, 237, Japan

*ISSIHIKI, Hiroshi, Technical Research Institute, Hitachi Zosen Corporation, 3-22 Sakurajima 1-Chome, Konahana-ku, Osaka 554, Japan

*INOUE, Reisaku, Ship Research Institute, Ministry of Transport, 6-38-1 Shinkawa, Mitaka-shi, Tokyo, Japan

INOUE, Yoshiyuki, Dept. of Naval Architecture and Ocean Engineering, Faculty of Engineering, Yokohama National University, 156 Tokiwadai Hodogaya-ku, Yokohama 240, Japan

*JEFFERYS, Richard, Dept. of Mechanical Engineering, University College, London, Torrington Place, London WC1E 7JE, United Kingdom

*KINOSHITA, Takeshi, Institute of Industrial Science, University of Tokyo, 7-22-1 Roppongi, Minato-Ku, Tokyo 106, Japan

LIGHTHILL, Sir James, Provost, University College London, Gower Street, London WC1E 6BT, United Kingdom

LOCKETT, Francis Peter, Mathematics Department, Coventry (Lanchester)Polytechnic, Coventry CV1 5FB, United Kingdom

LOVHAUGEN, Odd, Senter for Industriforskning, P.O.Box 350 Blindern, 0314 Oslo 3, Norway

*MAEDA, Hisaaki, Institute of Industrial Science, Univ. of Tokyo, 7-22-1 Roppongi, Minatoku, Tokyo 106, Japan

*MALMO, Oddbjorn, Norwegian Hydrotechnical Laboratotry, NHL, N-7034 Trondheim-NTH, Norway

MARQUES, Gil, Instituto Superior Técnico, Dept. de Engenharia Electrotécnica e de Computadores, Av. Rovisco Pais, 1096 Lisboa Codex, Portugal

*MASABUCHI, Masami, Dept. of Mechanical Engineering, Osaka University, 2-1 Yamadaoka, Suita, Osaka 565, Japan

MCCORMICK, Michael E., 1096 Sands Drive, Annapolis, Maryland 21401, USA

*MCPEAK, Francis, Civil Engineering Dept., David Keir Building, Stranmillis Rd, Belfast, N. Ireland

*MEHLUM, Even, Norwave A.S., Forskningsveien 1, P.O. Box 350, Blindern, 0314 Oslo 3, Norway

*MENDES, A.C., Dept. Engenharia Mecânica, Instituto Superior Técnico, Av. Rovisco Pais, 1096 Lisboa Codex, Portugal

MILOH, Touvia, Dept. of Fluid Mechanics, School of Engineering, University of Tel-Aviv, Ramat-Aviv 69978, Israel

*MOLLISON, Denis, Dept. of Actuarial Mathematics and Statiscs, Riccarton, Edinburgh EH14 4AS, United Kingdom

*NAITO, Shigeru, Dept. of Naval Architecture, Osaka Univer., Yamada-Oka 2-1, Suita, Osaka, 565, Japan

*NICHOLS, Nancy K., Department of Mathematics, University of Reading, P.O.Box 220, Reading, Berkshire RG6 2AX, United Kingdom

*NICOL, David A.C., Marchwood Engineering Laboratories, Marchwood, Southampton, United Kingdom

*NIELSEN, Kim, Department of Ocean Engineering, The Technical University of Denmark, Building 101 E, DK-2800 Lyngby, Denmark

NOREN, Anders, Inst. for Fluid Technology, Royal Institute of Technology, Stockholm, Sweden

OSKARSEN, Harald, Kvaerner Brug A/S, Box 3610 Gb, 0135 Oslo 1, Norway

*OLTEDAL, Geirmund, Division fo Experimental Physics, The Norwegian Institute of Technology, University of Trondheim, N-7034 Trondheim-NTH, Norway

*OLIVEIRA PIRES, Henrique N., Instituto Nacional de Metereologia e Geofísica, Rua C ao Aeroporto, 1700 Lisboa, Portugal

*PERDIGÃO, J.N.B.M. Azeredo, LNETI, Dept. de Energia Renovaveis, Estr. do Paço do Lumiar, 1699 Lisboa Codex, Portugal

*PESSANHA, Luis E. Vasconcelos, Instituto Nacional de Metereologia e Geofísica, Rua C ao Aeroporto, 1700 Lisboa, Portugal

PINA, H.L.G., Instituto Superior Técnico, Departamento de Engenharia Mecânica, Av. Rovisco Pais, 1096 Lisboa Codex, Portugal

*PONTES, M. Teresa V.S., Instituto Superior Técnico, Departamento de Engenharia Mecânica, Av. Rovisco Pais, 1096 Lisboa Codex, Portugal

REIS, L.M.M., Instituto Superior Técnico, Departamento de Engenharia Mecânica, Av. Rovisco Pais, 1096 Lisboa Codex, Portugal

*REITAN, Arne, Fysisk Institutt, UNIT/AVH, N-7055 Dragvoll, Norway

RETZLER, Chris-Heinz, Dept. of Mechanical Engineering, University of Edinburgh, King's Buildings, Mayfield Road, Edinburgh EH9 3JL, United Kingdom

*SALTER, S.H.
Wave Power Project, Dept. of Mechanical Engineering, University of Edinburgh, King's Buildings, Mayfield Road, Edinburgh EH9 3JL, United Kingdom

SANTANA, J., Instituto Superior Tecnico, Dept. de Engenharia Electrotécnica e de Computadores, Av. Rovisco Pais, 1096 Lisboa Codex, Portugal

*SARMENTO, A.J.N.A., Instituto Superior Técnico, Departamento de

Engenharia Mecânica, Av. Rovisco Pais, 1096 Lisboa Codex, Portugal

*SCHER, Robert M., 3371 S. Stafford St., Arlington, Virginia 22206, USA

*SIMON, Michael J., Dept. of Mathematics, The University, Manchester MB 9PL, United Kingdom

SOARES, C.G., Instituto Superior Técnico, Dept. Eng. Mecânica, Av. Rovisco Pais, 1096 Lisboa Codex, Portugal

TAKEZAWA, Seiji, Department of Naval Architecture & Ocean Engineering, Yokohama National University, Tokiwadai 156, Hodogaya-ku, Yokohama, Japan

THOMAS, G.P., Department of Mathematical Physics, University College, Cork, Ireland

VALENTE, F.P., Instituto Superior Técnico, Dept. Matemática, Av. Rovisco Pais, 1096 Lisboa Codex, Portugal

*VANTORRE, M., Dienst voor Scheepsbouwkunde, State University of Ghent, B-9710 Gent-Zwijnaarde, Belgium

WATTS, K. Christopher, P.O.Box 1000, Halifax, Nova Scotia, Canada B3J 2X4

*WHITTAKER, T.J.T., Dept. of Civil Eng., The Queen's University of Belfast, David Keir Building, Stranmillis Rd., Belfast, N. Ireland

*YAMAZAKI, Tetsuo, Ryokuseisha Co. Ltd., 9-11 Narita nishi, 3-chome, Suginami-ku, Tokyo, Japan

*YEUNG, R.W., Dept. of Naval Architecture and Offshore Eng., University of California Berkeley, Bekeley, California 94720, USA

Contents

The Hydrodynamic Efficiency of Wave-Energy Devices

D. V. EVANS

School of Mathematics,
University of Bristol, Bristol, BS8 1TW.

Summary

A description is given of theories leading to expressions for the mean power which can be extracted by one or more devices absorbing energy from a long-crested monochromatic wave. Extensions to constrained motions and various approximate methods are described and comparisons with more accurate numerical methods or experimental results made.

1. Introduction

Despite its all-embracing title, I shall only be concerned with one aspect of the hydrodynamics associated with extraction of energy from ocean waves. Nonetheless it is an important aspect and one which first comes to mind in considering the conversion of wave energy to mechanical energy. Can we develop a theory which will tell us how good a specific device is or how efficient is the primary conversion process, and how much confidence can we put in such predictions? I shall show how classical linear water wave theory can be used to good effect to provide the answers we need with generally excellent agreement with experiment. Furthermore I shall show how judicious approximations can lead to powerful simple results for such quantities as maximum efficiency or capture width ratio (defined later) thus providing the inventor with a guide as to how his latest wave-energy device might be improved.

Much of the material which follows will be familiar to the older workers in the field, and can be found in various publications, including my own, from 1976 onwards. It is inevitable in a rapidly expanding field that similar if not identical results are discovered by research workers in different countries at about the same time. Let me apologise, therefore, in advance, if, in dwelling on the development of my own ideas on wave-energy, I omit to acknowledge the fine work which has been and continues to be done by theoreticians in Trondheim, Oslo, Tokyo, Lisbon, Cambridge, U.S.A., and elsewhere as well as my colleagues in the U.K. Most of all I am indebted to my good friend Dr Brian Count, a familiar name in wave-energy research, who has produced a steady stream of illuminating papers on theoretical and experimental aspects as well as field studies, and with whom I have had the pleasure of collaborating on joint work in the area for a number of years.

2. Simple Devices

Although Masuda in Japan had already developed his wave-powered navigation buoy, and McCormick [1] had provided a hydrodynamic theory for the air flow through the turbine, it was not until after Salter [2] published the results of narrow-tank tests on his 'duck' showing how unsymmetric rocking cylinder sections could be highly efficient in absorbing wave energy that researchers in the U.K., U.S.A., Norway and Japan began to produce general hydrodynamical theories. With one or two exceptions such theories have been based on linear water wave theory used with considerable success in ship hydrodynamics. In fact many of the ideas from this field carry over to wave energy theory; in particular the ideas of added mass and damping which play a key rôle in what follows.

Despite the complexity of the sea surface and the structure being used to extract the energy of the incident waves, it is possible, with reasonable success, to model the problem by considering the effect of a single long-crested monochromatic wavetrain on the device. The net effect of the more complicated wavefield actually present can then be determined by superposition of the effects of a large number of such wavetrains each of different amplitude, wavelength and angle of incidence. This procedure requires a knowledge of an energy spectrum appropriate to the region of installation of the wave-energy devices. One of the most commonly used is the Pierson & Moskowitz [3] spectrum suitably modified to allow for directionality effects. In this paper we shall concentrate on the fundamental problem of the interaction of a device with a single infinitely long-crested wavetrain of given wavelength and direction.

Wave-energy devices come in many different shapes and sizes. Single devices often comprise a rigid body having relative motion between adjacent parts. Examples include the floating pontoon initially developed by Cockerell [4] in the U.K. and later by Hagen [5] in the U.S.A., and the Salter 'duck' with its outer part in contact with the waves and rotating about the inner spine. Notable amongst other types is the oscillatory water column (OWC) device where the structure itself is often fixed and the induced oscillations of an internal water surface forces air through a turbine to generate power. In the former case power is developed through the wave-induced force and velocity of the moving parts, whereas in the latter case it is the induced pressure on and displacement of the internal free surface which gives rise to absorbed power. More complicated devices such as Kaimei [6] can consist of a combination of rigid-body motions and internal free surface pressure distributions. Falnes and McIver [7] provide a theory applicable to this situation. Again the surface in contact with the waves can be flexible as in the Clam device a description of which along with many other devices can be found in Shaw [8].

It is clearly not possible to embrace all these different types in one simple theory although we shall attempt to go some way towards this. Thus a notation will be used which enables extension to a wide class of devices to be made and which enables a number of important results to be derived in a simple manner. To fix ideas we consider first the time-harmonic motion of a Salter 'duck' in a narrow wave tank oscillating in one mode of motion characterised by the complex-valued (angular) velocity U. The actual angular velocity is Re U $\exp(i\omega t)$ where $\omega/2\pi$ is the incident wave frequency. Implicit already in this is the assumption of linearity since the wave-frequency $\omega/2\pi$ is assumed to give rise to a response at that frequency only. A treatment of non-linear power take-off systems has been given by Count [9] and Sarmento and Falcão [10]. Aspects of this problem are also considered by a number of authors in the present symposium. The treatment given here may provide upper bounds on performance since it is generally accepted that non-linear effects will tend to reduce efficiency.

The rocking motion of the duck is produced by wave-induced pressure forces, the hydrostatic restoring force, and the external opposing forces modelling the power take-off.

Thus we have

$$X_e + X_w + i\omega^{-1}CU = -i\omega IU \tag{2.1}$$

where X_e, X_w are the time-independent complex external force and wave-induced pressure force respectively, C is a real constant characterising the linearised hydrostatic restoring force and I is the inertia of the section. For this two-dimensional case all forces are per unit width of the device. The modulus of the terms $-i\omega U, i\omega^{-1}U$ describe the maximum acceleration and displacement of the section respectively.

The wave force is conveniently separated into a term X describing the magnitude and phase of the exciting force on the section assuming it remains at rest, and a term $+i\omega UM - UB$ where M, B are the frequency-dependent added mass and damping coefficients for the section, which describes the induced fluid pressure force on the body due to its own motion with velocity U in the absence of the incident waves. The functions M, B are well-known in ship hydrodynamics theory and depend solely on the geometry of the particular shape in question. In general $B \geq 0$, for all frequencies since in forced motion the net work done over a cycle in radiating waves away from the body is $\frac{1}{2}B|U|^2$.

Equation (2.1) can now be written

$$X_e + X = ZU \tag{2.2}$$

where

$$Z \equiv B - i\omega(I + M - C\omega^{-2}). \tag{2.3}$$

It is convenient to express the external force X_e in terms of the velocity by writing

$$X_e = -\Lambda U \tag{2.4}$$

so that the external force opposes the motion of the section with a force proportional to the velocity. In practice Λ will most likely be real and positive although the use of a sophisticated power take-off may allow Λ to be complex. In air turbine systems Λ may be slightly complex to model compressibility effects.

The mean rate of working of the wave forces over a cycle is

$$P = \tfrac{1}{2}\text{Re}\ \bar{X}_w U = -\tfrac{1}{2}\text{Re}\ \bar{X}_e U \tag{2.5}$$

where the bar indicates complex conjugate.

It follows from (2.4) that

$$P = \tfrac{1}{4}(\Lambda+\bar{\Lambda})|U|^2$$

$$= \tfrac{1}{8}\frac{|X|^2}{B}\left\{1 - \frac{|\Lambda-\bar{Z}|^2}{|\Lambda+Z|^2}\right\} \tag{2.6}$$

from (2.2)-(2.4). Displayed in this form, the mean power can be seen to have a unique maximum given by

$$P_{max} = |X|^2/8B \tag{2.7}$$

and achieved when $\Lambda = \bar{Z} = B + i\omega(I + M - C\omega^{-2})$.

In two-dimensions the quantities $|X|$ and B are connected by the formula (see, for example Newman [11])

$$|X|^2/B = 8P_W\gamma \tag{2.8}$$

where P_W is the mean power incident upon the section, and

$$\gamma = |A_+|^2/(|A_+|^2+|A_-|^2) \tag{2.9}$$

where $A_+(\omega)$, $A_-(\omega)$ are the complex wave amplitudes of the waves generated upstream and downstream respectively by the forced motion of the section in the absence of the incident waves. It follows from (2.7) to (2.9) that the maximum efficiency E_{max} is

$$E_{max} = P_{max}/P_w = |A_+|^2/(|A_+|^2+|A_-|^2) \qquad (2.10)$$

a result obtained independently by Mei [12] Evans [13] and Newman [11].

Equation (2.10) illustrates the desirability of designing a shape like the Salter duck which generates little wave motion downstream compared to upstream so that $|A_-/A_+| \ll 1$. The conclusion for two-dimensional sections is that a section which is a good uni-directional wave generator is also a good absorber, the extreme example being a vertical piston wave maker for which $\gamma = 1$ and $E_{max} = 1$, at a particular frequency.

In practice a particular real value of Λ must be chosen which will in general be fixed for all wave frequencies. In this case it is easily shown that, with Λ real

$$P_{max} = \frac{|X|^2}{4(|Z|+B)} \qquad (2.11)$$

when

$$\Lambda = |Z| \qquad (2.12)$$

so that

$$E_{max} = \frac{2\gamma B}{|Z|+B} \qquad (2.13)$$

If we now choose a predominant wave frequency ω_0 at which to tune our device, thus fixing $\Lambda \equiv \Lambda_0 = |Z(\omega_0)|$ we can determine the efficiency at any other frequency ω from

$$E = \frac{2\gamma B}{|Z|+B}\left(1 - \frac{(\Lambda_0-|Z|)^2}{|\Lambda_0+Z|^2}\right). \qquad (2.14)$$

This expression has been computed by Count [14] for four different devices, a Salter duck, a Cockerell raft pair with just a single hinge, an NEL oscillating water column and a simple OWC type device using a shallow draft section with an enclosed water column in its front section. In order to model the last two devices he assumed the free surface could be modelled by a thin weightless horizontal piston and used the above rigid-body theory.

The results which appear in Count [14] are shown in Fig. 1 to 4 together with the results of narrow tank experiments, and are reproduced with permission of the Marchwood Engineering Laboratories, UK Central Electricity Generating Board where the work was carried out.

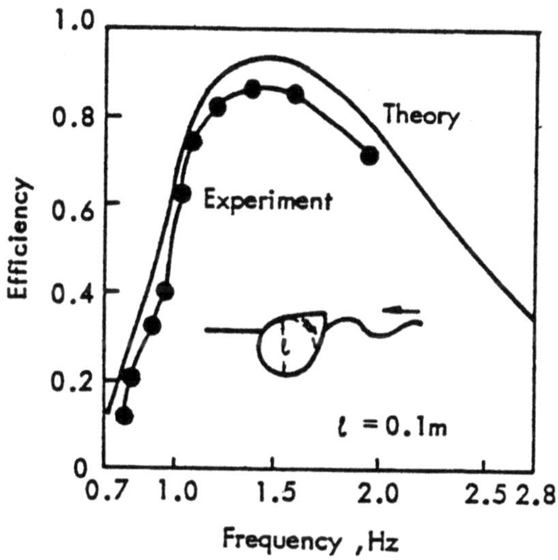

Fig.1 The performance of a Salter duck operating in a single mode of motion - theory and experiment.
(From Count [14], reproduced by permission of the UK Central Electricity Generating Board).

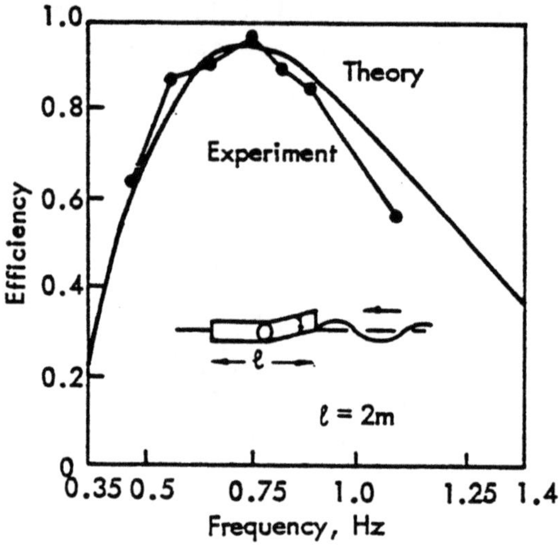

Fig.2 The performance of a raft pair with the rear section held fixed - theory and experiment.
(From Count [14], reproduced by permission of the UK Central Electricity Generating Board).

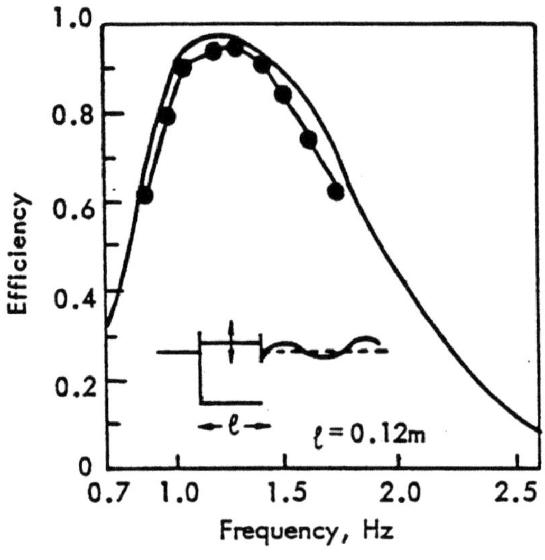

Fig.3 The performance of an oscillating water column tested by the National Engineering Laboratory - theory and experiment.
(From Count [14], reproduced by permission of the UK Central Electricity Generating Board).

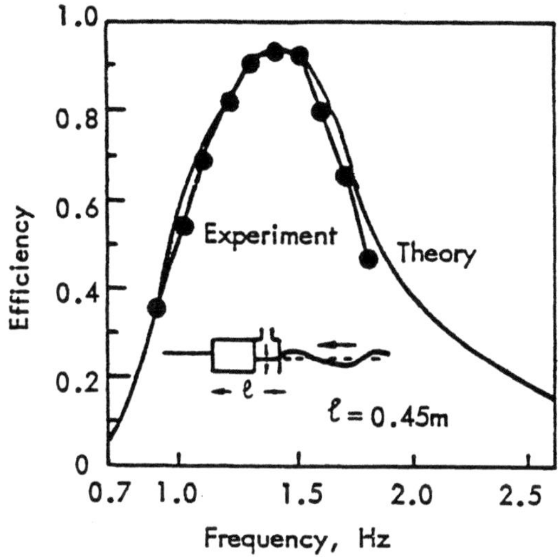

Fig.4 The performance of a shallow draught oscillating water column operating in a single mode of motion - theory and experiment.
(From Count [14], reproduced by permission of the UK Central Electricity Generating Board).

Two things emerge from the Figures. First it is clear that linear wave theory is entirely adequate to describe the wave energy absorption phenomenon, giving very good agreement with experiment. Secondly, it appears that there is very little difference in the behaviour of the three different devices over the complete range of frequencies.

3. Multi-degree of Freedom Devices

The previous theory applies only to single degree of freedom device operating in a narrow wave tank. In practice many floating devices will have more than a single degree of freedom. For example the Salter duck is to be mounted on a long spine and allowance has to be made for.its motions as well as for the motion of the duck itself. The previous theory can be modified to take care of this. To begin with we suppose that power is absorbed from all possible degrees of freedom which are characterised by the vector velocity U having components U_i ($i = 1,2,..n$). The integer n may also include all possible degrees of freedom of any number of independent devices. The quantities X_e and X also become n-vectors whilst Z, B, M, C are each $n \times n$ matrix with complex elements describing the forces in one direction due to motion in another.

Thus now

$$P = -\tfrac{1}{2}\mathrm{Re}\ X_e^* U \tag{3.1}$$

where * denotes conjugate transpose and

$$X_e + X = ZU \tag{3.2}$$

as before. Thus from (3.1), (3.2)

$$P = \tfrac{1}{2}\mathrm{Re}\ X^*U - \tfrac{1}{2}\mathrm{Re}\ U^*Z^*U \tag{3.3}$$

$$= \tfrac{1}{2}\mathrm{Re}\ X^*U - \tfrac{1}{2}U^*B_1U \tag{3.4}$$

where $$B_1 = \tfrac{1}{2}(Z + Z^*) \quad (= B \text{ if } Z = Z^T \text{ from (2.3)}) \tag{3.5}$$

This can be re-arranged in the form

$$P = \tfrac{1}{8}X^*B_1^{-1}X - \tfrac{1}{2}(U - \tfrac{1}{2}B_1^{-1}X)^*B(U - \tfrac{1}{2}B_1^{-1}X)$$

provided $\det B_1 \neq 0$. Thus

$$P_{max} = \tfrac{1}{8}X^*B_1^{-1}X \tag{3.6}$$

provided $$U = \tfrac{1}{2}B_1^{-1}X. \tag{3.7}$$

To find whether the requirement (3.7) is feasible we assume that each mode is controllable by a linear combination of components λ_{ij} of the $n \times n$ matrix Λ where

$$X_e = -\Lambda U \tag{3.8}$$

so that $$(\Lambda + Z)U = X. \tag{3.9}$$

It is clear from (3.5), (3.9) that the condition (3.7) is satisfied by $\Lambda = Z^*$. Notice that here we have not assumed that Z is symmetric since although both B and M are symmetric we must allow the possibility that C may not be. Although $B_1 = \tfrac{1}{2}(Z + Z^*)$ is complex the expression (3.6) for P_{max} is still real since we can write using (2.3), (3.5)

$$B_1 = B + i\,D, \text{ say, where } D^* = -D \qquad (3.10)$$

and the complex conjugate of P_{max} is

$$\bar{P}_{max} = \overline{\tfrac{1}{2}U^* B_1 U} = \tfrac{1}{2}U^T \bar{B}_1 \bar{U} =$$

$$= \tfrac{1}{2}U^* B_1^* U \quad \text{(after taking the transpose)}$$

$$= \tfrac{1}{2}U^* B_1 U = P_{max}$$

since $B_1^* = B_1$ from (3.10).

The result (3.6) generalises to unsymmetric Z the result first proved for symmetric Z independently by Falnes [15] and Evans [16].

A simple illustration of this result in two dimensions is given by the Bristol Submerged Cylinder Wave-energy device (Clare et al. [17]). For a submerged cylinder in deep water restrained horizontally and vertically by opposing springs and dampers, the added mass and damping coefficients are the same in heave as in sway (Ogilvie [18]) and the effect is to produce an efficiency

$$E = 1 - \frac{|\Lambda_o - \bar{Z}|^2}{|\Lambda_o + Z|^2}$$

where here $\Lambda_o = \begin{pmatrix} \lambda_{10} & 0 \\ 0 & \lambda_{20} \end{pmatrix}$ and Z is diagonal also.

Fig.5 shows the variation of E with non-dimensional wavelength and illustrates the good agreement with experiment for a cylinder of radius a immersed to a depth $\frac{5}{4}a$ in small amplitude incident waves. See Evans et al. [19].

It is clear that the capacity to absorb energy in more than one mode is advantageous. In most cases extra modes of motion exist in which power is not being absorbed and this can reduce performance unless special care is taken in tuning the device.

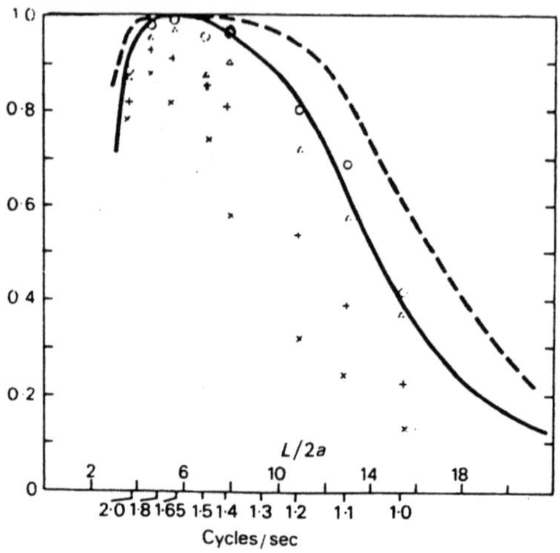

Fig.5 The efficiency of a model of a horizontal submerged cylinder wave energy device operating in two modes of motion-theory and experiment. Cylinder of radius .05m, with axis submerged to depth .0625m. L/2a - wavelength to diameter ratio. Solid line theory, dotted line - theory for neutrally buoyant cylinder. Wave amplitude to cylinder radius A/a = .033, o; .054, +; .096 x.
(From Evans et al. [19], reproduced by permission of C.M.L. Publications).

4. Three-dimensional Devices

The simplest and perhaps most remarkable result which can be deduced in three dimensions is for an isolated device oscillating vertically along a vertical axis of rotation, where it is known (Newman [11]) that

$$2\pi|X|^2 = 8B\lambda P_w \qquad (4.1)$$

so that from (2.7)

$$\ell_{max} = \frac{P_{max}}{P_w} = \frac{\lambda}{2\pi} \qquad (4.2)$$

where here now P_w is the mean power per unit crest length of the incident wave-train. The quantity ℓ is called the capture width for the device. What is remarkable about (4.2) is that the device appears capable of capturing all the power in a wavecrest of about one-sixth of a wavelength, independently of the diameter of the device. In reality of course the motion of the device will be restricted and (4.2) will never be achieved except for large diameter devices or for short wavelength waves.

Motion constraints

To illustrate this it is useful to consider the maximum power which can be achieved when a device or group of devices is constrained in the amplitude of the motions it can make. Thus we might assume

$$|U_m| \le \omega A \alpha_m = \beta_m \quad \text{say } (m = 1,2 \ldots n) \tag{4.3}$$

where A is the incident wave amplitude, and α_m are constants, although it is easier to replace these by the global constraint

$$U^*U = \sum_{m=1}^{n} |U_m|^2 \le \sum_{m=1}^{n} \beta_m^2 = \beta^2, \quad \text{say} \tag{4.4}$$

since matrix methods can still be used to maximise the mean power P subject to the constraint (4.4). The 2n-dimensional function

$$P(U) = \tfrac{1}{8}X^*B^{-1}X - \tfrac{1}{2}(U - \tfrac{1}{2}B^{-1}X)^*B(U - \tfrac{1}{2}B^{-1}X) \tag{4.5}$$

regarded as the closed surfaces P(U) = const. in the 2n-dimensional U space has a maximum at $U = \frac{1}{2}B^{-1}X$ given by $\frac{1}{8}X^*B^{-1}X$ (assuming $Z = Z^T$) when U is unconstrained. The appropriate maximum when U satisfies (4.4) occurs when the surfaces P(U) = const. just touch the sphere $U^*U = \beta^2$. The required constrained maximum P^A_{max} (A for amplitude constraint) can be obtained using Lagrange Multipliers in a straightforward manner when it is found that

$$P^A_{max} = \tfrac{1}{8}X^*(B + \mu I)^{-1}X + \tfrac{1}{2}\mu\beta^2 \tag{4.6}$$

where μ satisfies

$$\tfrac{1}{4}X^*(B + \mu I)^{-2}X = \beta^2 = \omega^2 A^2 \alpha^2, \text{ say} \tag{4.7}$$

For a single body oscillating vertically in one degree of freedom along the vertical axis of rotation all vectors and matrices become scalars and it can be shown that

$$\ell^A_{max} = \frac{P^A_{max}}{P_w} = \frac{\lambda}{2\pi}\{1 - H(1 - \delta)(1 - \delta)^2\} \tag{4.8}$$

where H is the Heaviside step function and

$$\delta = B/(B+\mu) = 2\omega A\alpha B|X|^{-1}$$
$$= \alpha(Ka)^{\frac{3}{2}}\lambda_{33}^{\frac{1}{2}}(2M/\rho a^3)^{\frac{1}{2}} \qquad (4.9)$$

where $\lambda_{33} = B/M$ is the non-dimensional damping coefficient, $K = \omega^2/g$ is the wave-number, and a the radius of the body. Thus Fig.(6) shows how $\ell^{A}_{max}/2a$ varies with $Ka(= 2\pi a/\lambda)$ for a vertically oscillating sphere constrained to oscillate at a maximum of 0.5, 1, and 2 times the incident wave amplitude. For example if the amplitude of motion of the sphere cannot exceed the wave amplitude, the maximum capture width never exceeds 70% of the sphere diameter. (See Evans [20]).

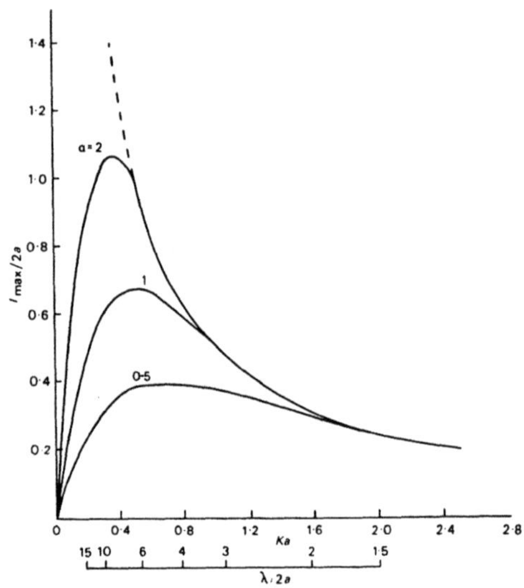

Fig.6 Variation of maximum capture width/diameter ratio with wavelength for half immersed energy-absorbing sphere for different motion constraints, ---, unconstrained sphere. (From Evans [20], reproduced with the permission of C.M.L. publications).

A different constraint limiting the efficiency may occur in devices which consist of a number of sections each of which responds separately to the adjacent wave field but which drive a fixed internal air volume through the device. For example certain designs of the Clam device being developed in the UK worked on this principle. In this case the total internal volume is fixed, a condition which translates to

$$I^T U = 0 \tag{4.10}$$

where U is the complex n-vector describing the velocity of each section, and I is an n-vector all of whose elements are unity. Again the maximum mean power subject to (4.10) can be found using Lagrange Multipliers. Thus, (with v standing for volume constraint)

$$P^v_{max} = \tfrac{1}{8} Y^* B^{-1} Y \tag{4.11}$$

where

$$Y = X - \mu I \tag{4.12}$$

and the complex number μ is given by

$$\mu = (I^T B^{-1} X)/(I^T B^{-1} I) \tag{4.13}$$

An illustration as to how this constraint reduced the maximum capture ratio is shown in Fig.(7) taken from Evans [21] for a simple device consisting of three horizontally pulsating boxes of small width and draft equal to half the box length. The curves show the variation of the constrained ℓ_{max} - non-dimensionalised with respect to the total box length a - with angle of incidence. Here $\beta = 0°$ is head seas and $\beta = 90°$ is beam seas. The reason for the drop to zero as $\beta \to 90°$ can be explained by the approximation used to estimate the exciting force on each of the boxes. Since the diffracted field has been neglected the phase of the exciting force on each box is the same and the volume constraint thereby prevents any movement at all.

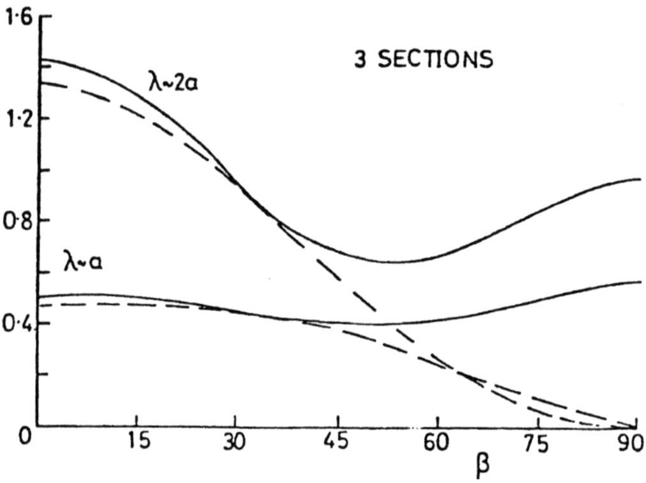

Fig.7 Maximum capture width ratio $\ell_{max}(\beta)/a$ with (-----) and without (———) the constant volume constraint, as a function of incident wave direction β, for three pulsating box sections at two different wavelengths $\lambda \sim a, 2a$.
(From Evans [21]).

Absorption for m(<n) modes of motion

It proves possible to extend the results above to the case when, as is inevitable in practice, not all the modes of motion are energy-absorbing, whilst still allowing the appropriate motion constraints to apply. This was first done by Count and Jefferys [22] and extended to motion constraints in a later paper by Count and Miyazaki [23]. The method is a good example of the power of matrix techniques but the details are complicated. Here for simplicity we illustrate the idea for the unconstrained case of a group of devices or a single device oscillating in n degrees of freedom, m of which can be controlled by a linear controller $(m \le n)$. As before we shall extend existing results by allowing Z to be unsymmetric.

The external force vector can be written

$$X_e = -R\Lambda R^T U \qquad (4.14)$$

where U is a complex $n \times 1$ vector, Λ is $m \times m$, R is $n \times m$ and $R_{ij} = 0,\ i \neq j;\ 1,\ i = j$. For example if $n = 3$, $m = 2$, then

$$X_e^T = -(\lambda_{11}U_1 + \lambda_{12}U_2, \lambda_{21}U_1 + \lambda_{22}U_2, 0)$$

and we have control over U_1 U_2 only, through the elements of the 2×2 matrix Λ.

As before the mean power absorbed is

$$P = -\tfrac{1}{2}\mathrm{Re}\, X_e^* U$$

$$= \tfrac{1}{4}U^* R^T (\Lambda + \Lambda^*) RU \tag{4.15}$$

from (4.14)

where U satisfies

$$(R\Lambda R^T + Z)U = X. \tag{4.16}$$

We wish to maximise P as a function of the elements of the matrix Λ. By considering a variation $\delta\Lambda$ in Λ we obtain after some algebra the condition

$$I_m = (\Lambda + \Lambda^*) RAR^T \tag{4.17}$$

where I_m is the $m \times m$ identity matrix, and

$$A = (R\Lambda R^T + Z)^{-1} \tag{4.18}$$

If $m = n$, $R = R^T = I_n$ and the condition reduces to

$$I = (\Lambda + \Lambda^*)(\Lambda + Z)^{-1}$$

whence $$\Lambda = Z^* \tag{4.19}$$

which extends (2.12) to n power-absorbing modes. Now using results from the theory of partitioned matrices, it can be shown that

$$(RAR^T)^{-1} = \Lambda + (R^T Z^{-1} R)^{-1}$$

so that (4.17) can be solved for Λ to obtain

$$\Lambda = (R^T Z^{*-1} R)^{-1} \tag{4.20}$$

as the optimal value for Λ which clearly reduces to $\Lambda = Z^*$ when $n = m$.

In both (4.19) and (4.20) it is clear that the elements λ_{ij} of Λ for maximum power absorption need to be complex implying a reactive power take-off characteristic. In addition the optimal external force from (4.14) involves knowledge of the velocities of all controllable modes. A simpler system more easily implemented in practice would be to use the same pure damping power take-off for each controllable mode. Then $\Lambda = \lambda I_m$ a real positive scalar and I_m is the $m \times m$ identity matrix. It is then straightforward to maximise P as a function of the single variable λ. The required condition for this is in the case $n = m$ is

$$U^*\{I - \lambda(\lambda I + Z)^{*-1} - \lambda(\lambda I + Z)^{-1}\}U = 0 \qquad (4.21)$$

with

$$P_{max} = \tfrac{1}{2}\lambda U^*U. \qquad (4.22)$$

For $n = 1$ a little algebra shows that these reduce to $\lambda = |Z|$ and $P_{max} = \frac{1}{4}|X|^2/(|Z|+B)$ in agreement with (2.11). Count and Miyazaki [23] give the more general condition corresponding to (4.21) when $m < n$ and go on to apply the theory to both a bottom and side opening set of oscillating water columns as possible models of the Kaimei system.

5. Approximate Methods

We have shown how it is possible using matrix methods to obtain general expressions for the maximum hydrodynamic power which can be captured by a wave-power system under various conditions, in terms of hydrodynamic properties such as added mass, damping and exciting force coefficients. For a complicated system involving a number of modes of motion these quantities are not easily computed and it is desirable to seek good approximations.

Point absorbers

The first example relates to the expression

$$P_{max} = \tfrac{1}{8}X^*B^{-1}X \qquad (5.1)$$

for an unconstrained set of absorbers oscillating in all modes with $Z = Z^T$. A general result connects X the exciting force vector with the damping matrix B. Thus (Newman [11])

$$B = \frac{1}{8\lambda P_W} \int_0^{2\pi} X(\theta)X^*(\theta)d\theta \qquad (5.2)$$

where $X(\theta)$ is the vector of exciting forces as a function of the angle θ of the incoming wave. Clearly, knowledge of the exciting force as a function of θ (and frequency) is sufficient to determine P_{max}. Of particular interest is the value of P_{max} for n identical equally-spaced devices on a line. A simple approximate result can be obtained by treating the devices as point absorbers so that the exciting force on the mth device is just proportional to the phase of the incident wave potential at the position of the device. The result is (Evans [16])

$$\ell_{max}(\beta) = \frac{\lambda}{2\pi} L^*J^{-1}L \qquad (5.3)$$

where $$J_{mn} = J_o\{Kd(m-n)\} \qquad (5.4)$$

and $$L_m = \exp\{iKd_m \sin\beta\} \qquad (5.5)$$

Here d is the spacing between adjacent devices, $K = 2\pi/\lambda$, λ the wavelength, and β is the angle made by the incident wave with the line of devices. A comparison with a full numerical treatment has been made by Count [14] and Simon [24] and results from the former reference are shown in Figs.(8),(9). Here

$$\ell_{max} = \frac{\lambda}{2\pi}.n.q$$

so that the q-factor or interaction factor indicates the improvement or otherwise in capture width over n single devices. It is clear that the approximate expression is close to the exact values over all spacings for two devices and for the limited exact numerical values for three devices, and that the theory agrees well with experiment for two devices over various angles of incidents. An appropriate spacing for two devices appears to be about two-thirds of a wavelength.

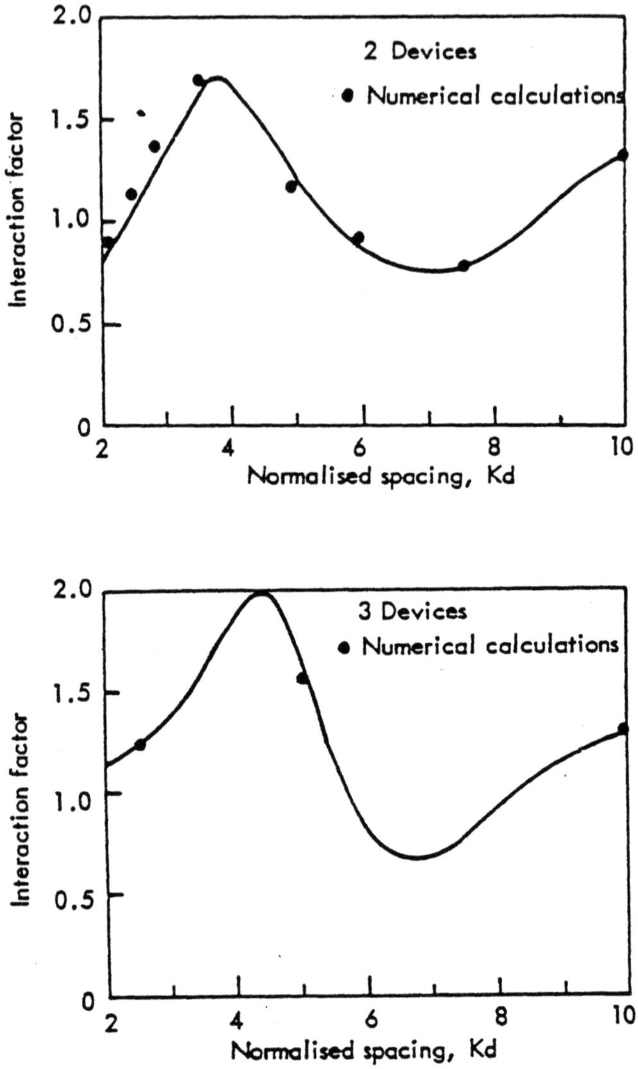

Fig.8 A comparison between Evans' approximate theory and numerical calculations for arrays of two and three point absorbers.
(From Count [14], reproduced with the permission of the UK Central Electricity Generating Board).

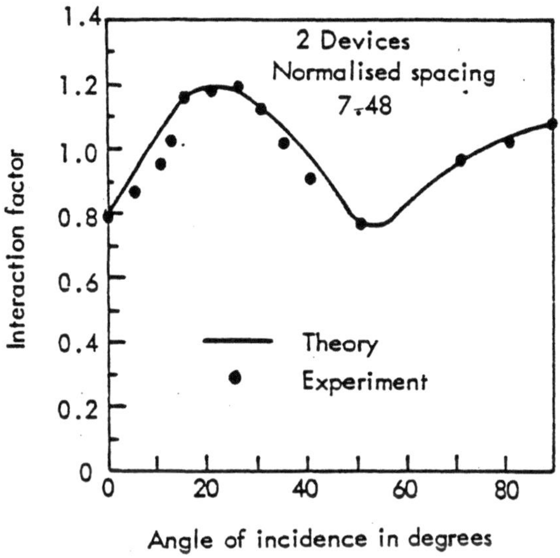

Fig.9 A comparison between theory and experiment for a pair of point absorbers.
(From Count [14], reproduced with the permission of the UK Central Electricity Generating Board).

Resonant devices with projecting side-walls

The idea of building out side-walls from a wave-energy device in order to create an additional resonance with the incident wave-field was first suggested by Ambli et al. [25]. Notable amongst the theoretical papers tackling various aspects of the device is the work of Reitan and Malmo [26],[27] in Norway and Evans [28], [29], and Count [30], [31] in the U.K. It is clearly of interest to be able to calculate the possible improvement in performance this idea produces. The expression (2.6) will still apply although the estimation of X, B is further complicated by the new geometry of the device including projecting side-walls. A simple approximate theory (Evans, [28]) is possible enabling results to be calculated using solely information about the two-dimensional performance of the device. A typical device is shown in Fig.(10). Although in practice the projecting side-walls are less than a wavelength in length we shall assume that an obliquely-incident wave in the open ocean gives rise to a plane wave of amplitude A travelling down the 'harbour' formed by the side-walls, there being no local effect felt at the device

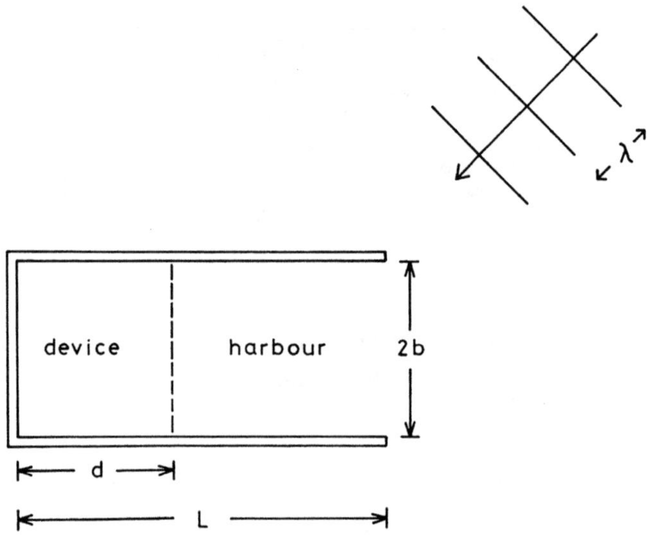

Fig.10 Device with extended side-walls in the open sea.

which responds as if to a plane wave in a narrow wave-tank. This wave is reflected as a wave of (complex) amplitude Ar which in turn is reflected from the open end as a wave of amplitude ArR back down towards the device. Here r, R are reflection coefficients corresponding to the device and the open end respectively. This process when repeated indefinitely gives rise ultimately to a wave of amplitude A/(1 - rR) travelling towards the device and Ar/(1 - rR) travelling away. The capture width ratio, based on the width of the device 2b is then clearly

$$\ell/2b = (1 - |r|^2)/|1 - rR|^2$$

which may be re-written

$$\frac{\ell}{2b} = (1 - |R|^2)^{-1}\left(1 - \frac{|r - \bar{R}|^2}{|1 - rR|^2}\right).$$

It follows that

$$\ell_{max}/2b = (1 - |R|)^{-1} \text{ when } r = \bar{R}. \tag{5.6}$$

If we now choose |R| the expression

$$|R| = e^{-2Kb} \tag{5.7}$$

(Noble [28]) for plane waves travelling towards the open end of a semi-infinite duct, we see that

$$\ell_{max}/2b = (1 - e^{-2Kb})^{-1} \tag{5.8}$$

as the maximum capture width of an isolated device with projecting side-walls. Notice that in long waves $\ell_{max} \sim K^{-1} = \lambda/2\pi$ as $Kb \to 0$ the point absorber result, whilst in short waves $\ell_{max} \sim 2b$.

If the device is operating in the middle of a wave tank the same arguments can be used but now we need the reflection coefficient for waves travelling down and towards the open-end of a semi-infinite duct bounded by parallel infinite walls a distance 2a apart. It can be shown Evans [33] that

$$|R| = 1 - b/a \tag{5.9}$$

so that the maximum efficiency of devices with side-walls of width 2b operating in the middle of wave tanks of width 2a is

$$E_{max} = (2 - b/a)^{-1} \tag{5.10}$$

Now

$$E_{max} \to \tfrac{1}{2}, \quad b/a \to 0$$

in agreement with Srokosz [34] for point absorbers in channels, and

$$E_{max} \to 1, \quad b/a \to 1$$

since now the device spans the entire tank.

The same simple multiple-reflection arguments can be used to derive an expression for the capture width ratio in terms of Λ, the power take-off characteristic Z, the <u>two</u>-dimensional impedance term given by (2.3), and r_o the two-dimensional reflection coefficient for the device in a wave-tank when power is not being absorbed.

Thus

$$\ell/2b = \frac{|\Lambda + Z|^2 - |\Lambda - \bar{Z}|^2}{|(\Lambda + Z) - r_o(\Lambda - \bar{Z})|^2} \tag{5.11}$$

obtained by using the relations

$$Z^h = (Z + r_o R\bar{Z})/(1 - r_o R) \tag{5.12}$$

and

$$X^h = X/(1 - r_o R) \tag{5.13}$$

between the three-dimensional impedance Z^h and exciting force X^h for the 'harbour' formed by the device plus side-walls and the two-dimensional hydrodynamic characteristics r_o,R,Z,X.

The advantage of using (5.11) to determine the capture width ratio is that the qualities r_o,R,Z require the solution of problems in two space dimensions, not three. The accuracy of (5.11) can be judged from Fig.(11) where a comparison is made

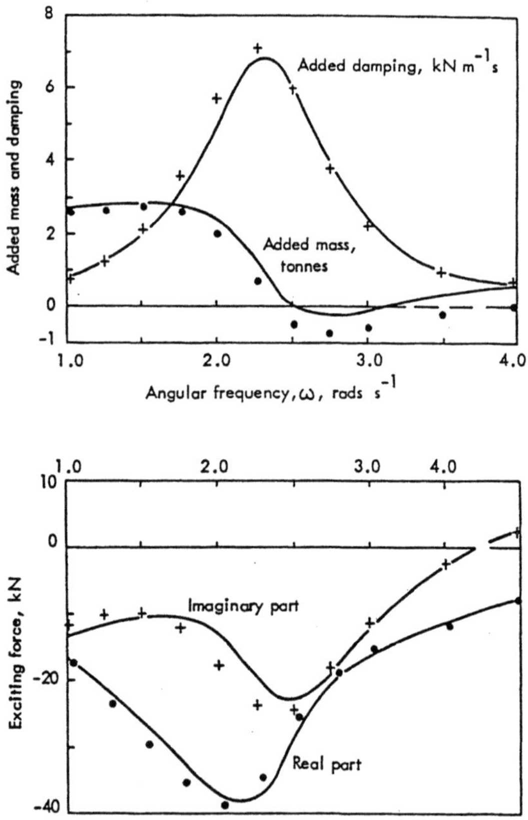

Fig.11 A comparison between theory of Evans (solid lines) and numerical calculations (•,+) for L = 2.4m, d = 1.2m and b = 1.0m. (From Count [31], reproduced with the permission of the UK Central Electricity Generating Board).

between the added mass, damping and exciting force derived from (5.12) and (5.13) and a full numerical three-dimensional scheme for computing these quantities (Count [31]). It is also clear from Fig.(12) that the projecting side-walls can improve performance markedly. The case L = 1.2m corresponds to no side-walls.

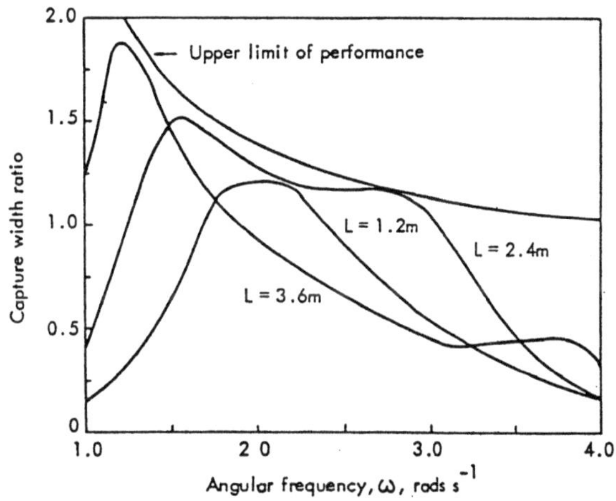

Fig.12 The performance of an idealised oscillating water column with and without a harbour with d = 1.2m, b = 1.0m. (From Count [31] reproduced with permission of the UK Central Electricity Generating Board).

When one or more such devices are embedded in a cliff-face as with the single Kværner prototype device being built off Bergen in Norway, it is appropriate to make a different approximation involving matched asymptotic expansions. If the dimensions of the devices are small compared to a wavelength we may consider an "outer" approximation in which the devices appear to be point sources of unknown strength on a wall. In the near field an "inner" solution can be derived which is valid near the mouth of the device and requires the solution for a potential flow most easily determined using conformal mapping techniques. This solution is matched to the outer source solution and to another "outer" solution valid between the side-walls but away from both mouth and the device itself. Matching this to the solution for the device gives a composite solution everywhere and in particular determines the strength of the sources in the outer region.

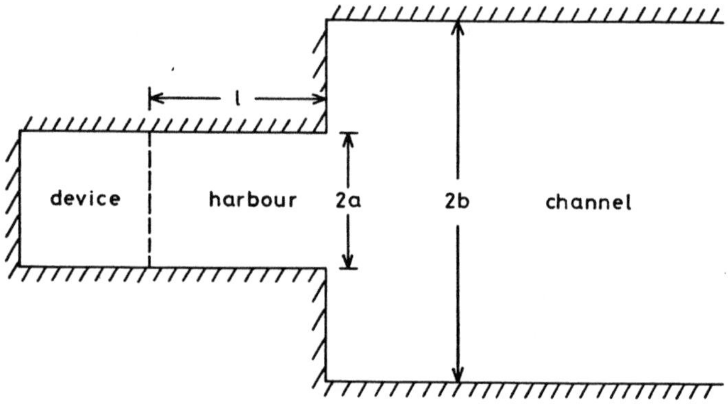

Fig.13 Device with extended side-walls operating in a channel. (From McIver and Evans [36]).

For example Evans and McIver [35] have applied the technique to a side-wall device operating in a narrow wave-tank as in Fig.(13). There are two different near field solutions depending on whether the channel walls are included in the inner solution or not. Typical results are shown in Figs.(14), (15) where it can be seen that the agreement is good in the two cases shown provided the appropriate wide or narrow channel approximation is used. In both cases $\ell/a = 1$ but for the wide channel $b/a = 5$ whilst for the narrow channel $b/a = 1.2$.

By a simple extension of this approximate method it is possible to determine the performance of any number of side-wall devices built into a vertical cliff. In this case the outer field will consist of the totally reflected incident wave plus a sum of pulsating sources of unknown strength at the position of each of the devices. These strengths will be determined by suitable matching to inner solutions valid in the vicinity of each device. Details can be found in McIver and Evans [36].

For example if we denote M_{ij} as the source strength at the jth harbour mouth due to the oscillations of the ith device it can be shown that, in the notation of equations (5.1) - (5.4)

$$X_i(\beta) = cM_{ij}L_j \tag{5.14}$$

where c is a constant.

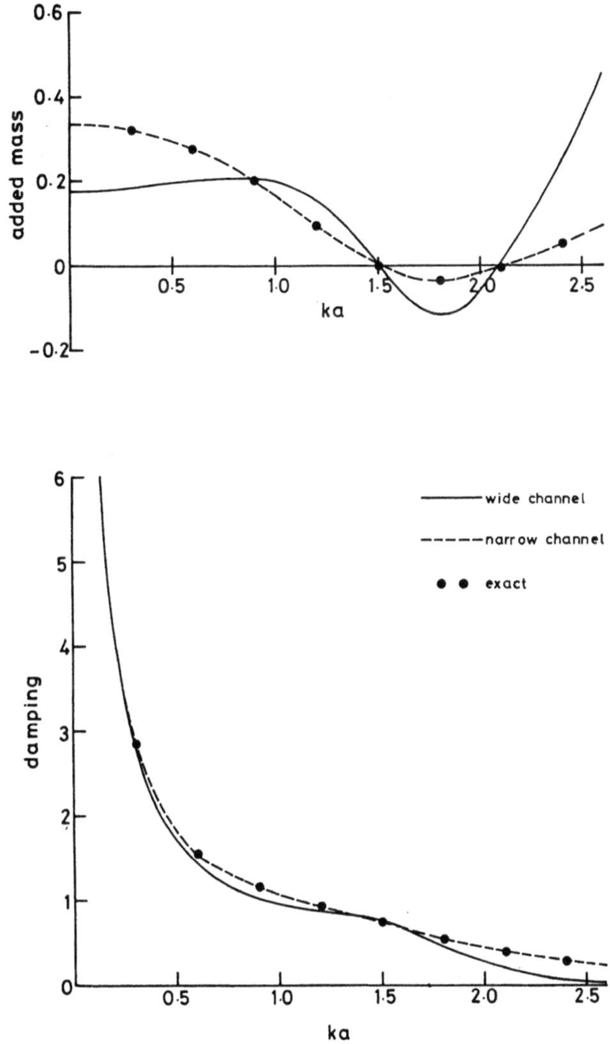

Fig.14 For explanation see text.

Notice that in contrast to point absorber theory, the exciting force on the ith device here depends on the presence of all others through the source strengths M_{ij}. Despite this the expression for the maximum capture width remains the same, namely (5.3), as substitution of the matrix form X = cML into (5.2) and (5.1) easily confirms. This fact helps to explain the success of the point absorber result (5.3) in giving good agreement with more accurate numerical approaches in which the interaction is taken fully into account.

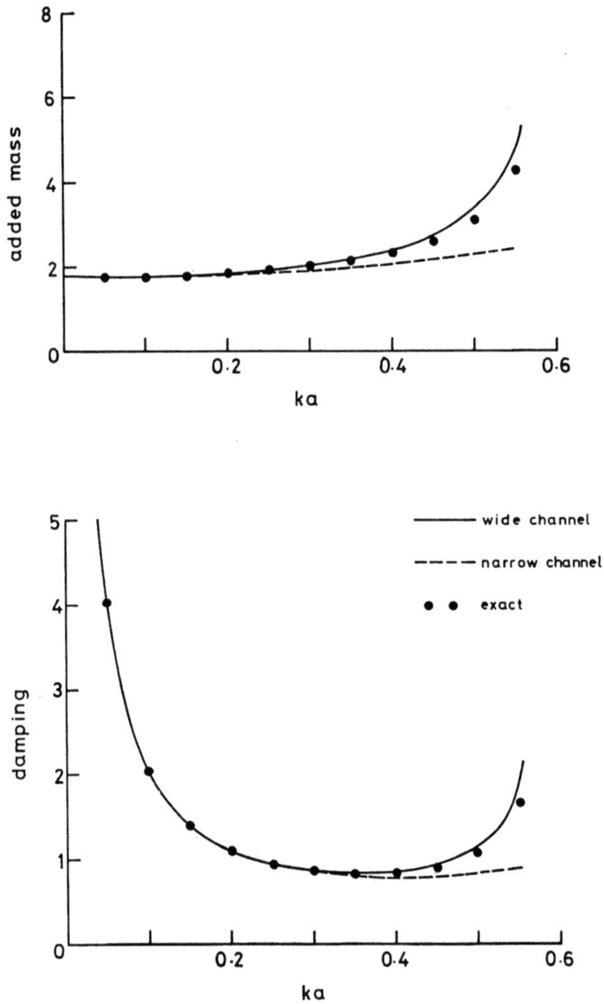

Fig.15 For explanation see text.

A typical curve based on the above theory is shown in Fig.(16) where the ratio of mean power to maximum power for four identical equally-spaced devices built into a cliff face, is plotted against dimensionless wave-number. The spacing is that which gives maximum possible power at ka = 0.5. The mean power P is based on a real damping coefficient λ. The solid line corresponds to optimal λ at each frequency obtained by numerical optimisation from (4.21) and (4.22) and equivalent to solving for λ from (4.21). The dotted line corresponds to λ fixed at its optimal value for ka = 0.5. The curve is for a/h = 1, $\ell/a = 1$, $\beta = 0$ where h is the water depth and β the angle of incidence of the waves. More accurate studies of these type of problems have been made by Malmo and Reitan [26], [27] and their paper in this Symposium.

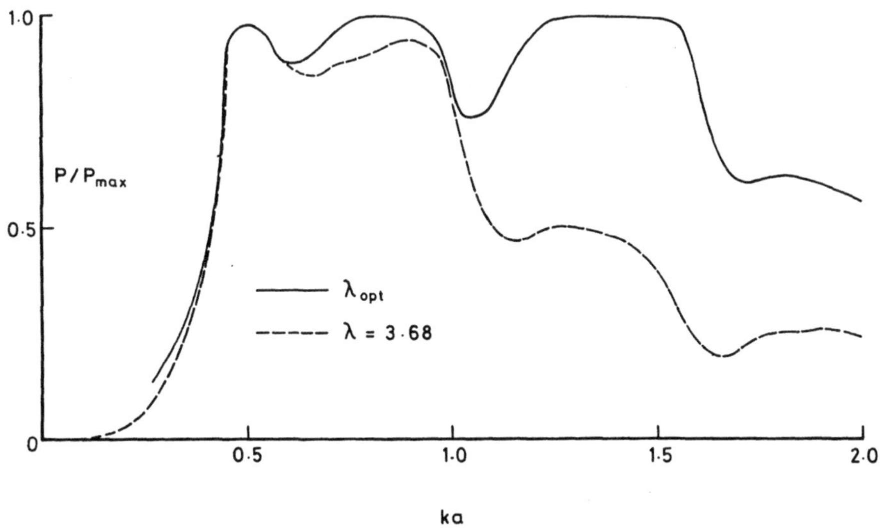

Fig.16 For explanation see text.

6. Surface Pressure Distributions

Many devices including oscillating water columns operate on the principle of resonating a mass of water which in turn drives a trapped volume of air through an air turbine. Mean power is absorbed through the product of the increased surface pressure and volume flux which in turn is related to the integral of the vertical component of the surface velocity over the free surface. The extension of the theory for rigid-body motion to the case of internal pressure distributions has been considered in specific cases by Sarmento and Falcão [10] and Masuda et al. (quoted in Falnes and McIver [7]) and a general theory for an arbitrary number of internal pressure distributions, assuming all structures are fixed, has been given by Evans [37]. This has recently been extended by Falnes and McIver [7] to the case of any number of oscillating rigid-bodies and internal pressure distributions.

The main conclusion of these studies is that the maximum absorbed power can still be written in the form (3.6) but now X, B take on different meanings. For example if all structures are fixed, the correspondence is made by identifying the exciting force on, and velocity of, the rigid body with the volume flux across the free surface, and the surface pressure on the free surface respectively. The usual diffraction problem when the rigid-body is held fixed has, as its counterpart, when the pressure is zero i.e. the device is open to the atmosphere, whilst the forced radiation problem when the rigid-body has given velocity is replaced by an imposed pressure on the surface. This is not the only correspondence that can be made.

Conclusion

This paper is not intended as a review of the hydrodynamics of wave-energy devices. A review by the author in 1981 (Evans [38]) contained nearly 80 references almost all concerned with theoretical aspects. A similar review at the present time might contain four times that many, including a majority from Japan.

Instead I have concentrated on just one aspect which I have been concerned with over a number of years, namely, the maximum hydrodynamic efficiency to be expected from different types of devices. As devices become more complicated it becomes increasingly difficult to predict their behaviour without resort to extensive and costly numerical procedures. However it has proved possible to provide simple accurate expressions in many cases of interest using suitable approximations. Although the full linear theory is just an extension of ship hydrodynamic theory, the various approximations have called upon a wide variety of tools from mathematical theory, including matrix methods, matched asymptotic expansions, constrained optimisation, and complex analysis.

The interest expressed in wave energy devices shows no sign of flagging. There is therefore a continuing role for the theoretician to translate the physics into a manageable mathematical model and subsequently interpret the consequences of that model in clear language to the design engineer.

Acknowledgments

Many of the approximate results presented in this paper were validated using full numerical theory by Dr Brian Count of the UK Central Electricity Generating Board (C.E.G.B.) who has until recently been at the forefront of both theoretical and experimental work in wave-energy in the U.K. since its renaissance in 1974. I would like to acknowledge a most fruitful collaboration with Dr Count since that time as well as thanking the Marchwood Engineering Laboratories, U.K. C.E.G.B. for permission to reproduce Figures from Technical Reports. Similar thanks are due to C.M.L. Publications for permission to use Figures from papers of mine published in

Applied Ocean Research. Finally I would like to thank the Marine Technology Directorate of the U.K. Science and Engineering Research Council for providing research grants over a number of years to enable wave-energy research to proceed at Bristol University.

References

1. McCormick, M.E.: Analysis of a wave energy conversion buoy. J. Hydronautics 8 (1974) 77-82.

2. Salter, S.H.: Wave power. Nature 249 (1974) 220-84.

3. Pierson, W.S.; Moskowitz, L.: A proposed spectral form for fully developed wind seas based on the similarity theory of S.A. Kitaigorodski. J. Geophys. Res. 69 (1964) 581-90.

4. Cockerell, C.; Platts, M.J.; Comyns-Carr, R.: The development of the wave-contouring raft. 1978 Proc. Wave Energy Conference. London-Heathrow, H.M.S.O. London, ed. P. Quarrell.

5. Hagen, G.: U.S. patent application No.4077213, Official Gazette, U.S. Patent and Trademark Office, March 7, 1978, filed Feb.13, 1976.

6. Masuda, Y.: Experimental full scale result of wave power machine KAIMEI in 1978. Proc. First Symp. on wave energy utilisation, Gothenburg, Sweden 1979.

7. Falnes, J.; McIver, P.: Wave interaction with oscillating bodies and water columns. Paper presented at this Symposium 1985.

8. Shaw, R.: Wave energy - a design challenge. Ellis Horwood, p 49 1982.

9. Count, B.M.: The theoretical analysis of wave power devices with nonlinear mechanical conditioning. CEGB Marchwood Rep.No.R/M/N1008, 1978.

10. Sarmento, A.J.N.A.; Falcão, A.F. de O.: Wave generation by an oscillating surface-pressure and its application in wave-energy extraction. J. Fluid Mechanics 150 (1985) 467-485.

11. Newman, J.N.: The interaction of stationary vessels with regular waves. Proc. 11th Symp. Naval Hydrodynamics, London (1976) 491-501.

12. Mei, C.E.: Power extraction from water waves. J. Ship Research 20 (1976) 63-66.

13. Evans, D.V.: A theory for wave-power absorption by oscillating bodies. J. Fluid Mechanics 77 (1976) 1-25.

14. Count, B.M.: On the physics of absorbing energy from ocean waves. Ph.D. thesis, Dept. of Physics, University of Exeter 1982.

15. Falnes, J.: Radiation impedance matrix and optimum power absorption for interacting oscillators in surface waves. Applied Ocean Research 2 (1980) 75-80.

16. Evans, D.V.: Some analytic results for two and three dimensional wave energy absorbers, in *Power from Sea Waves* ed. B. Count, London/NY, Academic (1980) 213-250.

17. Clare, R.; Evans, D.V.; Shaw, T.L.: Harnessing sea wave energy by a submerged cylinder device. Proc. Instn. Civ. Engns. Part 2,73 (1982) 565-585.

18. Ogilvie, T.F.: First- and second-order forces on a cylinder submerged under the free surface. J. Fluid Mechanics 16 (1963) 451-72.

19. Evans, D.V.; Jeffrey, D.C.; Salter, S.H.; Taylor, J.R.M.: Submerged cylinder wave-energy device: theory and experiment. Applied Ocean Research 1 (1979) 3-12.

20. Evans, D.V.: Maximum wave-power absorption under motion constraints. Applied Ocean Research 3 (1981) 200-203.

21. Evans, D.V.: A comparison of the relative hydrodynamic efficiencies of attenuator and terminator wave energy devices. School of Mathematics, Univ. of Bristol, Rep.No. AM-82-05. See also Proc. 2nd Int. Symp. Wave Energy Utilisations, Trondheim, Norway (1982) 137-154.

22. Count, B.M.; Jefferys, E.R.: Wave power, the primary interface. Proc. 13th Symp. Naval Hydrodynamics, Tokyo. (1980).

23. Count, B.M.; Miyazaki, T.: Study on floating alternator wave energy devices. J. Soc. Naval Arch. Japan 155 (1984) 164-171.

24. Simon, M.J.: Multiple scattering in arrays of axisymmetric wave-energy devices. Part 1. A matrix method using a plane-wave approximation. J. Fluid Mechanics 120 (1982) 1-25.

25. Ambli, N.; Bønke, K.; Malmo, O.; Reitan, A.: The Kvaerner multi-resonant OWC. Proc. 2nd Int. Symp. on Wave Energy Utilisation, Trondheim, Norway (1982) 275-297.

26. Malmo, O.; Reitan, A.: Wave-power absorption by an oscillating water column in a reflecting wall. Preprint, University of Trondheim, Norway (1984(a)).

27. Malmo, O.; Reitan, A.: Wave-power absorption by a finite row of oscillating water columns in a reflecting wall. Preprint, University of Trondheim, Norway (1984(b)).

28. Evans, D.V.: Wave-power absorption within a resonant harbour. Proc. 2nd Int. Symp. on Wave Energy Utilisation, Trondheim (1982) 371-378.

29. Evans, D.V.; Count, B.M.: Approximate impedance methods for wave-energy absorption by devices in harbours. Proc. Int. Workshop on ship and platform motions, Berkeley, Calif. U.S.A. (1983) 163-182.

30. Count, B.M.; Evans, D.V.: The influence of projecting sidewalls on the hydrodynamic performance of wave-energy devices. J. Fluid Mechanics 145 (1984) 361-376.

31. Count, B.M.: Theoretical hydrodynamical studies on harbour systems for wave energy absorption. Central Electricity Generating Board Rep. TPRD/M/1334/N83 (1983).

32. Noble, B.: *Methods based on the Wiener-Hopf technique*. Pergamon 1958.

33. Evans, D.V.: The influence of projecting sidewalls on the hydrodynamic performance of wave-energy devices. University of Bristol School of Mathematics, Rep.No. AM-83-01 (1983).

34. Srokosz, M.A.: Some relations for bodies in a canal, with an application to wave power absorption. J. Fluid Mechanics 95 (1980) 717-741.

35. Evans, D.V.; McIver, P.: A hydrodynamic theory for wave-energy devices with projecting sidewalls in harbours. Proc. Ocean Space 85, Nihon University, Tokyo (1985).

36. McIver, P.; Evans, D.V.: In preparation (1985).

37. Evans, D.V.: Wave-power absorption by systems of oscillating surface pressure distributions. J. Fluid Mechanics 114 (1982) 481-499.

38. Evans, D.V.: Power from water waves. Ann. Rev. Fluid Mech. 13 (1981) 157-187.

Progress on Edinburgh Ducks

S. H. Salter,

Department of Mechanical Engineering,
University of Edinburgh,
Scotland.

Summary

This paper describes the work of the Edinburgh group on their duck. It reviews the techniques used to improve device productivity by controlling both power take-off and mounting-movement. It outlines the results of a test programme on long compliant spines and discusses costing methodology. Some future developments are suggested.

The Evolution of Control Techniques

The first 1973 duck models (1) rotated about a fixed axis. Their shape was intended to mimic water displacement on the waveward side while not generating new waves on the leeward side. The only variable was the damping coefficient which determined the ratio of torque to velocity presented to the model by its power take-off dynamometer. The best coefficient could be determined in a few minutes.

In 1974 we built models with different water line lengths which could accept a range of different ballast weights. We found that increasing the water line length increased the added inertia and so allowed good operation in longer waves. It delayed the onset of non-linear effects and let the models absorb larger amplitudes. However, if the water line length was made too great, the torque requirement of the power take-off became very large.
We found that the most important variable was the position of the ballast weights. The centre of gravity had to be at the highest point which still allowed the model to recover from capsize. The natural nodding period had to coincide with the most useful parts of the spectrum. It took several hours of weight adjustments to tune a new model.

In 1975 we found that, if the power take-off were modified by the addition of a phase-shifted term which acted like a negative spring, performance in longer waves was improved (2) (3). A negative inertia term could help in short waves. The combination provided a useful wide-band characteristic which simultaneously helped all parts of the spectrum. The values of these reactive terms were affected by ballast position so that it took several days to optimise a new model.

At this stage we believed that fixed mountings would be expensive for full scale Atlantic conditions and so we built an elaborate mounting which allowed separate control of the mechanical characteristics in heave and surge. It was fitted with strain gauges which we used to measure fluid loading. We also believed that departures from rigidity, necessitated by practical considerations of full-scale construction, would spoil performance. However, David Evans encouraged us to explore the effect of compliance variation and we discovered that the right mounting movement could add a good deal to productivity (4). For short waves, say eight times the draught of the duck, a rigid mounting is good. But as wavelength increases there is more water motion below the duck and so energy can escape underneath it. If the mounting is allowed to move with the right phase and amplitude, the rear surface of the duck will create waves behind which cancel those that pass beneath. There will therefore be no loss of energy to leeward. The best value of rigidity falls as wavelength increases. For wavelengths 25 times the duck diameter the correct rigidity can double the output.

In addition to model-shape and ballast there were now five control variables: spring, damping and inertia of the duck nodding motion, and the two spring-rates of the mounting. By now it took several weeks to harmonise the settings for a new model.

By 1976 we had decided to concentrate our efforts on close-packed spine-mounted devices (5). From our extensive series of force measurements we knew that the bending moments on a long rigid spine would be unacceptably large. We decided that some yielding mechanism would be necessary. A non-intelligent yielding mechanism would move without damage when the local bending moment exceeded a preset value. But an *intelligent* yielding mechanism (based on high-pressure oil-hydraulics) could provide the required mounting compliance to enhance duck performance. The hardware which was necessary to ensure protection from fatigue damage could therefore enhance productivity in smaller seas. It was natural to try damping the motion of the mounting to see if it too could produce useful power in the manner of the highly efficient Evans cylinder (6).

Our first attempts to do this were not encouraging. Power could easily be generated from the mounting but most of it was robbed from the nodding motion of the duck. Our problem was that the number of variables was by now at the limit of manual optimisation techniques. We decided to assign the optimisation task to a computer.

The experiments were conducted by Henry Young. His first software used a hill-climbing optimisation technique known as the 'parrots and monkeys method'. It exploited the fact that the repeatability of absorbing wave-makers is excellent. The stiffness and damping of the model and its mounting are implemented as electronic commands to servo-motors on the various axes. It made very little difference whether the electronic drive signals were produced by the electronic position-transducers or were played back 'parrot-fashion' from a data store recorded from a previous run of the same pseudo-random wave sequence. It was therefore possible to make random 'monkey-like' changes to a proportion of numbers in the data store and then repeat the experiment with a modified time series to see if the random changes were beneficial or not. Good changes were retained. Bad ones were replaced by the original values for the next run.

The method produced some useful performance gains relative to manual techniques but was clearly not suitable for the general case of a genuinely random sea. We therefore developed a second technique (7) which varied the values of control coefficients. The best values were determined for each of the 46 seas specified for the official UK assessment procedure.

The results for a typical run are shown in Figure 1. We can see a rapid initial rise for the first 80 runs but the computer failed to find improvements after 240. In this case the computer 'learned' how to improve performance by nearly 13%. We found that the biggest gains could be achieved in the seas with large wavelengths and amplitudes. The best improvement was a gain of 30% for a sea with Hrms = 1.62 metres and a Te of 13.26 seconds. We found that mounting stiffness was much lower. In some cases it was one sixth of that which we had previously expected. For a 14 metre duck 37 metres wide the optimum stiffness in a 10 second sea is only

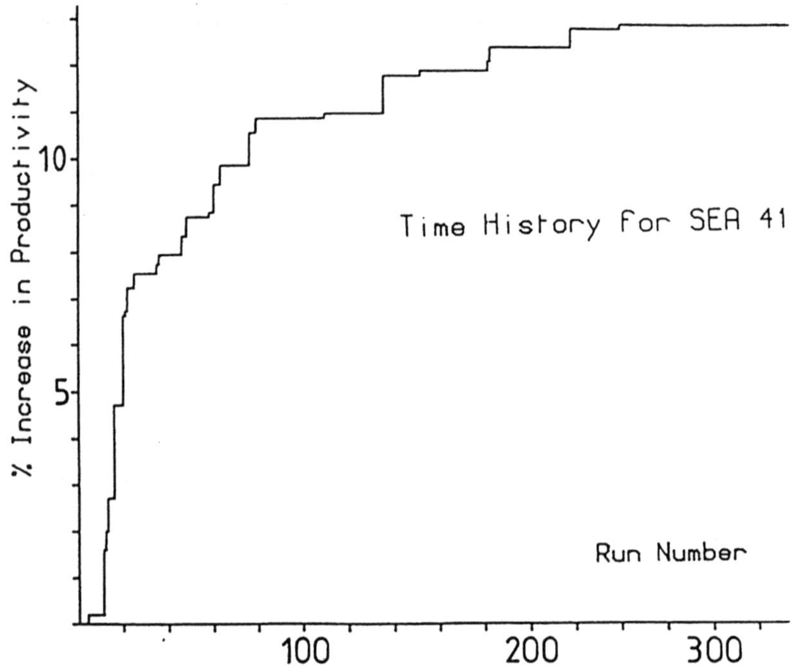

Fig.1. Optimisation with variable coefficient control.

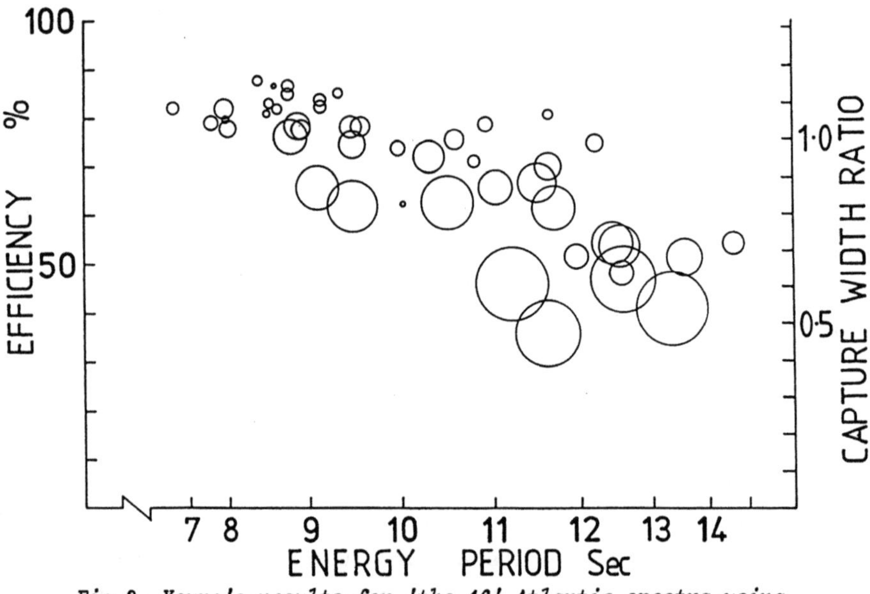

Fig.2. Young's results for 'the 46' Atlantic spectra using seven coefficients. Circle diameters are proportional to output power. Duck diameter is 14 metres.

5MN/metre in both heave and surge. The best stiffness falls linearly to zero at periods of 15 seconds. There is less of an obvious trend for the preferred spine damping coefficients but figures of 7MN sec/m for surge and 1.6MN sec/m for heave may be taken as typical. Factors of two either way do not seriously reduce performance.

The maximum force applied by the mounting was limited to the equivalent of 8MN. The model yielded to any load greater than this, thereby creating a stern-wave and reducing its mooring force. The computer program placed equal value on spine power and duck power. It was therefore open to it to tell us to get rid of the ducks. In small seas the duck did most of the work, but as wave amplitude increased the mounting produced a larger fraction of the total. In the very largest seas its output equalled or even exceeded that of the duck.

The computer reached nearly the same optimised settings from different initial conditions and, in one case, learned how to avoid a capsize. We do not yet understand the reasons for all the final coefficient values but it was encouraging to discover that, for common seas in the middle of the scatter diagram, a global coefficient based on pooled coefficients produced *better* results than those from individually optimised ones. This was not true for 'unusual' seas from the perimeter of the scatter diagram. Coefficient calculation has outstripped the power of our first computer and a more powerful one is being installed which will be able to handle more coefficients. The present narrow tank mixed-sea results scaled up to a 14 metre diameter duck are shown in Figure 2. The horizontal axis is energy period in 'expanded South-Uist seconds'. The intervals between axis marks are proportional to the annual input energy in that part of the spectrum with amplitudes of the larger seas reduced to 1 metre rms. We find this much more useful than the usual frequency axis which compresses the more valuable low-frequency region of efficiency graphs. The model occupied only 0.76 of the tank width, equivalent to the expected full-scale packing fraction. We have yet to apply the technique to a solo duck in a wide tank but we are indebted to Brian Count of the Central Electricity Generating Board for calculations which show that a peak capture-width of 3 should theoretically be possible. A value of 1.85 on rigid mountings has been measured experimentally by Bisceglia (8) at MIT.

It is useful to inquire why a narrow tank model does not reach 100% efficiency at all frequencies. The most obvious losses are bearing friction and viscosity, which we estimate at about 4% for our tests and which should reduce at full scale. Other losses must be from reflected or transmitted waves. Our next experiments will measure the phase, frequency and amplitude of all energy radiated from the model. It should then be possible to calculate combinations of heave, surge and nod movements such that a cancelling wave is transmitted. If the energy is lost as higher harmonics then clearly non-linear terms will have to be included. Calculating them may be increasingly difficult as amplitudes increase but it should not be impossible. If we can call on an infinite level of intelligence then the only limits will be those of swept volume, force and power rating.

The History and Philosophy of Spines

All wave energy devices require a reference frame against which to react. Our decision to use the crest-spanning spine configuration was a response to the official policy laid down at the height of the energy panic of the mid seventies. British teams were required to design for 2 GW installations. The Edinburgh group wanted a general solution which could be used with any sea bed geology. We wanted to maximise the use of sea space and avoid collisions between neighbours. We knew that long spines allowed easy power accumulation and we expected that their mooring behaviour would be excellent. We were however much concerned about bending moments and in particular the fatigue that they would induce. It was clear that we would need to know a great deal about the fluid loading on long compliant spines. Unfortunately the very best theoreticians told us that the problem was beyond the limit of either analytical or numerical techniques, particularly if non-linear effects were present. There were no tanks in which we could control a wide test region with a realistic spectrum. It was therefore necessary to build a wide tank. This was done in 1977. The design is described in references (9) (10) and has proved popular elsewhere.

We spent 1978 and 1979 building wide tank models with invaluable help from school-leavers and vacation students. It was necessary to develop new methods for representing the scaled elasticity of the prototype structure and the yielding behaviour of the joints. We split the spine into a large number of separate sections, each connected to the next by a quick-release joiner-ring. Each section contains a bearing mechanism which allows relative rotation between it and the next. A transducer at each joint senses the local bending moment and passes the information to a central computer which calculates the appropriate angular deflection for the beam characteristics selected for the experiment. The model behaves like a python with a variable degree of rigor mortis. We believe that it is the world's first computer-controlled beam. It has been under test by Jamie Taylor and Chris Retzler since 1979 and the results for bending-moment, joint-angle and mooring force are now known (11) (12) for a wide range of different lengths, stiffnesses, sea-states and fixed appendages. The main conclusions are as follows.

There seems to be no obvious relationship between length and bending moment for long compliant spines. Length-squared and length-cubed rules do not apply. Bending moments are nearly independent of energy period over the central part of the useful spectrum and fall for the very shortest periods. In medium amplitude seas, surge bending moments are about twice heave moments. Heave bending moments are proportional to the first power of wave amplitude. Surge bending moments for a low-freeboard circular spine rise with wave amplitude to the power 0.8. For engineering purposes it is safe to assume that there is a 90° phase shift between heave and surge moments. Bending moments rise gently with stiffness. A logarithmic relationship fits our test range. Reducing stiffness makes the effects of period change even smaller. Changing the stiffness in one direction produces little effect on the bending moments in the other so that we cannot run away from surge moments by yielding in heave. Seas with a mean approach direction of 90° to the length of the spine produce a bending moment pattern which oscillates about a mean level in the central regions and which is raised at points approximately a wave-length in from each end. Seas which approach the spine from an oblique direction produce smaller moments at the up-wave end and much larger ones at the down-wave end. The down-wave enhancement factor in monochromatic waves can be as high as twenty-five, so that obliquity is by far the most powerful variable. It is a relatively narrow effect, changing over one or two degrees. We believe that enhancement happens when the velocity of a flexural wave along the length of the spine is close to the velocity of an intersecting wave crest and flexural energy is reflected from a free end. As flexural propagation velocity is a function of spine rigidity the effects can be controlled by the use of intelligent joints and perhaps even turned to advantage for power generation. The maximum joint angle in the steepest, largest possible wave is only just over 4°, so that the 12° design allowance is generous: the result is unaffected by any attempt to stiffen or loosen the joint. Broken joints reduce their own bending moment and raise the moments at adjacent joints. Low free-board models have lower mean mooring forces as wave amplitude rises. They can sometimes move forward. High free-boards reverse this effect. Short spines have larger and much less steady mooring forces. The addition of Norwegian-style harbours *reduces* surge moments slightly but increases the smaller heave ones.

Moving from the Narrow to the Wide Tank

We believe that we will soon be quite close to the ultimate efficiency limit for a narrow-tank duck on an active mounting. We will know what torque should be applied to the nodding motion and what forces should be applied between the mounting and the fixed reference of the tank walls in any sea state. We recognise that achieving the same performance with a free-floating spine model in a wide tank will be a challenging task. The argument that it should be possible runs as follows.

Imagine that there exist a large number of adjacent narrow tanks all fitted with ducks on correctly controlled mountings. Next imagine that the wave-makers in each tank are driven so that the wave pattern in front of each model corresponds to a full-scale, multi-directional spectrum. We could record sets of time-series of the force command to each mounting. We could then write a set of new time-series of the differences in forces between each adjacent pair. This set would be equivalent to the time series of the shear forces at each joint along the length of the equivalent spine. It is well known from elementary beam theory that the bending-moment in a beam is the integral of the shear force. This means that we know what the pattern of bending-moments in the equivalent spine should be. If we can provide a mechanism which applies this pattern of bending moments to the joints of a full-scale spine, then all the ducks on it will behave in the wide tank in the exemplary manner of their counterparts in the bank of narrow tanks. It would, of course, not be wise to apply a bending moment greater than the safe structural limit. The spine would have to move more than the optimum amount if this requirement arose.

It is important to realise that the bending moments at the joints in our spine have absolutely nothing to do with the sea state. They are merely the result of oil-pressures in the controlling hydraulic rams. This in turn is a function of the state of the controls of the axial-piston machines which absorb or supply their oil. Bending moments can be very large in a calm sea if the spine is swimming to its station, or zero if no resistance is offered to oil flow. The hardware which was necessary for limiting the bending moments of the spine to the safe endurance limit can perform the function of applying the correct pattern of bending moments. The task of supplying the necessary software may appear formidable today. It will appear trivial by the time that the energy crisis returns.

Costing Methods

The task of predicting the cost of future projects is extremely difficult and the process is open to abuse either from self-deceiving enthusiasts or defensive rivals. The present official estimates of capital costs of wave power devices are based on an elaborate schedule of standard rates suggested by a group of leading consultants and contractors who could draw on their experience of oil-related offshore work (13). This document is one of the most useful to emerge from the British wave programme and is invaluable to the newcomer to civil engineering. It lists the costs of steel as a function of diameter or sheet thickness, painting and fabrication complexity. It separates bulk concrete costs from those of form work and shuttering of different curvatures. It distinguishes between different lengths of post-tensioning tendon and lists the hire charges for different kinds of welding plant. It provides labour costs for different skills together with details of insurance, sick-benefit and holidays. Even the costs of 'office ablutions' and 'field ablutions' are separated. The rates given appear to be in good agreement with a random sample of independent estimates which we have obtained elsewhere.

The data from this document, supported by numerous commercial quotations, were applied to the reference design separately by the Consultants to the United Kingdom Department of Energy (14) and the John Laing costing department (15). The results were close. For all hardware up to the beach, the 'ex-works' Laing figure was £982 per kilowatt while the official figure in the Consultants' 1983 report was £1141. After additions for construction facility (usable for future production) and the costs of installation, land cable, switchgear etc. the totals were £1290 and £1520 per kilowatt.

I am sorry to report that this admirable costing method may be replaced by a much simpler one. In future all structural concrete would be deemed to cost £400 per tonne, all ballast to cost £50 per tonne and all our steelwork, whether the pressed sheet of our flywheels, the ship-like shell of the power canister or the most elaborate hydraulic pumps, is costed at

£10,000 a tonne. These data give reasonably accurate results when applied to wind turbines: while we would have preferred numbers which work for ships and harbours we congratulate ourselves that we are not required to design for the weight ratings of military aircraft.

Costs per kilowatt are quite meaningless without a value for the load factor. This can be calculated from a knowledge of the wave climate, the conversion efficiency predictions of the tank tests, calculations of the mechanical and electrical losses of the hydraulics, generators, transformers and transmission cables and finally some estimate of equipment availability. The consultants calculated that in the 48 kw/metre net power input of the Atlantic wave climate each 37 metre wide duck would convert a mean 1.1 MW to raw mechanical power. The hydraulic and electrical conversion processes had an effective mean efficiency of 82% although an unresolved dispute about the possible double accounting for generator losses might reduce this slightly. If all the equipment worked perfectly its annual load factor would be 40.4% - unusually high for a renewable energy device. If we use the official capital cost of £1520 per kilowatt, the 5% test discount rate set by the Treasury, assume that the plant becomes instantaneously valueless at the end of 25 years and ignore for the time being the vexed question of maintenance, we would require an annual payment of £107.85 for each kilowatt of capacity. With a 40.4% duty factor the annual output from each kilowatt of capacity would be 3537 kilowatt hours and so each kilowatt hour would cost 3.05 pence.

This figure would be lower if we used a longer repayment period or assumed that the plant retained some value at the end of 25 years. For example, a 50 year repayment period (not impossible for hydro-electric schemes and harbour installations) would imply 2.35 pence per kilowatt hour. The extreme assumption that (because of careful maintenance and proper understanding of fatigue) the plant retained its full value at the end of 25 years would imply a cost of 2.15 pence per kilowatt hour. The greatest reduction would, however, result from the use of the return on investment reported by the Central Electricity Generating Board (16). For the years 1981-84 (before the miners' strike) this has averaged 2.28%. At CEGB repayment rates and a 25 year life the cost would be 2.27 pence. Over 50 years it would be a dazzling 1.45 pence! These figures are of course mere fantasy because there is no question of the Treasury treating the early phases of renewable energy schemes as favourably as established methods.

In reality a wave scheme would sometimes be out of action. Energy would be lost and the costs of repairs and maintenance would have to be added to those of capital repayment. The saga of duck availability is an extraordinary one involving classified information sources, bogus and untraceable references, unrepeatable calculations, lost data, destroyed documents, banned communications, bizarre assumptions about salt, sand and dust in the hermetically sealed power canisters, acceleration levels one hundred times greater than those in the roughest seas and unexplained discrepancies between failure-rate data used for identical components in the hydraulics of wind and wave devices (17) (18). Our confidence in the Consultants suffered a severe blow when they refused to answer questions and denied us access to a publicly-funded computer program written to predict device availability. However, the Energy Technology Support Unit took over and ran the program on the Harwell computer (19). They found, to our surprise, that equipment reliability was much less important than assumptions about maximum sea-states for changing failed items.

The reliability program takes as input the failure rates for various parts of the plant. It allows the selection of a weather window for each task in terms of the duration for which the sea-state must be continuously below a particular threshold for the critical phases of the exchange. Next it uses Monte Carlo methods and the known statistics of the Atlantic to predict the values of plant availability for any number of repair crews. It assumes accurate forecasting (an increasingly reasonable expectation) and makes an allowance for the observed ability of healthy, close-packed devices to take most of the energy from adjacent, failed neighbours provided that their own power-ratings have not been exceeded.

Figure 3 shows a sample of the output. Failure-rate assumptions for the case shown were that vacuum canisters, duck-to-spine cables and joints all suffered one failure a year, that the magnetic bearing, the generators and the internal spine electrics for a group of six ducks suffered 0.2 failures a year and that the rate for all other auxiliaries, communications etc was 0.1. This leads to a total annual failure rate of 4.23 per duck. Readers can judge the severity of this assumption for themselves. It was assumed that each task required a weather-critical phase of 12 hours for sea states below 2.5 metres significant. Increasing the critical time to 24 hours makes little difference but results get very much worse if the sea-state threshold falls much below 2 metres significant. It was assumed that each repair crew, its tools and spares cost £2m a year and that they worked from a base which cost £20m a year.

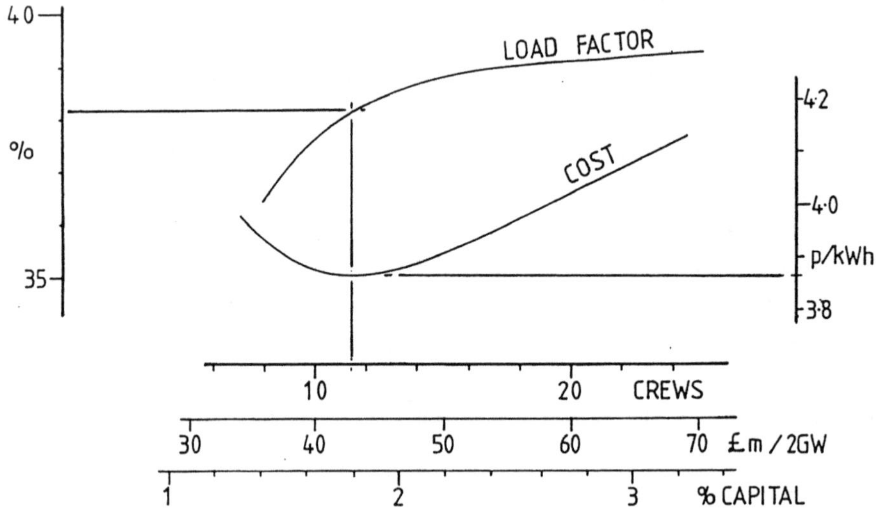

Fig. 3. Sample output derived from the official reliability program.

Values for load factor and electricity cost are plotted against the number of repair crews. The cost of maintenance is expressed both as an annual charge and as a percentage of the officially calculated capital cost of marine-side items. For the failure rates, repair times and sea-state access specified above, it can be seen that the best number of repair crews is 12, that optimum annual maintenance costs are about 2% of capital, that the availability-corrected load-factor is 38% and that the electricity cost is 3.87 pence per kilowatt hour.

Working inside our spines will be more comfortable than working in the engine room of a ship because there is more space and considerably less violent motion. The root-mean-square acceleration is only 0.007g per metre significant wave height, almost irrespective of period. The difficult tasks are the separation of power canisters from duck bodies and the transfer of personnel. Narrow tank tests show that duck nodding can be limited to ±2° in sea-states up to 2.5 metres significant by the method of overdamping the power conversion mechanism of one of the two power canisters. If both power canisters have failed, the duck motion can still be reduced to ±2° by feeding a signal proportional to duck angular velocity into the heave axis. Above 2.5 metres significant motion becomes much more violent because force-limits and torque-limits are exceeded. We are confident that the safe separation of the power canister from the duck body can be achieved by means of 'Vetter bags'. These are inflatable cushions used for lifting bulldozers and road rollers. Uninflated bags would be inserted into slots between canister and duck body. Air pressure would be applied and the retaining straps released. Bag pressure would provide the separating force and protect the two bodies from the forces due to differing accelerations. The procedure will be reversed for canister replacement with tension winches pulling the new canister into place against the Vetter bags. Squeeze film effects limit the speed of the operation.

While any figure below 5p (suitably index-linked) would have been welcome at the height of the energy crisis, present predictions are still not low enough to be attractive to the United Kingdom. Our next cost reductions may come from the continuation of the improvements in control technique. We should also take special note of Even Mehlum's siting methods developed by Norwave (20) for choosing places for the early installations. Good sites can double input energy without increasing extreme loading. However, we want to concentrate particularly on materials for duck bodies: while the Laing post-tensioned concrete design is exactly right for the spine sections, it is too strong and too heavy for the ducks. Most of the expense goes into resisting hydrostatic pressure over the air compartments. Proposals for ice/wood pulp construction have not been well received and, though thermodynamically excellent, ice might be troublesome for initial work. Expanded polystyrene is not quite strong enough and too light. We would like to average its properties with those of concrete to make solid ducks.

Solo Ducks

However ingenious we become at controlling model spines, shaving full-scale design costs and improving reliability, there is little immediate prospect for the deployment of large British installations because the United Kingdom is saturated with depletable energy. It is therefore interesting to look at the smaller initial step of a solo duck. Comparisons with the present design are interesting.

The disadvantage of wasting sea-space will not apply immediately. The duck-to-spine bearing and spine joints are eliminated. A considerable weight saving - several thousand tons of concrete per device - a doubling of input energy by 'hot-spot' siting and a possible doubling of capture-width by point absorber effects will all help the economics.

The surface area between duck and spine is removed and the asymmetric ellipsoidal shape is close to that of an egg shell which is very effective at resisting external compressive forces - much better than the concave bearing section of spine-mounted ducks. The combined volume of duck-beak plus the previous spine space is available for power conversion mechanisms. These could use sealed gyros or very large diameter ring-cams reacting against a harbour. We are also looking at methods of using duck motion directly to drive a vapour-compression desalination plant. A duck could produce more than 1000 tonnes of fresh water a day in a very moderate sea.

A solo duck would obtain its reference from 4 tension-leg rodes tied to two sea-bed attachment points, one forward and one aft of the duck. By using long Evans cylinder-like rams, or the Jim Platts tube-pumps (21) in the rodes we may be able to provide the desirable mounting compliance and some extra output.

The major change is that we have replaced the high bending moments of the spine by high tensions in the rodes. For the first time in the Edinburgh project we are facing a force without a means to limit it. We need tensions of the order of 10 meganewtons which will be subject to much more variation than the rodes of spine counterparts. Rodes for tensions of this size are not as difficult as the sea-bed attachment. Some leading mooring consultants have been employed by the Atomic Energy Authority to advise wave power teams on mooring techniques. I believe that their proposals can be improved in several ways.

The designs (22, 23) on offer contain examples of acute stress-concentrations in some parts combined with under-used steel in others. Water-exposed welds are subject to peeling forces and tensile fatigue. Pile-heads are exposed to unnecessary offset loads and torsion. The volume of rock to be drilled is higher than necessary while the contact area to the surrounding rock is undesirably low. A successful solo duck would need sea-bed attachments designed with much more thought. I hope to describe the results to a future conference.

Conclusions

The emerging vocabulary of wave energy allows an exact, if highly dense, description of our work. It should not be read too rapidly: each word represents three man-months of effort.

The duck is an asymmetric, close-packed, hard-skinned, surface piercing, low-freeboard terminator with unlimited rotation, mounted on chlorine-dosed, contactless, magnetic squeeze-film bearings about a long, compliant circular, active, vertebrate, crest-spanning, post-tensioned concrete spine. Duck nodding provides a large swept-volume to self-volume ratio and is resonant at the central parts of the spectrum. Bandwidth is widened by reactive phase-control of the non-manifolded, hydrostatic, vacuum-sealed power-conversion mechanism which reacts against a gyro-stabilised reference and contains bulk front-end energy-storage. Control strategies of arbitrary complexity can be used. No mechanical penetrations are required. Spine survival is ensured by ram-controlled joint-yielding below the fatigue limit. Joints contribute to output. Long wave performance is improved by calculated joint movements which deliberately generate stern waves to cancel energy transmission. Electrical generation is by mean-rated synchronous AC machine locked to the grid-frequency. Power collection is by enclosed, constant-tension, wet, spool-wrapped tape conductors between duck and spine, followed by dry air-spaced parallel cable inside the spine, to multi-wound transformers which step voltage up to 132kV. Sea-bed transmission lines use low-capacitance cross-linked polyethylene insulation. Single faults can be tolerated without energy loss. Mooring is by constant-tension variable-direction rodes with high redundancy. Mooring forces are reduced by stern transmission of excess energy. Installations have low dependence on sea-bed geology but require water-depths of 80 metres or more. The design minimises the exposure of corrodible materials. Hydrodynamic performance seems to be little affected by severe external surface fouling. Access to the spine for maintenance is allowed but duck power canisters can only be repaired by exchange. Cost of installed Atlantic plant including land-transmission and switch gear is about £1500 per kilowatt with a load factor approaching 40%, leading to electricity costs between 3.5 and 4.5p per kW/h. Hot-spot sited solo-ducks offer an attractive first step provided that careful attention is paid to their heavily-loaded sea-bed attachments.

Acknowledgements

Wave energy research at the University of Edinburgh is supported by the British Department of Energy through the Energy Technology Support Unit of the United Kingdom Atomic Energy Authority.

References

1. Salter, S. H.; Wave Power. Nature vol 249 pp720-724 June 21, 1974.
2. Edinburgh Wave Power Project; 1st Year Interim Report. September 1975.
3. British Patent C2224/76. January 1976.
4. Edinburgh Wave Power Project; 4th Year Report vol 1. July 1978.
5. Salter, S.H.; Jeffrey, D.C.; Taylor, J.R.M.; The Architecture of Nodding Duck Wave Power Generators. The Naval Architect pp21-24, January 1976.
6. Evans, D.V. et al; Submerged Cylinder wave energy device: Theory and Experiment. Applied Ocean Research vol 1 pp3-12. 1979.
7. Young, H.E., Pollock, J.; Variable Coefficient Control of a Wave Energy Device. Edinburgh Wave Power Project report. March 1985.
8. Bisceglia, S.; Economic Feasibility of Cam-type Wave Power Generators. MSc dissertation. Department of Ocean Engineering M.I.T. June 1978.
9. Salter, S.H.; Absorbing Wavemakers and Wide Tanks. ASCE Directional Wave Spectra Applications Conference pp185-202. Berkeley. 1981.
10. Salter, S.H.; Physical Modelling of Directional Seas. Symposium on Description and Modelling of Directional Seas ppB1.1-B1.9. Copenhagen June 1984.
11. Retzler, C.H.; Long Spines in Multi-directional Seas. Ibid ppB5.1-B5.10.
12. Edinburgh Wave Power Report; Bending Moments in Long Spines. 1984.
13. Rendel, Palmer and Tritton; Analysis of Civil Construction Costs. Consultant's Working Paper Number 36. July 1981.
14. Rendel, Palmer and Tritton; Technical Appraisal of the Edinburgh Duck Volume 3D (ETSU WV 1570P26). June 1983.
15. Edinburgh, Merz and McLellan, Laing; Report to Department of Energy. October 1981.
16. CEGB Report and Accounts p10. March 1984.
17. Y-ARD; Reliability Study of the Salter Duck. Report YR 2814/81 appendix 3 fig A3.8. 1981.
18. Y-ARD; Comparison of Mechanical and Hydraulic Transmission Systems for Aerogenerators. Report YR 2788 Y 2407 pA1.3.3. December 1982.
19. Taylor, R.; A Coda to the 1982 Duck Assessment WESC(85)P702. 1985.
20. Mehlum, E.; The Wavetrack Software Package. Norwave A.S. - undated technical literature.
21. Platts, J.; Spirally Wound Elastic Mooring and Tube-Pump. WESC(79)MA72.
22. Single Buoy Moorings (DBV) Ltd.; A Study on the Assessment of Costs for Installing and Maintaining Mooring Systems for Wave Energy Devices. WESC(80)MA115. 1980.
23. Wimpey Laboratories Ltd.; Anchorages for Wave Energy Devices. Report number ST51/82 WESC(82)MA135. March 1982.

TAPCHAN

Even Mehlum

NORWAVE A.S
Forskningsveien 1
0371 Oslo 1, Norway

ABSTRACT

The TAPCHAN Wave Power Plant combines the following features:

- The wave energy is converted to potential energy in an on-shore water reservoir

- The generation of electricity is carried out by standard hydroelectric power plant technology

- The conversion device is entirely passive and has no moving parts

- High energy conversion efficiency is maintained over a broad range of wave heights, frequencies and directions.

The first full scale TAPCHAN power plant is now under construction on the west coast of Norway. It is expected that this plant will deliver electricity at a cost of 5-6 US cents pr kWh.

Significant cost reductions are expected when we get beyond the prototype phase. The prototype will be connected to the Norwegian electrical network in October 1985. Most of the lecture at the conference will be used to describe the actual prototype, whereas the enclosed paper describes working principles.

TAPERED CHANNEL WAVE POWER PLANTS

In many parts of the world the energy content of the ocean waves is considerable and the cost of traditional electricity is high. This makes wave power an interesting proposition. The critical problem is the development of efficient and reliable methods for converting wave energy into electrical energy. In general, such energy conversion devices must meet the following requirements:

* The conversion efficiency should be insensitive to variations in wave height, frequency and direction.

* The construction itself should be able to cope with extreme weather conditions.

Most of the devices proposed for wave energy conversion involve the exposure of moving parts and machinery to adverse off-shore conditions. As a result, production and maintenance costs for these devices are high.

To overcome these problems, NORWAVE A.S and the Central Institute for Industrial Research, Oslo, Norway, have developed a novel method for wave energy conversion: the TAPERED CHANNEL WAVE POWER PLANT(TAPCHAN). TAPCHAN POWER PLANTS are now offered by NORWAVE A.S on a commercial basis.

THE TAPCHAN POWER PLANT

The TAPCHAN POWER PLANT combines the following features:

* The wave energy is converted to potential energy in an on-shore water reservoir.

* The generation of electricity is carried out by standard hydroelectric power plant technology.

* The conversion device is entirely passive and has no moving parts.

* High energy conversion efficiency is maintained over a broad range of wave heights, frequencies and directions.

The principle of operation of the NORWAVE TAPCHAN can be explained by dividing the system into the following sub-systems: (see fig. 1.)

* A <u>collector</u> designed to concentrate the wave energy and optimize collection efficiency for a range of frequencies and directions.

* An <u>energy converter</u> in which the energy of the collected waves is transformed into potential energy in a water reservoir.

* A <u>water reservoir</u> for storage of the converted energy. The reservoir also serves to average out short-term fluctuations in the wave energy level.

* A <u>conventional hydroelectric power plant</u> for electricity generation.

Fig. 1

(a) The wave energy converter

This stage is the unique part of the TAPCHAN POWER PLANT. It consists of a gradually narrowing channel with wall heights equal to the filling level of the reservoir (typical heights 3-7m). The waves enter the wide end of the channel and, as they propagate down the narrowing channel, the wave height is amplified until the wavecrests spill over the walls. The gradual narrowing of the channel causes a continuous sideways spill-off as the wavecrests move along. As a result, the wave energy is gradually transformed into potential energy in the reservoir. The advantage of this conversion method is that it is entirely passive, and almost completely insensitive to variations in wave height and frequency.

Various models of converters have been thoroughly tested in a wave flume as well as in a wave tank and shown to behave according to specifications. Typical figures for the conversion efficiency are in the range of 65-75 %.

(b) The collector

The collector serves the dual purpose of concentrating the incoming waves before they enter the converter, and matching the converter to the local wave climate in an optimum manner. The horn shaped collector is designed so as to optimally collect the wave energy over a range of incoming wave frequencies and directions.

In some cases it is desireable to modify the local wave climate (i.e. increase the energy density) by using offshore lens elements for wave focusing. Such wave lenses act as a pre-collector, and must be adapted to the collector stage.

(c) The reservoir

The main function of the reservoir is to provide a stable water supply for the turbines. The reservoir must be large enough to smooth out the fluctuations in water spill-off due to the individual variations in the waves propagating down the tapered channel. This is achieved with a small reservoir. Storage capacity for less than one hour's operation of the turbines is ample. The reservoir should be as small as possible, the lower limit being determined by the technical requirements of the water turbines.

(d) The hydroelectric power plant

Well established techniques are used for the generation of electric power. The water turbine driving the electric generator is of a low head type such as a Kaplan, or a tubular turbine. It must be designed for salt water operation, and should have good regulation capabilities. Otherwise, there are no special requirements for the power station and its equipment.

DESIGN OF A TAPCHAN POWER PLANT

The TAPCHAN POWER PLANT is based on a concept which is adaptable to a large spectrum of external conditions. Plants of sizes between 0,5 MW and 300 MW can be constructed.

A TAPCHAN POWER PLANT is individually designed to ensure an optimal configuration. The capital costs of the plant and the unit cost of power, i.e. average production cost per kWh. in a year, should be minimized.

Normally, NORWAVE will start with a feasibility study in order to evaluate the possibilities for an economical utilization of the available wave power in a region. The economy of any wave power plant is highly dependent upon local variations in the wave energy. Key parameters, such as capital cost per installed kW and unit cost, vary with installed machine capacity, dimensions and shape of converter etc. Thus, optimization of plant parameters is a very important part of the design.

NORWAVE and partners have developed the theoretical basis and adequate design tools for optimization, design, and performance analysis of TAPCHAN POWER PLANTs. With our Wavetrack computer program, the local wave climate is analysed. The Wavetrack results are used together with results from well established procedures for site selection to determine the most advantageous plant location.

Computer programs are important tools in the optimization and design of the power plant and its sub-systems. One example is a program for design of optimum size and shape of the collector stage of the system. With a given spectrum of sea waves as input, this program calculates the detailed wave patterns and the overall transmission efficiency of converters of various shapes. The output data from the program are important parameters used in a computer-aided optimization procedure in which plant parameters (dimensions of converter and reservoir, installed machine capacity, etc.) are determined to give optimum overall performance of the TAPCHAN POWER PLANT.

NORWAVE has found that great cost reductions can be achieved by a thorough and intelligent optimization and design. Optimization of potential TAPCHAN POWER PLANTs has resulted in reductions in unit costs of power by factors of between 2 and 4. The unit cost of power from some actual TAPCHAN POWER PLANT propositions has been calculated to less than 2 new pence per kWh. (about 3 U.S. cents).

This figure compares favourably with most other energy costs, and indicates that our TAPCHAN WAVE POWER PLANT in many cases might be the right device for the supply of electric power.

PARTICIPATION BY NORWAVE A.S IN WAVE POWER DEVELOPMENT

NORWAVE A.S is a company formed by former staff members of the Central Institute for Industrial Research, (SI) to develop and market products based on wave phenomena. The NORWAVE employees and their SI colleagues have carried out a major research programme on ocean waves since early in the seventies. One notable result is the renowned focusing wave power plant, in which specially designed water wave lenses are used to convey the wave energy into a focal area, where the converter is placed. This concept is well suited for large wave power plants, producing several hundred megawatts.

During the later years SI and NORWAVE A.S have concentrated on the development of smaller wave power plants. The result of this effort is the TAPCHAN WAVE POWER PLANT which now can be constructed on a commercial basis.

NORWAVE A.S and its partners in Norwegian industry combine some of the world's most advanced scientific expertise on wave power, with extensive international experience in the design and construction of hydroelectric power plants, as well as large offshore structures.

NORWAVE A.S can assist in connection with any activity related to the development of wave power. This assistance can range from cooperation with local companies on special activities, to taking full responsibility for any part of the design and construction of a TAPCHAN POWER PLANT. This includes turnkey delivery of the complete power plant.

REFERENCES

First Symposium on Wave Energy Utilization, Proceedings, Chalmers University of Technology, Gothenburg, Sweden 30 October-1 November 1979

Wave Energy Utilization, Proceedings, The Norwegian Institute of Technology, Trondheim, Norway 22 June-24, 1982

Development of the Kvaerner Multiresonant OWC

O. MALMO

Norwegian Hydrotechnical Laboratory,
Trondheim, Norway

A. REITAN

Dep. of Physics, College of Arts and Science,
University of Trondheim, Norway

Summary

This work is a contribution to a research and development activity carried out at the Norwegian Hydrotechnical Laboratory on contract from and in close co-operation with Kværner Brug A/S. Our main contribution to the development of cost-effective wave energy converters has been the introduction of a socalled harbour in front of the absorber which increases the bandwidth. We are also able to bring the first results from the prototype tests of an oscillating water column built at Toftestallen west of Bergen.

1. Introduction

The Norwegian Hydrotechnical Laboratory (NHL) has been working with the development of a wave energy converter based on the principle of an oscillating water column since 1978. Oscillating water columns as wave energy converters are well-known from a large number of publications in recent years. In modern times the earliest description known to us is given in a patent held by Electricité de France Service National, registered in 1953 [1].

Our contribution to the exciting task of turning the wave energy into useful power is the introduction of the harbour concept, which we presented at the conference on Wave Energy Utilization in Trondheim 1982 [2]. We had then been working with the idea both experimentally and theoretically since 1980.

2. The harbour concept

The harbour is formed simply by a pair of walls protruding from the front of the absorber, thereby partly enclosing a rectangular basin. In this basin the phenomenon of harbour resonance occurs. The waves in the basin are amplified by the partial reflection from the harbour mouth of outgoing waves in the harbour. The principle is illustrated in figure 1.

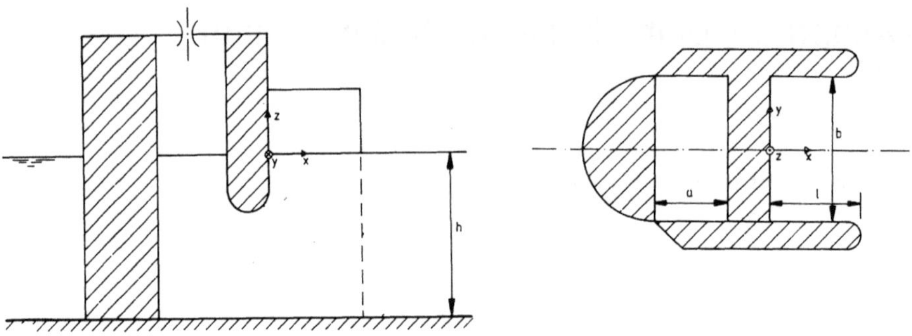

Fig. 1. Model.

The development of the harbour was triggered by the wish to keep the dimensions of the absorber small compared to the wave length. Thereby the absorber could be tuned to work as an antenna, and we would have a large absorption width ratio, which may be a cost-effective way of using the construction material. A difficulty with small absorbers is that the natural bandwidth tends to be narrow, and what is gained in increased absorption efficiency at the resonant frequency is more than lost because of the narrow bandwidth. Could there be a passive way of broadening the bandwidth without mechanical components and the wave prediction as needed for phase control? The harbour offers such an opportunity. To explain how the harbour works, we consider a simplified mathematical model of a wave energy absorber with a harbour. The absorber need not be an oscillating water column.

3. A simple mathematical model

In linear theory, the equation describing the movement of the absorber can be written

$$(Z + \Lambda)v = F_e \tag{1}$$

where v is the velocity of the absorber, Λ the load impedance, Z the radiation impedance, and F_e the excitation force. Details of the mathematical model are given in [3]. An alternative approach for the case of a semi-infinite harbour is described by Evans in [4] and [5].

The radiation impedance Z is conventionally separated in a real and an imaginary part

$$Z = i\omega A + B \tag{2}$$

where ωA is the reactance and B the radiation resistance.

If an oscillating water column is considered, Z includes the mass and stiffness terms due to "the column itself" in addition to the contributions from the harbour. If the harbour width is small compared to the wave length and the water can be considered shallow, the exact shape of the velocity distribution at the virtual boundary between the column and the harbour does not need to be matched. The total flux must, however, be conserved [3 and 5].

In analogy to pure mechanical systems we then write

$$\omega A = \omega M + \omega M_a - S/\omega \tag{3}$$

where M is the mass of the absorber (which may be the oscillating water column), M_a is the added mass defined so that ωM_a is the out-of-phase radiation force acting on the absorber at the (virtual) boundary to the harbour. S is the spring constant due to gravity.

Assuming that the applied load impedance is real so that $\Lambda = C$ (C real), the equation of motion can be written

$$(M + M_a)\ddot{z} + (C + B)\dot{z} + Sz = F_e \tag{4}$$

where $\dot{z} = v$ is the velocity of the absorber.

4. Optimal system design

It now remains to calculate M_a, B and F_e. In general these quantities can be computed numerically by a 3-dimensional boundary element method as applied for instance by Count [5]. For the problem of an absorber in a channel or in a reflecting wall a method of series expansions in vertical and horizontal eigenmodes is successfully applied by the authors [7, 8 and 9].

As we now intend to study separately the influence on the absorber of 1) the device itself, 2) the harbour and 3) the boundary conditions outside the harbour, we simplify the problem further by introducing a virtual rigid boundary at the harbour mouth. Again this is a good approximation if the harbour width is small compared to the wavelength. We note that in this approximation the maximum performance at complete impedance matching only depends on the radiational properties of the virtual boundary at the harbour mouth. The dynamics of the harbour and the absorber itself do not enter the problem until we start considering how impedance matching can practically be achieved.

Assuming a uniform water depth in the outer region we apply a 2-dimensional boundary element method to calculate the radiation impedance felt by the virtual rigid boundary at the harbour mouth. For each mode in the vertical expansion of the waves generated by the absorber the potentials and their derivatives are matched at the harbour mouth.

In figure 2 we show the calculated radiation resistance B of a single model-sized absorber for different harbour lengths ℓ. The details of the calculation are given in [3].

The radiation resistance is significantly increased by the introduction of the harbour as shown in figure 2. Note, however, that there will be a corresponding increase in the excitation force F_e so that the maximum absorption width W_{max} is not significantly changed. Local maxima in F_e and B occurs for

$$\ell + \ell' \simeq \frac{\lambda}{4}(1 + 2n) \; ; \quad n = 0,1,2... \tag{5}$$

and we recognize the well-known criterion for harbour resonance.

Resonance occurs when the total reactance is zero or from eq. 3

$$M + M_a - S/\omega^2 = 0 \tag{6}$$

The choice of parameters M and S in the first curve of figure 3 can be considered arbitrary. The second curve is obtained if the mass is just half as large and the spring constant is maintained.

We will find that for any constant S and M there is just a single resonant frequency in the extremal cases $\ell = 0$ (no harbour) and $\ell \to \infty$ (2-D case). The introduction of the harbour changes the added mass term. There is a moderate increase in the contribution from the evanescent modes, but the important change is caused by the contribution from the travelling wave mode M_o. In [3] it is shown that M_o undulates with the term $\mathrm{Im}\left\{(1 + r)/(1 - r)\right\}$ where r is the reflection coefficient at the harbour mouth. For small $|r|$ the leading term in M_o is $2\,|r|\sin\left\{2k\,(\ell + \ell')\right\}$.

M_o is zero when

$$(\ell + \ell')/\lambda = n/4 \;;\quad n = 0,1,2,\ldots \tag{7}$$

where ℓ' is the added length which depends on λ.

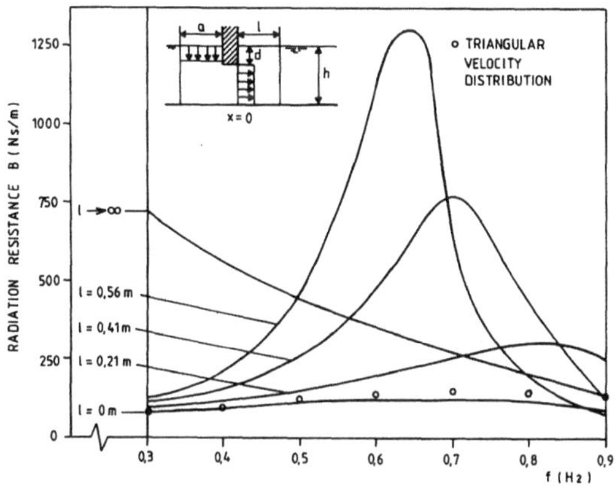

Fig. 2. The radiation resistance B of a single device for different harbour lengths. The resistance is related to the average chamber wave velocity for $a = 0.57$ m, $d = 0.25$ m and $h = 0.75$ m. The travelling wave mode component of the velocity distribution $V(z)$ at $x = 0$ is calculated under the assumption of $V(z) = 0$ for $-d < z < 0$ and the rectangular distribution $V(z) = V_o a/(h - d)$ for $-h \leq z \leq d$. The results marked by o are calculated by a triangular velocity distribution with the same total flux.

There will be a local maximum in M_o for $\lambda > 4(\ell + \ell')$ and a local minimum for $2(\ell + \ell') < \lambda < 4(\ell + \ell')$. For smaller λ corresponding to $n > 2$, the "undulation amplitude" of M_o which to the first approximation is proportional to the reflection coefficient $|r|$ becomes small for harbour widths of practical interest.

As a first estimate, the harbour length should therefore be chosen so that

$$\frac{1}{4}\lambda_{min} < \ell + \ell' < \frac{1}{3}\lambda_{min} \qquad (8)$$

where λ_{min} is the wavelength of the shortest waves from which energy is to be absorbed. We note that with a given frequency distribution of the wave energy flux, ℓ can be made smaller if the absorber is located in shallow water due to the shortening of the wavelengths at finite depths.

The two curves of $S/\omega^2 - M$ in figure 3 illustrate how ℓ, S and M can be chosen to optimize the performance of a wave energy absorber in a given wave spectrum.

For the first choice of $S/\omega^2 - M$ there will be just a single resonant frequency at f = 0.375 Hz while for the second choice resonance occurs at f = 0.44, 0.65 and 0.75 Hz, thus giving a much broader bandwidth.

By changing the geometry of the chamber we are thus able to match the system parameters S and M to the added mass term determined by ℓ and the reflection coefficient r at the harbour mouth. We can choose between a close match in a smaller frequency range, or an approximate match in a wider range of at least one octave.

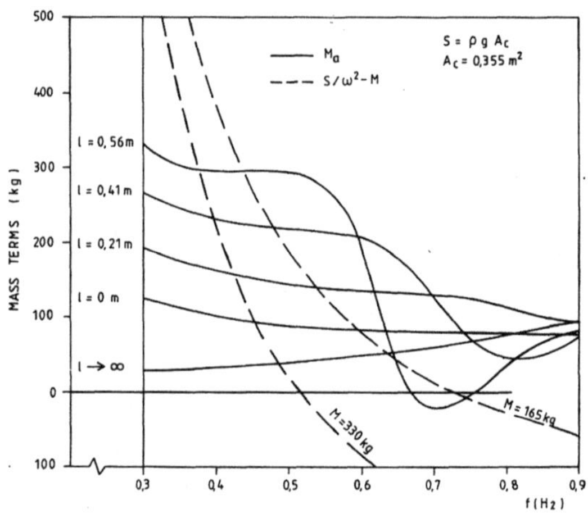

Fig. 3. Mass terms for different harbour lengths. The total reactance is $\omega A = \omega(M + M_a) - S/\omega$. Resonance occurs when the added mass M_a coincides with the spring/mass term $S/\omega^2 - M$ for some combination of harbour length ℓ and chamber geometry.

The important parameter regarding the energy absorption is not just the reactance term, but rather the phase angle characterized by the ratio $\omega A/(C + B)$. The ratio between the performance at optimal damping W'_{max}, and the performance at complete impedance matching W_{max} is given by the ratio $\beta = \omega A/B$ [3]. Figure 4 shows β for the no harbour case ($\ell = 0$), the 2-D case ($\ell \to \infty$) and for a finite harbour length ($\ell = 0.56$ m). The corresponding ratios W'_{max}/W_{max} are shown in figure 5.

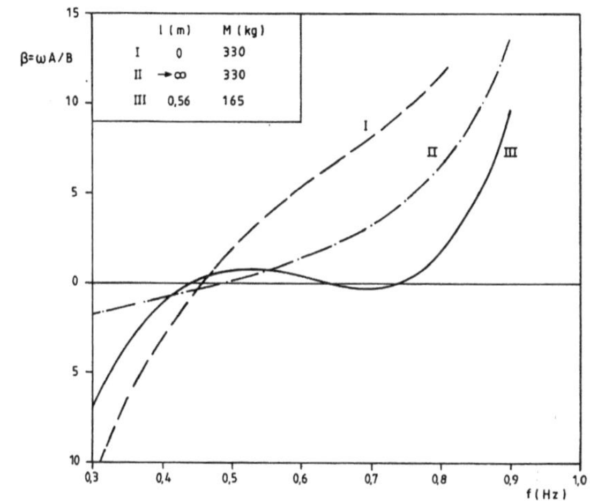

Fig. 4. The parameter $\beta = \omega A/B$ for different harbour lengths ℓ. Note that M is reduced for $\ell = 0.56$ m to maintain the lowest resonance in the same frequency range as for case I and II.

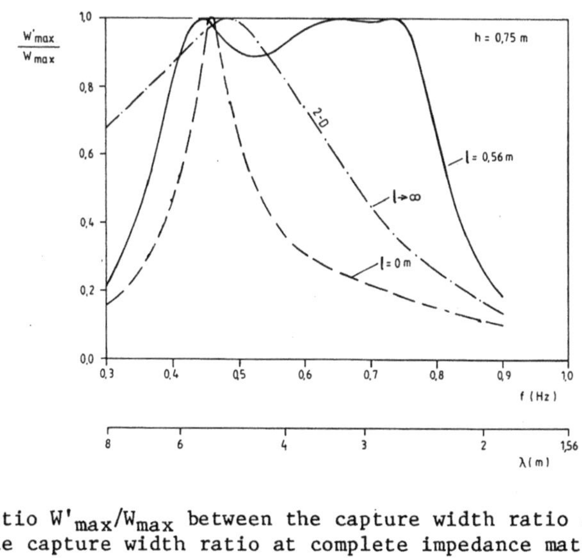

Fig. 5. The ratio W'_{max}/W_{max} between the capture width ratio at optimal damping and the capture width ratio at complete impedance matching.

There is a significant broadening of the frequency response of the absorber caused by the introduction of the harbour. This is not only because of the multiple resonances of the system thus obtained. The increase in the radiation resistance B relative to the reactance term ωA at the lowest harbour resonant frequency is also important in providing a better phase matching.

In the 2-D case ($\ell \to \infty$) we obtain a broader response than in the no harbour case because the radiation resistance is larger. The mass M and the spring constant S were chosen equal in the two situations.

In figure 6 we give the optimal load resistance C_o under the assumption of zero hydrodynamic losses and that the load impedance is real. In [3] C_o is derived as

$$C_o = B\sqrt{1+\beta^2} \tag{9}$$

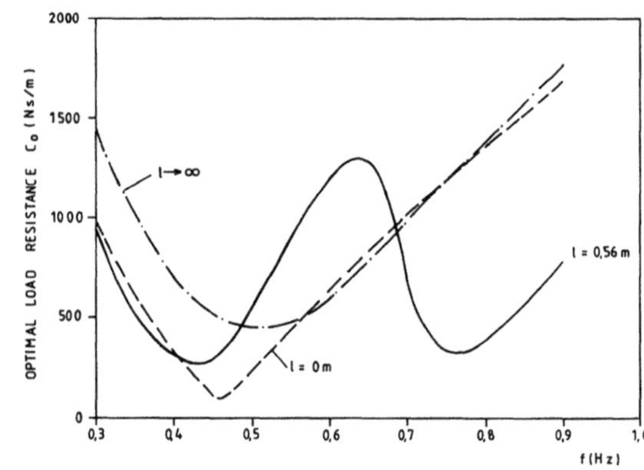

Fig. 6. The optimal load resistance C_o for different harbour lengths.

We observe a local minimum in C_o at the primary resonance (which we denote the column resonance) where $\beta = 0$ for all harbour lengths. For the finite harbour length, there is a local maximum at harbour resonance ($f \sim 0.65$ Hz) where B has a maximum, and a local minimum at $f \sim 0.76$ Hz where the rapidly decreasing B is balanced by the increase in β.

By choosing the secondary resonant frequencies so that $\beta = 0$ at harbour resonance, we obtain the minimal variation in C_o in the frequency range of interest for a given harbour length.

In an actual sea state, the load resistance C cannot be chosen optimal at every frequency simultanously. In figure 7 we give the relative performance for two fixed values of C compared to the performance when the damping is optimal at every frequency. Obviously there is an optimal choice of C for a given wave spectrum. The final optimization must, however, also include effects as the influence of hydrodynamic losses, amplitude restraints and the efficiency curve of the turbine and the generator.

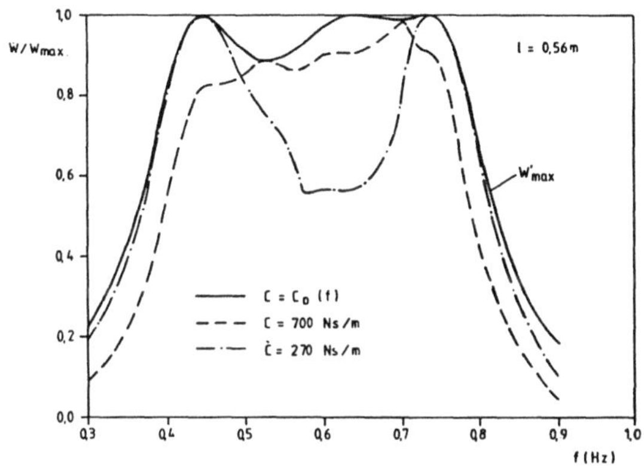

Fig. 7. The capture width ratio W at a fixed load resistance relative to the capture width ratio W_{max} at complete impedance matching. The capture width ratio at optimal damping W'_{max} is also shown. C = 700 Ns/m is about the mean value of the optimal load resistance in the actual range of frequencies, while C = 270 Ns/m is optimal at column resonance f = 0.44 Hz (see figure 2). Harbour resonance occurs at f = 0.64 Hz for this specific harbour length (ℓ = 0.56 m).

Figure 7 shows that the reduction in the absorbed power due to the non-optimal damping is not dramatic. The reduction in the range f = 0.6 - 0.7 Hz for C = 270 Ns/m is related to the harbour resonance. In this frequency range the damping is much lower than optimal ($\delta = C/C_o \simeq 0.2$), when the load resistance is optimized at column resonance.

5. Results from prototype tests

A prototype oscillating water column operated by Kværner Brug has now been at work at Toftestallen west of Bergen for approximately six months. Power takeout has been by means of an adjustable orifice. The production of electricity will, according to the plans, start in September 1985, when turbine and generator with a maximum capacity of 500 kW will be installed. Meanwhile the wave power input and the pneumatic power output has been recorded and absorption widths calculated. Some results are shown in figure 8.

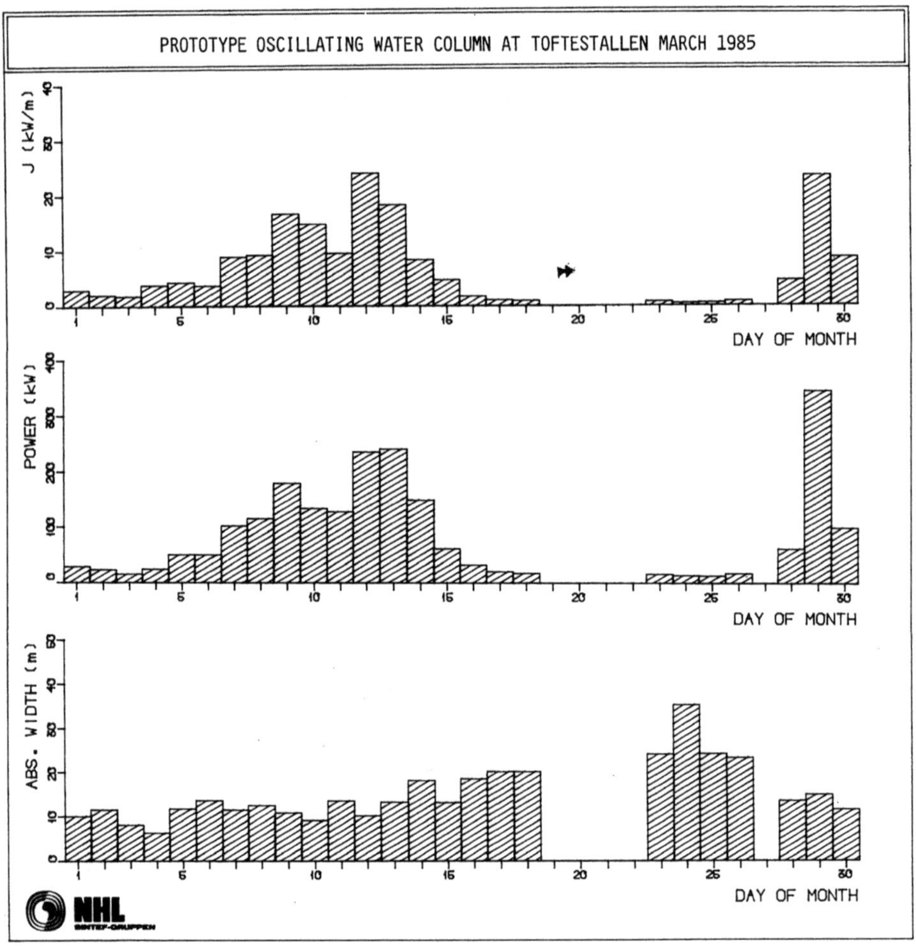

Fig. 8. Daily averages of incident wave energy, pneumatic power output and calculated absorption widths for the 10 m wide prototype oscillating water column at Toftestallen. The measurements and the analysis of results are performed by NHL in co-operation with the Oceanographic Company of Norway (Oceanor). The breaks in the power production are due to planned stops of the plant.

References

1. Electricité de France Service National: Improvements in or relating the Processes and Devices for Utilizing the Energy of Waves. Patent No. 9848153 April 10, 1953.

2. Ambli, N.; Bønke, K.; Malmo, O.; Reitan, A: The Kværner multiresonant OWC. Proc of the 2nd Int. Symp. on Wave Energy Utilization, Trondheim Norway (1982) 275-296.

3. Malmo, O.: A study of a multiresonant oscillating water column for wave-power absorption. Thesis, University of Trondheim, The Norwegian Institute of Technology, Department of Physics (1984).

4. Evans, D.V.: Wave-power absorption within a resonant harbour. Proc. of the 2nd Int. Symp. on Wave Energy Utilization, Trondheim, Norway (1982) 371-379.

5. Count, B.M.; Evans, D.V.: The influence of projecting side walls on the hydrodynamic performance of wave-energy deviced, J. Fluid Mech. 145 (1984) 361-376.

6. Count, B.M.: On the physics of absorbing energy from ocean waves. Ph.D. Thesis. Dept. of Physics, University of Exeter 1982.

7. Malmo, O.; Reitan, A.: Wave-power absorption by an oscillating water column in a channel, Accepted for publication in J. Fluid Mech.

8. Malmo, O.; Reitan, A.: Wave-power absorption by an oscillating water column in a reflecting wall. In preparation.

9. Malmo, O.; Reitan, A.: Wave-power absorption by a finite row of oscillating water columns in a reflecting wall. In preparation.

The Circular Sea Clam Wave Energy Converter

N.W. BELLAMY

Coventry Polytechnic and
Sea Energy Associates Limited
Coventry, England

Summary

The SEA Clam Wave Energy Generator has been developed to a prototype design suitable for supplying electricity to island communities and coastal communities. It consists of a 120 m long floating spine supporting six air bags breathing through self-rectifying Wells turbines into a common air duct. An alternative configuration of the SEA Clam has been model tested at Loch Ness and has achieved a dramatic improvement in efficiency. This paper describes the structural configuration of the Circular SEA Clam and explains the features which led to a breakthrough in performance. Predictions are made of the cost of power produced by a 60 m diameter full scale unit deployed in the North Atlantic.

Introduction

After eight years of research and development the current design of the SEA Clam wave energy device has reached a stage where further improvements would be difficult to achieve. The large 10 MW unit designed for a 2 GW scheme off the Hebrides was declared in 1982 to be the leading wave energy device arising out of the UK national wave energy programme and having the most potential for further development. Recent development has been towards smaller units aimed at the world wide market of supplying electrical power for small islands and coastal communities. Detailed design has now been completed for a 1 MW rated prototype for testing off the UK Atlantic coast. Unfortunately, with the current energy climate it is unlikely that a prototype SEA Clam will be built when the predicted cost of power generated is around 7½p/kWh.

Sea Energy Associates Limited and Coventry (Lanchester) Polytechnic have been involved in the UK national wave energy programme funded by the Department of Energy since its inception in 1975. Extensive design studies and sea trials have been carried out to evaluate structural designs, mooring configurations and power conversion systems which have resulted in an experienced team capable of pursuing all aspects of wave energy technology. By following the two basic themes of simplicity and overall cost effectiveness the SEA-Lanchester group have put together the most attractive concepts in wave energy to produce a successful device.

Although the design of the SEA Clam has changed little since 1981 a number of design variations have been explored and either rejected or shelved for further consideration in the future. The search has been for new ideas which would improve the productivity, reliability and cost-effectiveness of the total system without adding complexity or unproved technology. It appears that the present device configuration is a natural solution to the problem and that any further improvement in performance will come from the optimisation of the structural dimensions and the parameters of the power conversion system. However, there are alternative configurations of the SEA Clam which deserve investigation [1]. This paper deals with an alternative spine structure and power absorption system, which, for obvious reasons, is named the Circular SEA Clam.

The SEA Clam Wave Energy Generator

The large [2] and small [3] scale designs of the SEA Clam can be classified as spine-based pneumatic terminators. Devices utilising spines, that is long narrow structure, have been shown to be structurally efficient and have featured in the most cost effective devices developed in the UK. The use of air as a conversion fluid has many desirable properties including the ability to convert slow moving but large wave forces into high speed turbine driven electrical generation. Terminator is the name given to devices which face, or nearly face, the wave front and extract energy by terminating the wave in a matched load. Therefore, one can argue that a practical wave energy converter can be conceived as a floating spine terminator with a pneumatic wave absorbing face.

The design of the 1 MW SEA Clam, illustrated in Figure 1, utilises the displacement of air to extract energy from sea waves. Flexible air bags attached to the face of a floating spine breathe in response to wave forces. This causes air to be forced through self-rectifying turbines in and out of the hollow spine, allowing interchange of air between Clam bags. The randomness of sea wave patterns allows phased operation of the Clam elements enabling the spine to act as a stable reference body. For small scale wave energy deployment 1 MW units would feature six Clam elements on a 120 m long spine whereas larger 10 MW units, 290 m long, would be appropriate for 2 GW schemes.

Fig.1. Artists Impression of a 1 MW SEA Clam

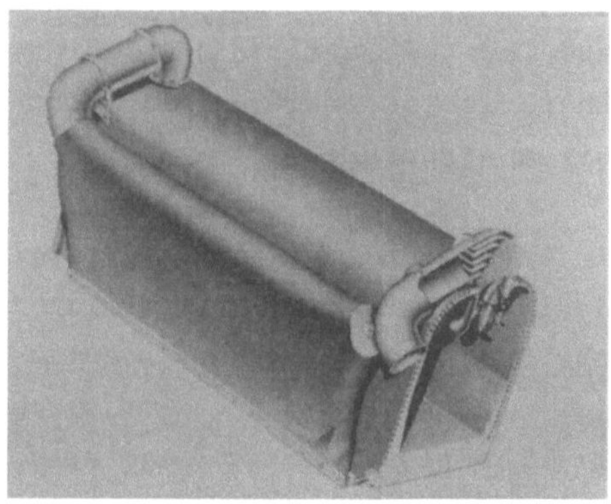

Fig.2. Cutaway View Showing Spine, Air Bag and Turbo-Generator

Figure 2 shows a cutaway section of the post-tensioned concrete spine with a flexible bag and its associated turbo-generator module. The air exchange between bag and spine is through ducting attached to one corner of the bag and the top of the spine. The turbo-generator module is contained in the top of the spine and can be sealed off by a butterfly valve in the ducting in the event of failure of the bag. The air turbines are of the self-rectifying "Wells" design with a single stage 6-bladed rotor directly coupled to an induction generator or alternator. The mooring is a self-aligning single point system with a V-yoke connection to the spine. Extensive testing of SEA-Clam models both in an indoor wide tank and in the natural waves of Loch Ness, Scotland has defined the structural behaviour and annual performance in the various sea states of possible wave energy sites. From this data, full scale designs have been produced, plant operation predicted and cost estimates made.

Limitations of the SEA Clam

Although the SEA Clam is a relatively inefficient wave energy device it does achieve a predicted low cost of power generated by virtue of its simple low cost structure and power conversion system. The overall annual efficiency of absorption per metre length of wave front for the earlier 10 MW design was around 30% and for the latter 1 MW design as low as 9% which shows that there is considerable room for improvement in performance of a given structure.

In the search for ways of improving the performance of the device, three fundamental limitations of the current design of the SEA Clam have been identified.

1. The flexible air bags used on the SEA Clam have a known high hydrogeometric spring rate which causes a severe mismatch of energy transfer from wave to air. A number of ideas have been explored to reduce or neutralise this spring rate but have had to be rejected because of engineering difficulties. Even if the bag spring rate problem was overcome, the resulting device would probably be hydrostatically unstable.
2. Experimental work has confirmed the predictable fact that the capture power of the device is roughly proportional to the total displacement of the air bags. This displacement is limited by the pitch restoring moment that can be provided by the spine which in turn is a function of the centres of gravity and buoyancy of the spine cross-section. Attempts to overcome this problem, such as attaching bags behind the spine, have not been effective.

3. The closed circuit air system of the SEA Clam has a finite air volume which impedes the common mode operation of the air bags. Any component of a wave which is parallel to the spine cannot be absorbed because the air has nowhere to go. The partial solution to the problem has been to align the spine at about 45° to the principal wave direction which gives a compromise between improving wave phase distribution and the loss of capture width. An alternative solution of providing a compensator to enable the air bags to effectively breathe to atmosphere has run into engineering problems.

These three fundamental limitations have impeded the further development of the SEA Clam. What is required is a new configuration of the device having efficient air bags of high capacity without incurring instability problems. The Circular SEA Clam appears to satisfy these requirements.

The Circular SEA Clam

By re-structuring the components, spine, air bag and turbo-generator modules, an alternative configuration of the SEA Clam can be evolved. The circular spine structure shown in Figure 3 is in the form of a rigid annulus 4 m in diameter, comprising of 12 identical sections. The separate sections of the model spine are held together by one compression joint and two tension links arranged in such a manner as to facilitate the measurement of wave-induced surge and heave bending in the ring. A continuous air duct around the inner top surface of the circular structure provides the common air connection between the 12 Clam elements. Each section has its own sealed buoyancy tanks for good sea-keeping and is fitted with membrane-type air bags (Figure 4) breathing through rotating vane air dampers (Figure 5) having the same characteristics as a Wells turbine. To activate the device the air system is pressurised to inflate the bags to a mean displacement in still water and during operation the air bags interchange air through the air dampers dissipating power in response to phased random waves.

Fig.3. Circular SEA Clam Model Ready for Launching

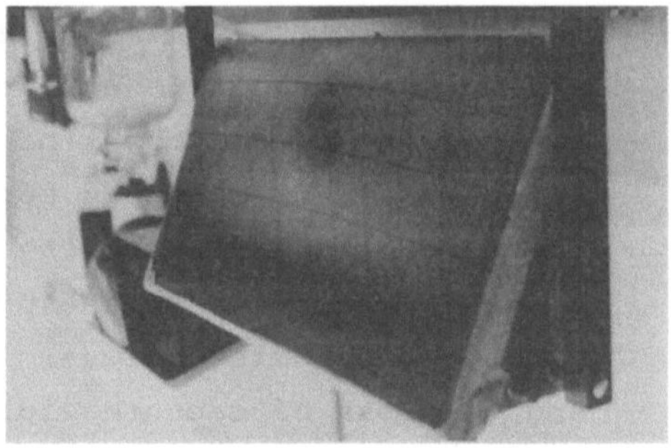

Fig.4. Profile of Membrane Bag

The key features of the Circular SEA Clam are:-

1. The circular frame of reference provides excellent stability due to its wave bridging properties in both the wave and crest directions. Pitch and roll motions are also minimised which helps to stabilise bag operation. A floating circular configuration has inherent stability which does not depend on the properties of section such as the centres of gravity and buoyancy.
2. The rigid annulus is well known for its inherent structural strength. In this application the heave, surge and torsional moments are roughly uniformly applied around the spine and hence the structural strength is efficiently used.
3. The stability of the structure can be enhanced by making its diameter about half a wavelength of the predominant waves. This has the effect of making the device a resonant absorber at the chosen wavelength and hence extending its bandwidth.
4. The modular structure of post-tensioned concrete sections or fabricated steel units is amenable to low cost production techniques and eases transportation problems.
5. The vertically inclined flexible membrane bag is considered to be the most ideal and efficient transformer of wave to air power discovered to date. Both experimental work and hydrodynamic analysis confirm this view. The properties of

an inclined bag are such that the controllable spring rate can be chosen to resonant with the hydrodynamic added mass to give a relatively low Q response. This results in a very efficient and nearly flat frequency response of the wave to air energy transfer provided the air turbine acts as a linear matched load. The Wells turbine running at constant speed is an ideal load for this system.

6. The circular spine with its stable hydrostatic characteristic can be designed to just compensate for the unstable buoyancy of the air system associated with low spring rate air bags. The resulting buoyancy forces on the nearly unstable combined bag-spine system are, therefore, reduced and hence the disturbance of the frame of reference by the buoyancy component of the waves is minimised. The corresponding effect of direct wave forces on the frame of reference is unclear at this time.
7. Inflation of the air bags on the pitch-stable circular spine raises the whole structure vertically rather than increasing the pitch angle as in the straight spine. Therefore, bag displacement is a function of water line area rather than restoring moment. This important difference in the bag inflation mechanism leads to a high ratio of air bag capacity to total device displacement which should improve performance and reduce costs.
8. The circular configuration of the device can extract energy from all components of a wave whatever angle they impinge on the device. Common mode waves cannot occur as with a straight spine and hence the air has always somewhere to go in spite of the closed circuit finite volume air system.
9. A separate 'power' air system from the 'buoyancy' air system is used which allows the structure to be partitioned for good sea keeping. Buoyancy air is the term used to describe the air which is only to keep the structure afloat and power air describes the air used to transmit power from bag to turbine.
10. It is essential for a wave energy device to be closed down irrespective of wave conditions for such reasons as operational requirements, maintenance, failure of the device

or transmission system, or the lack of consumer demand. Complete deflation of the air bags puts the Circular SEA Clam into a fail-safe mode where the flexible membrane bags are held inactive and under partial tension on the profiled spine surface. Inflation of the air system will restore power capacity progressively.

11. Being omni-directional, the energy capture is independent of device alignment and hence unaffected by currents and mooring resonances. The device is efficient in dealing with multi-directional seas.
12. Arguments within the wave energy community as to the relative merits of terminators, attenuators and point absorbers have continued for many years. This device can be classified as all three.

The provisional specification of a full scale Circular SEA Clam design for small scale applications in North Atlantic sea states would be:

Spine diameter	60 m
Spine depth	7 m
No. of sections and air bags	12
No. of turbines	8
Displacement	4000 tonnes
Capture air power (annual average)	1200 kW
Turbo-generator efficiency	65%
Transmission efficiency	95%
Availability	80%
Power landed (annual average)	593 kW
Annual energy output	5.2 GWh

Performance of the Circular SEA Clam

The capture efficiency of the Circular SEA Clam has been determined experimentally by testing a fully instrumented 1/15th scale model in Loch Ness (Figure 6). The model was fabricated from steel sheet and fitted with corded latex membrane air bags. Each section was numbered clockwise from 1 to 12 with the single point mooring attached between 12 and 1 o'clock.

Measurements were made of the pressure drop across each calibrated air damper during 6 minute test periods in order to compute the air power dissipated for each sea state.

Fig.5. Rotating Vane Air Dampers

Fig.6. Circular SEA Clam Tests on Loch Ness

Structurally bending moments and mooring forces were also recorded, but these will be reported in a later paper. Sea states were computed from the output of capacitance type wave gauge mounted on a mast near to the test area.

Two parameters were varied during the tests in order to seek optimum performance. First, the volume of air in the system was varied which, in effect, varied the mean operating profile of the air bags and hence their mean spring rate. Secondly, the speed of the rotating vane dampers was varied between 4000 and 10000 rev/min to give damping factors of 3636 to 9090 Nsm^{-5} for the linear pressure-flow characteristics. The model was tested in a range of sea states up to a significant wave height, H_S, of 0.5 m and energy periods, T_E, of 3 s. All results were processed by on-line computer and scaled to full scale.

The results of efficiency against energy period for the Circular SEA Clam model are plotted in Figure 7. The curve through the experimental points is the mean result of the 6 minute records taken with various air volumes and damping factors and represents an improvement by a factor of 4 over the efficiency of the straight 120 m long SEA Clam. When displacement of the devices is taken into account, this improvement in efficiency is equivalent to an improvement in power to weight ratio of 3 which

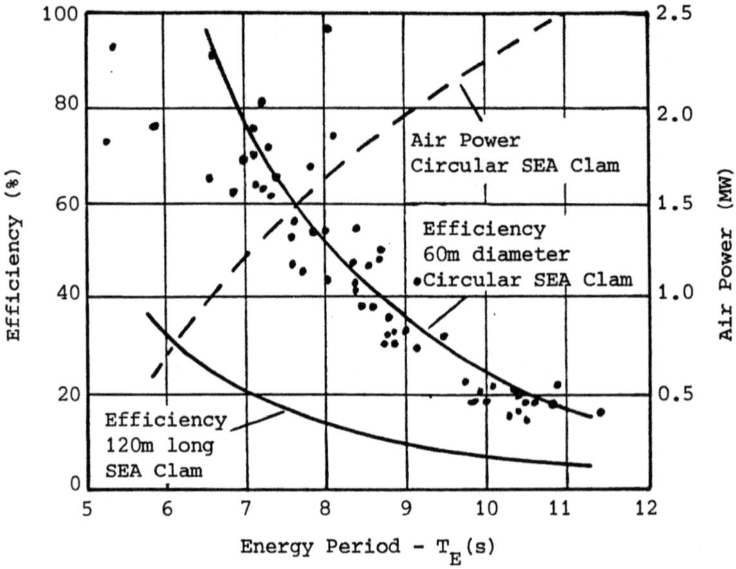

Fig.7. Scaled Air Power Results for 60 m Diameter Circular SEA Clam

should be reflected in the cost of power produced. Also shown on the graph is the air power curve for the Circular SEA Clam which has been derived from the efficiency curve.

Figure 8 shows samples of the distribution of air power dissipated by each damper numbered around the clock. As would be expected in a locally wind generated uni-directional sea like Loch Ness the distribution of power is reasonably symmetrical about the mooring line which is a check on the validity of the results. The major proportion of the total air power, P_T, is generated in the 8 front air bags and suggests the power captured by the 4 rear bags is not worth having.

Conclusions

As a result of investigating the hydrodynamic and structural limitations of the SEA Clam, a new device configuration has been tested which has produced a remarkable improvement in efficiency. After a series of model tests of the Circular SEA

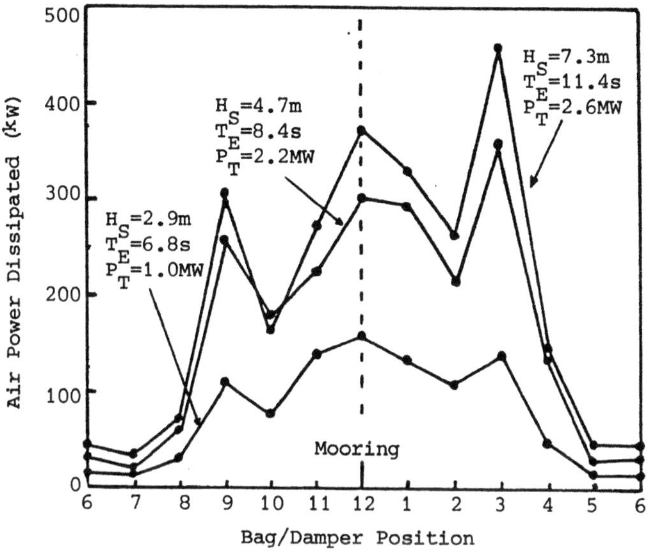

Fig.8. Scaled Air Power Distribution Around Circular SEA Clam

Clam in Loch Ness an improvement in productivity by a factor of 3 has been confirmed for the same structural displacement. This coupled with an improvement in the engineering features of the device is predicted to give a significant reduction in the cost of power generated from wave energy.

References

1. Bellamy, N.W., "An Alternative Design of the SEA Clam Wave Energy Converter", Proc. of the ASME Fourth International Offshore Mechanics and Arctic Engineering Symposium, Vol 1, Dallas, February 1985.

2. Bellamy, N.W., "The SEA Clam Wave Energy Converter", Proc. of the ASME Third International Offshore Mechanics and Arctic Engineering Symposium, Vol 11, New Orleans, February 1984.

3. Bellamy, N.W., "Further Development of the SEA Clam Wave Energy Converter", the IEE Fourth International Conference on Energy Options, London, April 1984.

Optimization of Wave Energy Absorber of an Attenuator OWC Device with an Air Turbine

H.Maeda*, T.Kinoshita*, K.Kudo** and K.Masuda***

*Professor, Institutue of Industrial Science, University of Tokyo, Minatoku,Tokyo, Japan; **Research Scientist, Japan Marine Science and Technology Center, Natsushima, Yokosukashi, Japan; ***Lecturer, College of Science and Technology, Nihon University, Narashinodai, Funabashishi, Japan

Abstract

The purpose of this paper is to establish the optimum design method of a combined system of a primary energy converter with a secondary energy converter of an attenuator OWC (Oscillating Water Column) device as a wave energy extractor which floats in waves, taking account of bottom plate and phase control.

The precise estimation of the characteristics of a primary energy converter depends only on the accuracy of the prediction of the hydrodynamics. Then, in order to get the accurate prediction of the hytdrodynamics, the authors carry out the experiments with regard to the hydrodynamics on each oscillating water column and on the whole device, and the authors develop the precise calculation method and the simple calculation method which correspond to the experimental results. Then the authors compose the mathematical model which represents the characteristics of the primary energy converter by the use of the above mentioned prediction method of the hydrodynamics. This mathematical model is applied to optimization of the system.

The conbination of the Wells turbine and generator is considered as the secondary energy converter. The published data of power charcteristics of a secondary energy converter is used in this paper. The authors analyze the characteristics of absorbed wave power of the conbined system of these kind of a secondary energy converter and a primary energy converter. Then the authors propose the optimum design method of the combined system of a primary energy converter with a secondary energy converter.

1. Introduction

In Japan, sea trials of "KAIMEI" have been and will be carried out as a national project, while sea trials of a private wave energy project named "KAIYO" have been done last summer. This paper corresponds to sea trials of "KAIMEI".

It is not easy to choose the better of OWC(Oscillating Water Column) devices and movable floating body type devices such as cam and raft, and the better of attenuator and terminator type, because the choice depends on sea conditions of a site, the field of wave energy utilization, evaluation basis of economics etc. But from the stand point of flux and reflux of tide, the maintenence, mobility of devices in various sites, practical capacity of mooring systems and so on, the authors think a floating attenuator type OWC device such as "KAIMEI" is one of the best promising ones.

In this paper the purpose is to develope mainly the analytical method for the optimization of wave energy absorbing systems which consist of a combination of a primary and secondary energy converter. The authors do not insist that the model adopted in this paper is the best one. According to the analytical method mentioned above, to design the best form and best system of wave energy absorbing device is the subject for the next study.

In order to understand the fundamental characteristics of optimization, a model of a device and surrounding conditions should be simplified as much as possible. Therefore the following assumptions are introduced: Linear terms are dominant for the prediction of behavior of a device in operation conditions except survival conditions. Then, since the behavior in irregular waves can be represented by linear superposition of behavior in regular waves, and since the authors have already developed the predicton method of absorbed wave power in irregular waves,[1)] incident waves are assumed to be regular sinusoidal waves. OWC can be replaced with the corresponding equivalent floating body.[2)] Compressibility of air in air chambers,[3)] however, are taken into accout, because the compressibility of air can not be ignored in large size of air chambers such as those of "KAIMEI". And also the effect of phase control of OWC is introduced.

2. Primary Energy Converter

The model shown in Figure 1 is a catamaran type floating body which has five air chambers.[4)] The scale ratio is considered as 1/50. Each air chamber has a circular hole (orifice) on the ceiling which simulates the load damping of secondary energy converter as an air turbine and generator. Dynamic pressure in each air chamber was measured by semi conductor type pressure gauge. Capacitance type wave gauge was set up just under the circular disk in order to measure the relative height of free surface inside the air chamber. Center of gravity is 0.116m above from the keel for model without a bottom plate and 0.083m for model with a bottom plate, and longitudinal moment of inertia M_3 is 0.3228kg·m·sec^2 for model without a bottom plate and 0.4225kg m·sec^2 for model with a bottom plate, respectively. The orifice gives load damping d_j to each OWC which is about 2.5 kg·m·sec^2.

In these tests, area ratio of orifice r (ratio of the area of an orifice to the area of the free surface of an air chamber) varies from 0 (air chamber closed completely-no hole on the lid), 1/100 to 1/4.5 (the lid or ceiling is removed). These numbers correspond to certain amount of load damping force. The wave height of incident waves is 4cm for whole tests, and the non dimensional frequency of transient water waves changes in the region of KL=1.5$\sim$15, where K is wave number and L is the length of the model. Waves progress in the longitudinal direction of the model. Only surge, heave and pitch mode motions are measured. Soft coil spring is attached to the model in surge direction in order to prevent drifting.

In both case of free and fixed floating model of catamaran hull, absorbing wave energy tests are carried out. Absorbed wave energy is defined as work done by free surface of air chamber. Absorbed wave energy coefficient is defined as abosorbed wave energy nondimensionalized by wave energy of incident waves with the width of the model length L . In order to obtain radiation forces, transient forced oscillation test in heave mode is carried out.

Right handed Cartesian coordinate systems are adopted. The origin is on the undisturbed free surface , the x-axis in the longitudinal direction and the z-axis positive downwards. The equation of motion, in which the equivalent floating body approximation is assumed, taking into account the impedance Z_{ij} written in matrix form, is shown as follows.

$$[Z_{ij}]\{\dot{\zeta}_j\}=\{E_i\} \quad ,(\ i,j = 1\sim 8\ ) \tag{1}$$

$$
\begin{aligned}
Z_{11} &= i\omega(M_1+m_{11})+N_{11} \\
Z_{1j} &= Z_{j1} = 0 && ,(\ j = 2\sim 8\ ) \\
Z_{22} &= i\omega(M_2+m_{22})+(N_{22}+\sum_{j=4}^{8} d_j)+\frac{1}{i\omega}(C_{22}+\sum_{j=4}^{8} a_j) \\
Z_{23} &= Z_{32} = -\sum_{j=4}^{8} d_j l_j-\frac{1}{i\omega}\sum_{j=4}^{8} a_j l_j \\
Z_{2j} &= Z_{j2} = i\omega m_{2j}+(N_{2j}-d_j)-\frac{1}{i\omega}a_j && ,(\ j = 4\sim 8\ ) \\
Z_{33} &= i\omega(M_3+m_{33})+(N_{33}+\sum_{j=4}^{8} d_j l_j^2)+\frac{1}{i\omega}(C_{33}+\sum_{j=4}^{8} a_j l_j^2) \\
Z_{3j} &= Z_{j3} = i\omega m_{3j}+(N_{3j}+d_j l_j)+\frac{1}{i\omega}a_j l_j && ,(\ j = 4\sim 8\ ) \\
Z_{jj} &= i\omega(M_j+m_{jj})+(N_{jj}+d_j)+\frac{1}{i\omega}(C_{jj}+a_j) && ,(\ j = 4\sim 8\ ) \\
Z_{ij} &= Z_{ij} = i\omega m_{ij}+N_{ij} && ,(\ i,j = 4\sim 8\ ,\ i \neq j\ )
\end{aligned}
\tag{2}
$$

where the subscript (i=1,2,3,4,····,8) corresponds to surge, heave, pitch, heave of No.1 OWC,······, heave of No.5 OWC, respectively. And ζ_j is displacement; E_i is wave excitation; M_i is generalized mass; m_{ij} is generalized added mass; N_{ij} is generalized wave damping coefficient; a_j and d_j are spring constant coefficient and load damping of orifice or secondary energy converter; C_{jj} is restoring force coefficient; l_j is distance between OWC and origin. Absorbed wave energy is calculated by the equation (19) in section 4.

The water depth is assumed to be deep. 2-D hydrodynamic radiation forces which are used in numerical calculation of strip method[5] are calculated by the hybrid FEM method. 3-D hydrodynamic radiation forces are calculated by the source distribution method.[6)7)] Wave exciting forces are derived by Haskind relations.

The summation of each absorbed wave energy coefficient from the first to fifth air chamber as the total absorbed energy coefficients is shown in Figure 2 in case of an attenuator type device with a bottom plate and without a bottom plate.

In case of attaching the bottom plate, the peak of absorbed wave energy coefficients shift to low frequency range and the coefficients are improved remarkably.

Wave making damping does not increase, while added mass coefficient increase when a bottom plate is attached to an attenuator type device, by comparing with the results of the no bottom plate device(Figure 3).

Wave exciting forces for pitch mode of the model with a bottom plate are remarkably reduced comparing with the results of the model without a bottom plate(Figure 4).

But motions of the external floating body with and without bottom plate are comparable with each other (Figure 5). The effect of natural frequency of catamaran hull with bottom plate and the difference between exciting force with and without bottom plate compensate for each other with regard to the motion. Then with regard to absorbed wave energy, the effect of a bottom plate may be appeared on optimum phase lag between the motion of OWC and a catamaran hull.

Absorbed wave energy coefficients of each OWC are shown in Figure 6 in case of the parameter of position of the air chamber. In this figure, calculated results by strip method cannot explain these experimental results enough .

The strip method is applicable to the motion prediction of the external floating body and is applicable to the total absorbed wave energy coefficient prediction. But it does not work well in case of the absorbed wave energy coefficient prediction of each OWC, while 3-D method is still held. In the following section, 3-D mathematical model is adopted.

3. Secondary Energy Converter

Let us consider that each OWC has an air chamber of 3.8m height, a combination of Wells turbines and a generator and that the generator is designed to have a torque characteristics which attain 1,000 rpm of the turbine . Principal particulars of the Wells turbines are determined by optimization. In this paper, optimization is carried out only for the primary energy conversion, but it would be better and it would be possible to optimize total energy conversion. Number of turbines instead of all principal particulars is determined, and the Wells turbines used at sea trial at Sanze by JAMSTEC[8)] is adopted as a standard, which are two and tandem positioned with an axis of a generator. Their blade section is NACA 0021, number of blade is 16, tip radius r_T is 0.67m and air flow area of each turbine A_T is 0.625 m. The turbine is made of stainless steel and its moment of inertia M_T is 7.926 $kg \cdot m \cdot sec^2$.

3.1 Equation of Motion of the Turbine.

Equations of motion of the turbine is as follows,

$$M_T \frac{dw_T(t)}{dt} = Q_0(t), \tag{3}$$

$$Q_0(t) = C_t(\alpha)\rho_a A_T r_T\{(r_T w_T(t))^2 + v^2(t)\}/2 - Q_L(t), \tag{4}$$

where $W_T(t)$, $V(t)$, $Q_0(t)$, ρ_a, $C_t(\alpha)$ and $Q_L(t)$ are an angular velocity of the turbine, axial velocity of air flow, torque of turbine axis, air density, torque coefficient, and load and loss torque of the turbine respectively. Relative flow angle at turbine tip α is given by

$$\alpha = \tan^{-1}\{(|v(t)|/(r_T w_T(t))\}, \tag{5}$$

Reaction force of the turbine to the oscillating water column $F_T(t)$ is given by

$$F_T(t) \equiv A_W\{p(t) - p_0\},$$
$$= A_W \rho_a \psi(\alpha)\{(r_T w_T(t))^2 + v^2(t)\} \frac{1}{2} \{v(t)/|v(t)|\}, \tag{6}$$

using pressure difference cofficient $\psi(\alpha)$ of the turbine, where A_W, $p(t)$ and p_0 are water plane area, pressure of the air chamber and atmospheric pressure, respectively.

The torque and pressure difference coefficients are shown in Figure 7, which are experimental results of a small model, given by Arakawa et al.[9] When the air can be assumed as adiabatic, energy conservation law gives the following equation,

$$p(t)\frac{d}{dt}\{V_a - A_W\zeta(t)\} + c_v\frac{d}{dt}[\{V_a - A_W\zeta(t)\}p(t)/R]$$
$$+ 2v(t)A_T p_e c_p/R = 0, \tag{7}$$

where c_p and c_v are specific heat at constant pressure and volume respectively and R is a gas constant, and $c_p/c_v = 1.4$, $c_p/R=3.5$ and $c_v/R=2.5$ for air. V_a and ζ_j are an initial volume of the chamber and heaving displacement of OWC, respectively. And

$$p_e = \begin{cases} p(t) & \text{for } v(t)>0, \\ p_0 & \text{for } v(t)<0. \end{cases} \tag{8}$$

Equations (5),(6) and (7) with a condition of $w_T(t)$ can be solved for p and v simultaneously. The load and loss torque $Q_L(t)$ is adjusted to 1,000 rpm of turbine.

3.2 Frequency Domain Calculations Linearized for Steady High Velocity Revolution. (Impedance of Secondary Energy Converter without Phase Control)

Since it is possible to linearize the equation (6) for steady high velocity revolution as

$$F_T(t) \equiv A_W\{p(t) - p_0\} \fallingdotseq \frac{1}{2} A_W \rho_a \psi_0 r_T w_{T_0} v(t), \tag{9}$$

where W_{T0} is the steady angular velocity of the turbine ψ_0 and is given by Figure 7 as about 6.03, the steady state can be estimated by linear frequency domain calculations.

At first we define d_T and a_T as

$$d_T + \frac{a_T}{i\omega} = F_T \,/\, \dot{\zeta}, \tag{10}$$

$$\begin{aligned} F_T &= \mathrm{Re}(F_T) \quad , & \dot{\zeta} &= \mathrm{Re}(\dot{\zeta}) \quad , \\ &= \mathrm{Re}\{F_{T_0} e^{i(\omega t + \theta)}\}, & &= \mathrm{Re}\{\dot{\zeta}_0 e^{i\omega t}\}, \end{aligned} \tag{11}$$

and further we put as follows,

$$v(t) = \mathrm{Re}\{v_0 e^{i(\omega t + \theta)}\}, \; p(t) - p_0 = \mathrm{Re}\{p_a e^{i(\omega t + \theta)}\}, \tag{12}$$

since pressure p(t) has the same phase as $F_T(t)$ and v(t) from equation (9).

Linearizing equation (7), we obtain

$$-A_w p_0 \dot{\zeta}_0 e^{i\omega t} + \frac{c_v}{R}\{i\omega V_a p_a e^{i(\omega t+\theta)} - A_w p_0 \dot{\zeta}_0 e^{i\omega t}\}$$

$$+ \frac{2c_p}{R} A_T p_0 v_0 e^{i(\omega t+\theta)} = 0 \quad . \tag{13}$$

$$\therefore \frac{2A_T v_0}{A_w \dot{\zeta}_0} e^{i\theta} = \frac{1}{1+i\Lambda} \quad . \tag{14}$$

where

$$\Lambda = \frac{\omega \rho_a \psi_0 r_T w_{T_0} V_a}{4A_T p_0} \cdot \frac{1}{\gamma} \quad , \quad \gamma = \frac{c_p}{c_v} \; , \tag{15}$$

$$\therefore d_T = \frac{A_w{}^2 p_0}{\omega V_a} \cdot \frac{\gamma\Lambda}{1+\Lambda^2} \tag{16}$$

$$a_t = \frac{A_w{}^2 p_0}{V_a} \cdot \frac{\gamma\Lambda^2}{1+\Lambda^2} = \omega\Lambda d_T \tag{17}$$

Here Λ can be named as an air compressibility factor of the air chamber. Figure 8 shows results for wave frequency $\omega = 0.6 \sim 1.0$ rad/sec and number of the set of the turbines $n = 1 \sim 4$.

3.3 Impedance of Secondary Energy Converter with Phase Control

In order to evaluate the effect of phase control on the impedance of secondary energy converter, a motion of the OWC must be considered in time domain, as follows

$$(M_i + m_{i\infty})\ddot{\zeta}_i(t) + \int_0^t K_{22}(t-\tau)\dot{\zeta}_i(\tau)d\tau + \rho_w g A_w \zeta_i(t)$$
$$= E_i(t) - A_w\{p(t) - p_0\}, \qquad (18)$$

where M_i, $m_{i\infty}$, ρ_w and g are mass of water column, added mass for an infinite wave frequency, water density and gravitational acceleration, respectively.

The kernel function $K_{22}(t)$ and heaving wave excitation $E_i(t)$ are obtained from linear water wave theory. Here 3-D effect, effect of the motion of the vessel and interaction between each OWC are neglected because otherwise we need too much computations for optimization. Equations (3) ~ (8) and (18) can be integrated simaltaniously. Details of this time domain simulation is given in Kinoshita et al.[10] One of the results of the simulation for regular incident waves of 1.5m wave height and ω = 0.7 rad/sec is shown in Figure 9. OWC moves even when a valve is closed due to a compressibility of air in the chamber. If the chamber height is changed to smaller, OWC moves a little bit as shown in Figure 9. The impedence is obtained by using of Fourier coefficients of the velocity of the displeamant of OWC, $\dot{\zeta}$ and the reaction force F_T. Results for 2.0m wave height are shown in Figure 8. Effect of wave height on the impedance is not serious in the range of 1.5m ~ 2.5m.

4. Optimization

By utilizing the impedance obtained in the previous section, we can estimate the optimum load condition for the attenuator in the frequency domain.

The attenuator considered here is 50 times of the model without a bottom plate, that is, particular dimensions are the length 78 m, the breadth 17.5 m (OWC breadth 7.5 m), and the draft 2.5 m. The number of the OWC's is five.

The equation of relative motions of OWC can be derived from equation (1). Then the absorbed power becomes

$$P = \frac{1}{2}\omega^2 \sum_{j=4}^{8} d_j |\zeta_j|^2 \qquad (19)$$

where we have to note that ζ_j is the complex relative amplitudes of the OWC's. The optimization is carried out as the maximization of P by changing the parameter d_j.[11]

Figure 10 shows the maximum absorbed wave energy coefficient. In case of the freely floating attenuator in longer waves, the absorbed wave energy becomes very low compared with that of the fixed condition. Moreover the spring constant caused by the compressibility of the air between the turbine and the OWC surface reduces significantly the absorbed wave energy. In these calculations, the external loads of the OWC are independently varied and optimally chosen. But the absorbed wave energy becomes same indistinguishably with the results of the uniform load distribution for every OWC.[12]

Figure 11 shows the optimum uniform load $d_T/\rho_w\omega V_w$, where V_w is the volume of the OWC. In this figure, three load dampings of secondary energy converter are also drawn. The load damping of the single turbine set (n=1) is found to be much larger in the moderate wave range. As the number of the turbine sets increases (n=2 or 3), the load damping tends to reduce and match well with the primary wave energy absorber.

The wave energy absorbing performances of attenuator with turbines are shown in Figure 12. Since the load of the single turbine set is in excess, it absorbs only the small part of wave energy. But two or three turbine sets for each air chamber can produce nearly optimum load and then improve the performance.

Finally assuming that we operate the phase control on the Wells turbines, the absorbed wave energy curves change to the dashed lines in this figure. In case of the single turbine set under phase control, the load damping increases while the spring constant does not change so much as shown in Figure 8. As the result, the performance reduces further in this case. With many turbines, on the contrary, the secondary system produces negative spring constant by the phase control and it can absorb more wave energy in the relatively longer wave range.

As shown in this paper, optimum secondary energy converter for a primary energy converter is obtained . Although number of turbine sets is determined , this result can be used for feedback to the design of turbine itself , namely torque and pressure difference coefficients and a size of turbines .

5. Conclusion

The characteristics of mathematical models of both 3-D and strip method for a floating type attenuator OWC as a primary energy converter device were made clear by comparing with the corresponding experiments and the remarkable effect of bottom plate under air chamber were shown.

Impedances such as equivalent load damping and spring constant of a secondary energy converter which consisits of Wells turbines and generator were derived from time domain simulation, taking account of the air compressibility and phase control.

Finally the characteristics of optimized system of above mentioned primary and secondary energy converter were revealed.

According to these analytical methods, to design the best system of a primary and secondary energy converter is the subject for the next study.

The authors wish to thank Messrs. H. Hayashi. and S. Miyajima for their cooperation in this investigation. This work was carried out partly with the support of the Grant-in Aid for Scientific Research of the Ministry of Education, Science and Culture , Japan.

REFERENCES

1) Maeda, H., Kinoshita, T., Kato, S. and Suzuki, F: Fundamentl Research on Absorbing Energy from Ocean Waves (3rd report) , J. of Soc. of Naval Arch. of Japan, Vol.150 (1981)

2) Maeda, H., Kinoshita, T., Masuda, K., Hayashi, H. and Kato, W. : Fundamental Research on Absorbing Energy from Ocean Waves (5th report) , J. of Soc. of Naval Atch. of Japan, Vol.152 (1983)

3) Maeda, H., Kinoshita, T., Masuda, K. and Kato, W.: Fundamental Research on Oscillating Water Column Wave Power Absorbers, 3rd International Symposium on Offshore Mechanics and Arctic Engineering, ASME (1984)

4) Maeda, H., Masuda, K. and Hayashi, H.: Fundamental Research on Attenuator Type OWC Wave Energy Absorber, ECOR International Conference (1984)

5) Kinoshita, T., Saijo, K. and Yokomizo, H. : Theoretical Analysis of Performance of Kaimei for Wave Power Absorber, J. of Kansai Soc. of Naval Arch. of Japan, Vol.187 (1982)

6) Kudo, K., Nakato, M. and Miyazaki, A.: Optimum Design of External Load Condition in the KAIMEI Type Wave Power Absorber, J. of Soc. of Naval Arch. of Japan, Vol.153 (1983)

7) Count , B.M. and Miyazaki, T.: Study on Floating Attenuator Wave Energy Devices, J. of Soc. of Naval Arch. of Japan, Vol.155 (1984)

8) Ishii, S., Masuda, Y., Miyazaki, T., Kudo, K., Hotta, H. and Washio, Y.: The Report on Operation and Sea Trial of the Shore Fixed Type Air Turbine Wave Power Generator, Tech. Reports of JAMSTEC, No.14, 1985.

9) Arakawa, C., Suzuki, M. and Kinoshita, T.: On the Fixed Type OWC Device with Wells Turbine, (1st report, Initial design method) , Proc. of Meeting of Japan Society of Mech. Eng., No.832-12, 1983.

10) Kinoshita, T., Masuda, K., Miyajima, S. and Kato, W.: Research of the System Simulation for a Fixed OWC Type Wave Energy Absorber, International Symposium on Ocean Space Utilization (Ocean Space '85), Tokyo, 1984.

11) Count,B.M.:On the Physic of Absorbing Energy from Ocean Waves, PhD Thesis (1982)

12) Kudo, K.:Optimum Design of Kaimei Type Wave Power Absorber, J. of Soc. of Naval Arch. of Japan, Vol.156 (1984)

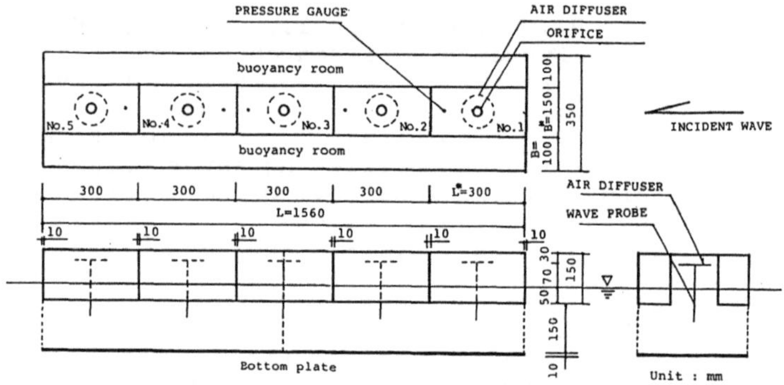

Fig.1. Model with bottom plate

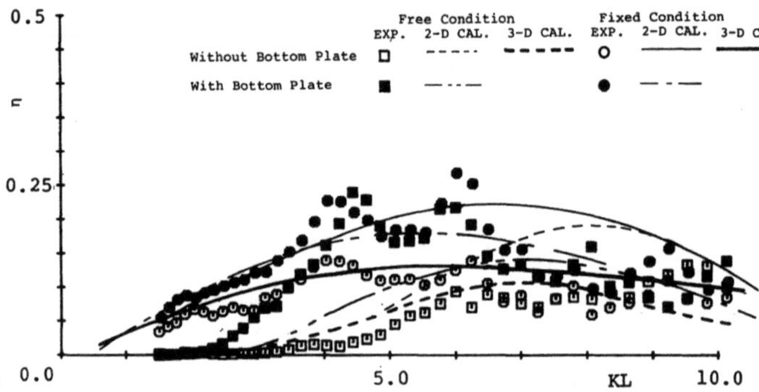

Fig.2. Absorbed wave energy coefficient in fixed and free condition of model with/without bottom plate (r=1/100)

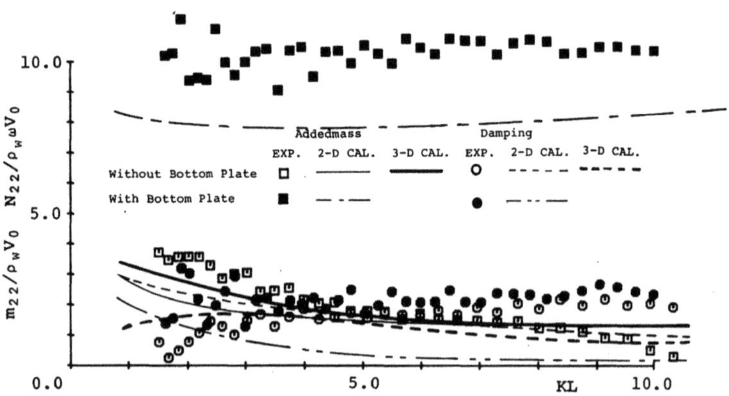

Fig.3. Added mass and damping coefficient of model with/without bottom plate (r=1/4.5)

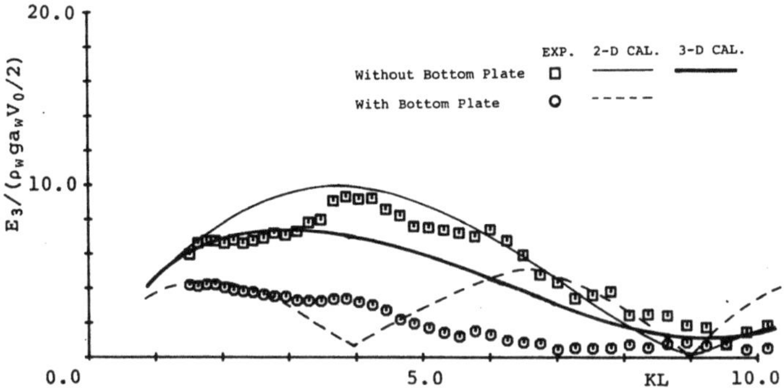

Fig.4. Wave exciting force of pitch mode of model with/without bottom plate (r=1/4.5)

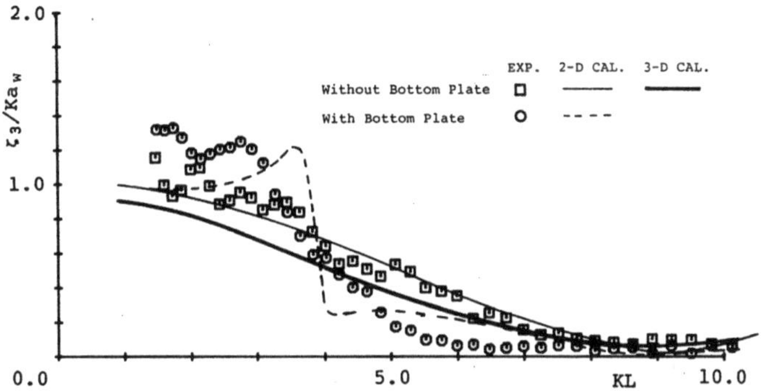

Fig.5. Pitch motion of model with/without bottom plate (r=1/4.5)

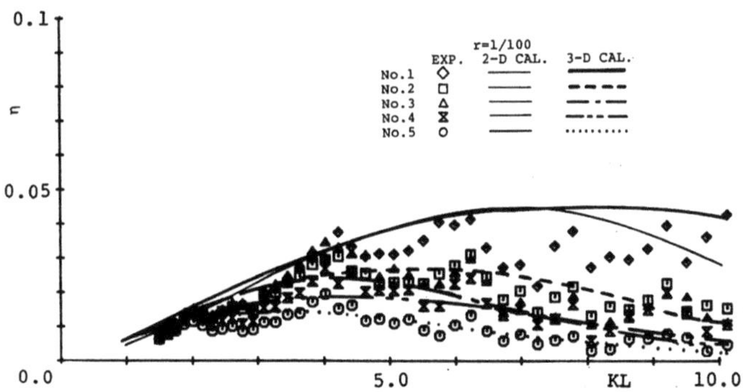

Fig.6. Absorbed wave energy coefficient of OWC in fixed condition of model without bottom plate (r=1/100)

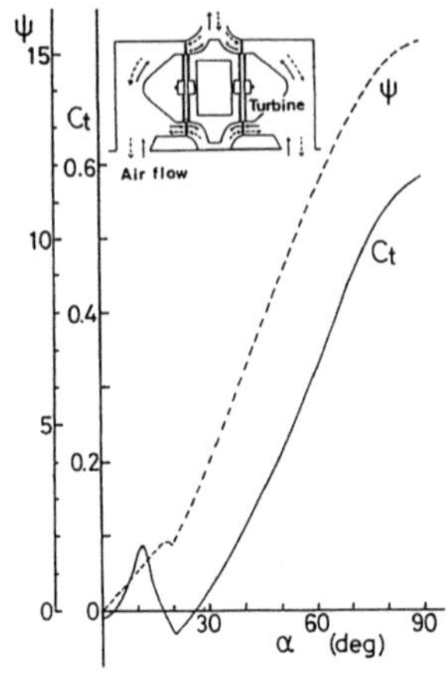

Fig.7. Performance of turbine

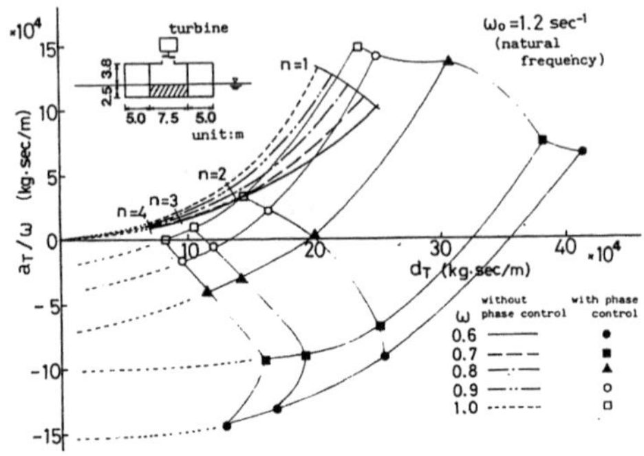

Fig.8. Equivalent load damping coefficient and spring constant of secondary energy converter

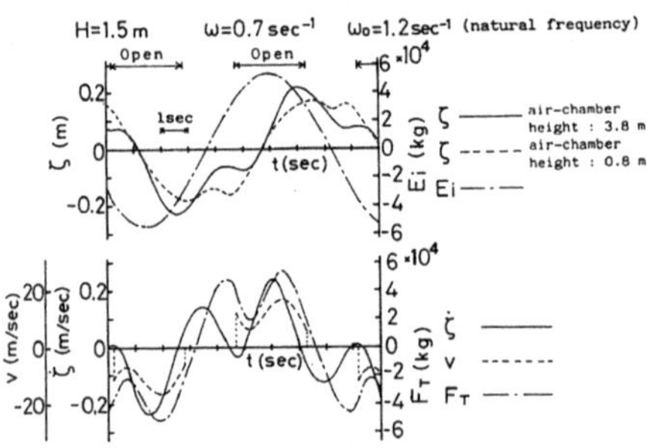

Fig.9. Time history of motion of OWC and air flow under phase control

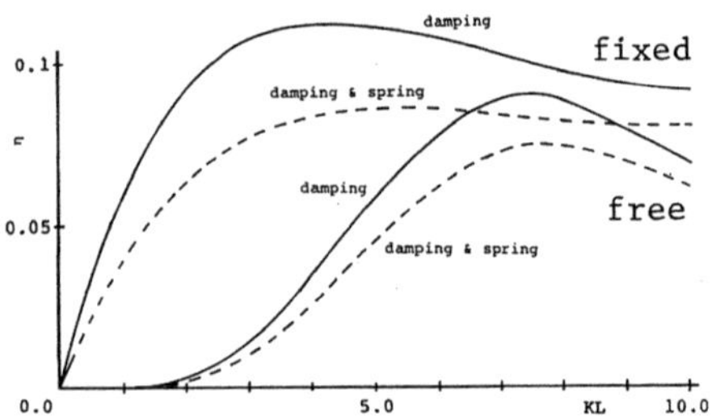

Fig.10. Maximum wave energy in fixed and free condition with different equivalent load damping

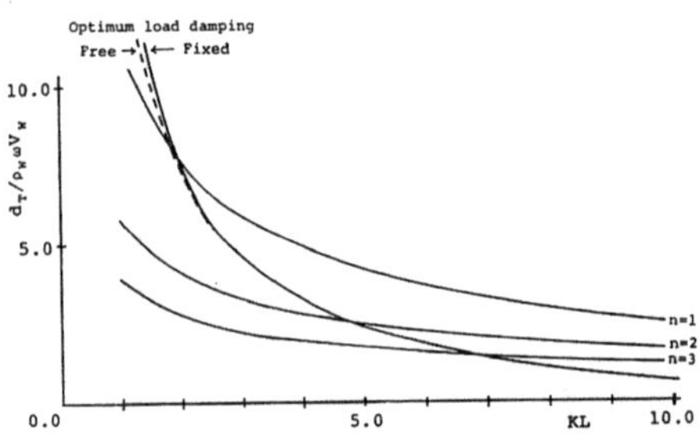

Fig.11. Optimum uniform load damping, and equivalent load damping provided by secondary energy converter

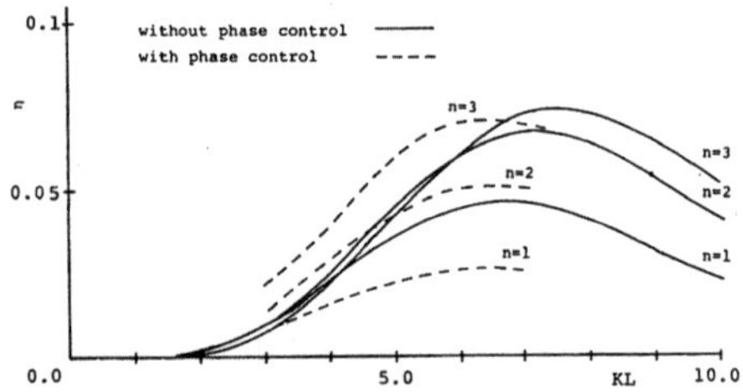

Fig.12. Optimum absorbed wave energy of freely floating attenuator with turbine, and with/without phase control

On the Experimental Investigation of a Wave Power Converter

KIM NIELSEN

Department of Ocean Engineering
The Technical University of Denmark
Lyngby, Denmark.

Introduction

Since 1978 utilization of ocean wave power has been investigated at the Department of Ocean Engineering, the Technical University of Denmark. A wave power converter has been developed after the KN principle from an idea to a small prototype with installed turbine and generator of 1kW. During 1985 the prototype will be on a sea trial in the Danish sound Øresund to demonstrate, how well both principle and design are able to deal with the forces of nature. If the construction stands the trial, future development can follow. The principal will be illustrated by the experimental research, presented in this paper. Presentation will involve experiments with: a) one buoy, dampened by two single working piston pumps, b) a manifolded system of four of these in an attenuator configuration, and c) test results from a five times larger model of the attenuator (the prototype due for sea trail).

The Origin of the Idea

The past seven years of research were inspired by the drawing and the formular presented in a textbook on hydraulics, Figure 1.

Fig.1. Power = $\rho g Q \Delta H$

It seemed a pity that the Danish countryside was so flat and without waterfalls. But Denmark is surrounded by plenty of ocean. Turbines could be placed on the seabed - the head ΔH (at most) equal to the waterdepth - the flow Q created by suction from pistonpumps driven by floats on the surface waves. Beyond a possible way of utilizing water turbines in Denmark a principle for utilization of wavepower had imerged.

The Experiment with a Single Buoy

The experimental set up is illustrated on Figure 2 and the experiments conducted in regular waves. For each wave period measurements of the head ΔH (see Figure 2) were made for different fixed adjustments (i) of a calibrated valve. For each valve adjustment the incoming wave-height was measured over a distance of 2 wave lengths in order to determine reflections.

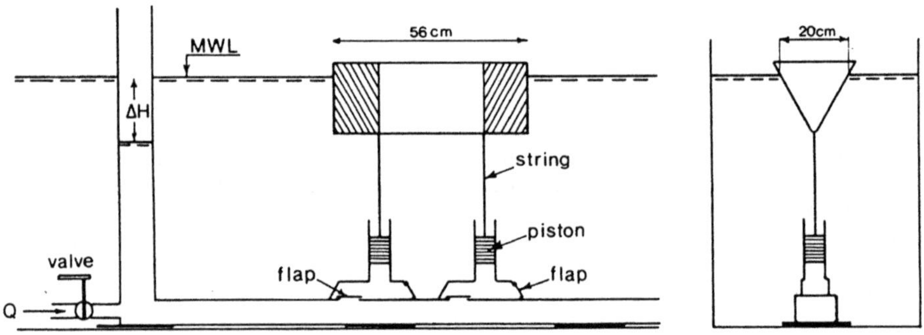

Fig.2. The experimental set up for the single buoy test in a 60cm. wide waveflume.

Geometry and Massdistribution of the Floating Body

The floating body investigated has a triangular 60° cross section and a rectangular waterplane. To obtain a large moment of inertia, waterballast is placed in stern and bow between bulkheads. Using two wires attached one below each bulkhead, both heave and pitch motion can be dampened in head sea.

The Piston Pumps

The single working piston pumps have two flap valves insuring a unidirectional flow. In the upward stroke the piston has to pull against the water head ΔH. On the downward stroke, when the flap toward the turbine side is closed, the weight of the piston has to press the water through the outlet valve. Design of flap valves are essential to ensure a good performence, as return losses should be minimized.

The Preparation of Measured Data

Power in regular waves was calculated as for deep water waves

$$P_{inc} = \frac{\ell}{32\pi} \rho g^2 H^2 T \qquad (1)$$

The power absorbed expressed in terms of the measured head ΔH and the flow Q

$$P_{abs} = \rho g Q \Delta H \qquad (2)$$

The flow through the calibrated valve can be written in terms of its effective opening area a(i), calibrated for certain adjustments (i):

$$Q = a(i)\sqrt{2g\Delta H} \qquad (3)$$

The measured head plotted for each valve adjustment is shown on diagram Fig.3a. The maximum power is absorbed, when the product of Q and ΔH has a maximum value.

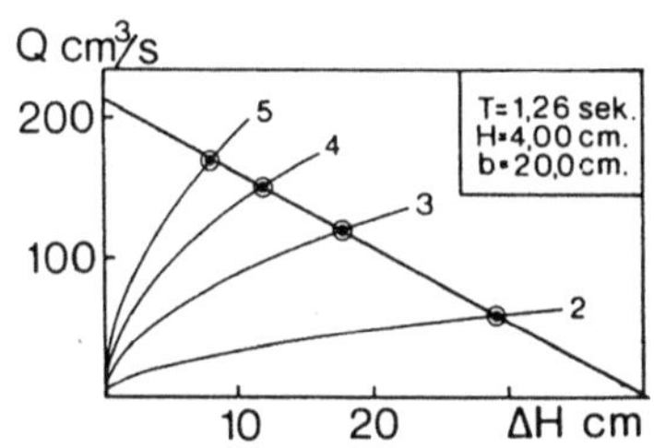

Fig.3a Measured Head and Flow.

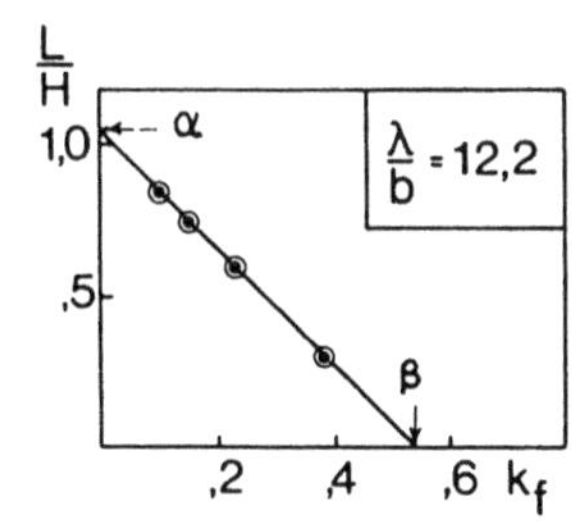

Fig.3b Non-dimensional.

To find this optimum the measured relation between Q and ΔH were represented by a straight line and the measurements plotted in a nondimensionalform, shown on Fig.3b. The parameters involved are defined as (symbols defined in list at the end of paper)

$$\frac{L}{H} = \frac{Q \cdot T}{A_r H} \qquad (4)$$

$$k_f = \frac{\Delta H A_r}{H A_w} \qquad (5)$$

$$\frac{\lambda}{b} = \frac{g T^2 \ell}{2\pi A_w} \qquad (6)$$

The α-β Diagram Presentation of Results

In the non-dimensional form the straight line in Fig.3b is expressed in terms of two parameters $\alpha(\frac{\lambda}{b})$ and $\beta(\frac{\lambda}{b})$ for each experiment of constant $(\frac{\lambda}{b})$. This way the average piston stroke response $(\frac{L}{H})$ can be expressed as a function of the damping ratio k_f as

$$\frac{L}{H} = \alpha(\tfrac{\lambda}{b})\left(1 - \frac{k_f}{\beta(\frac{\lambda}{b})}\right) \qquad 0 \leqq k_f \leqq \beta(\tfrac{\lambda}{b}) \qquad (7)$$

If $k_f = 0$ the unrestrained motion of the buoy occurs $\frac{L}{H} = \alpha(\frac{\lambda}{b})$, and if $k_f = \beta(\frac{\lambda}{b})$ the piston stroke becomes zero and there is no vertical motion of the buoy. When $k_f = \frac{1}{2}\beta(\frac{\lambda}{b})$ the maximum capture width ratio for each wave length can be expressed, using the formulas above as:

$$\frac{W}{\ell} = \frac{P_{abs}}{P_{inc}} = 4\,(\tfrac{\lambda}{b})^{-1}\,\alpha(\tfrac{\lambda}{b})\,\beta(\tfrac{\lambda}{b}), \quad \text{for} \quad k_f = \tfrac{1}{2}\beta(\tfrac{\lambda}{b}) \qquad (8)$$

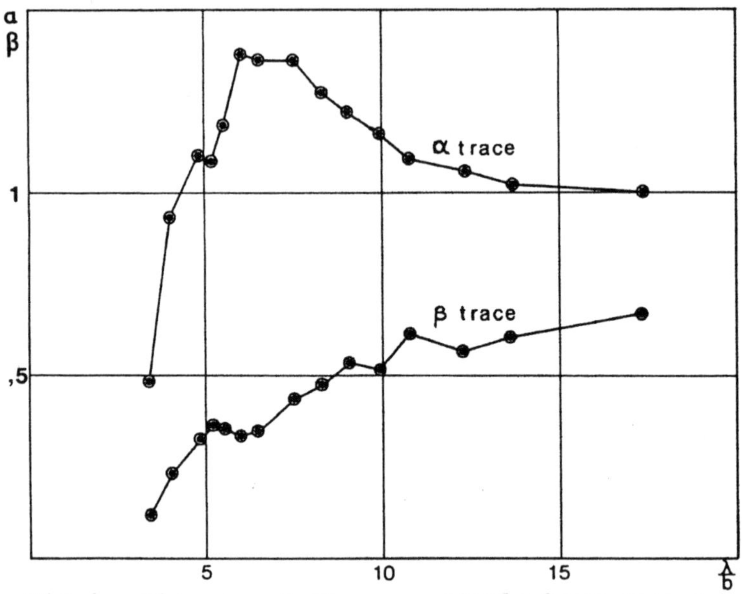

Fig.4. The α - β diagram single buoy system.

Experiments over a wide range of wave lengths are presented in Fig.4. The α-trace reminds you of the response of a vessel in waves and the β-trace of the non-dimensional hydrodynamic force on a vessel fixed in waves. The non-dimensional functions describes the behaviour of the buoy investigated.

The Attenuator Configuration

Testing the attenuator configuration on Figure 5a was inspired by new and exciting theories presented by English scientists. Pulsating "boxes" could possibly capture wave power from a wave crest equal to half the total length of the "boxes".

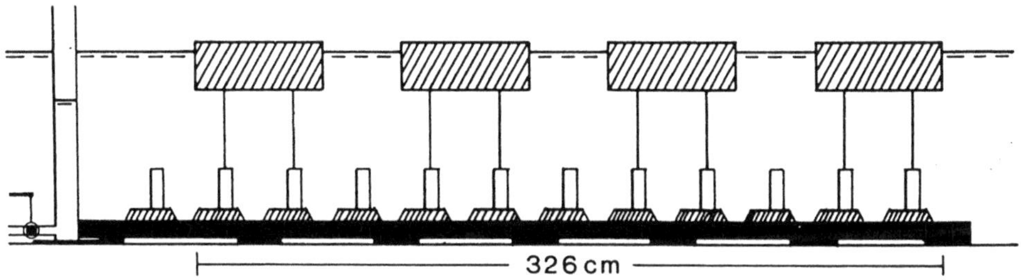

Fig.5a. The attenuator tested in a 400cm. wide waveflume.

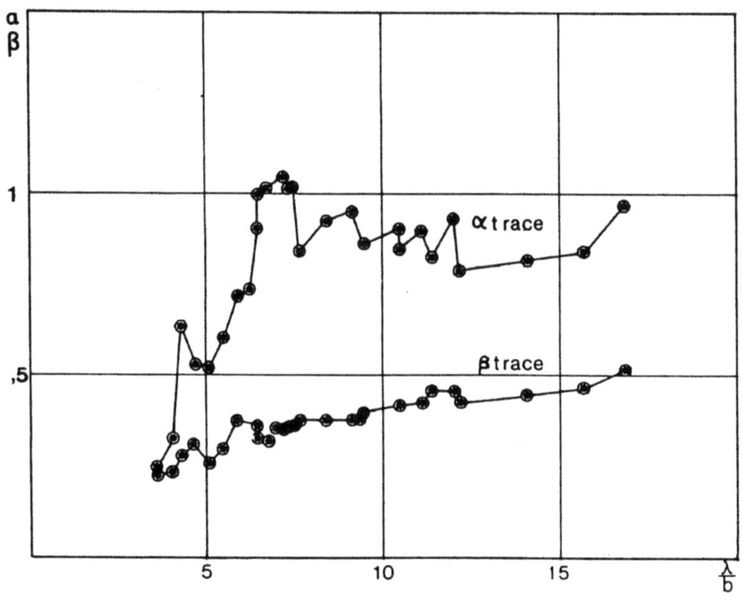

Fig.5b. The α - β diagram
4 buoy attenuator.

The first experimental results Figure 5b were not encouraging. Average power absorbed per buoy was approximately half the amount, compared with the single buoy system. The result was improved by placing rubberbands on the horizontal inlet flap valve. The valves made of plastic were too light to ensure fast closing.

Comparison of Small Scale Experiments

For the experiments conducted a comparison in terms of average capture width ratio (8) is shown in Fig.6. The two top curves have improved flap valves and are comparable within the experimental uncertaincy. The single buoy system is the most efficient one. The decrease in performance of the attenuator is probably caused by a combination of reduced wave height down the row and manyfolding effects increasing pressure fluctuations in the hydraulic interconnection. Introduction of buffers could improve performance but has not yet been investigated.

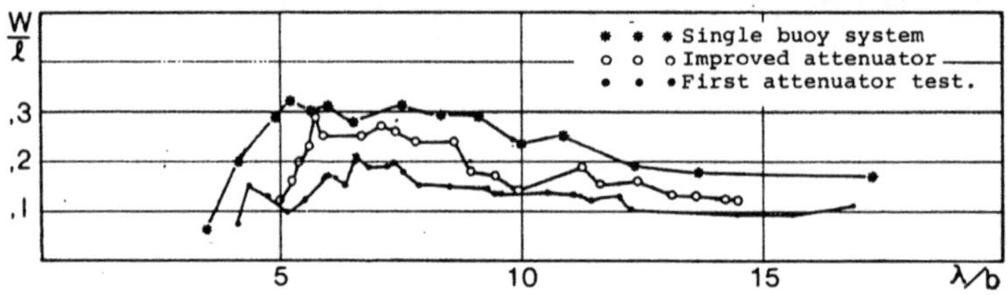

Fig.6. Comparison of test results in small scale.

The 1kW Prototype

To gain practical experience on how well the design is able to work at sea, it was decided to undertake a sea trial. Due to test site conditions the previous model attenuator was scaled up 5 times and now capable of working in waves up to 1 meter high.

The prototype illustrated in Fig.7 was initially tested in the 12 meters wide, 5.5 meters deep and 240 meters long wave flume of the Danish Maritime Institute.

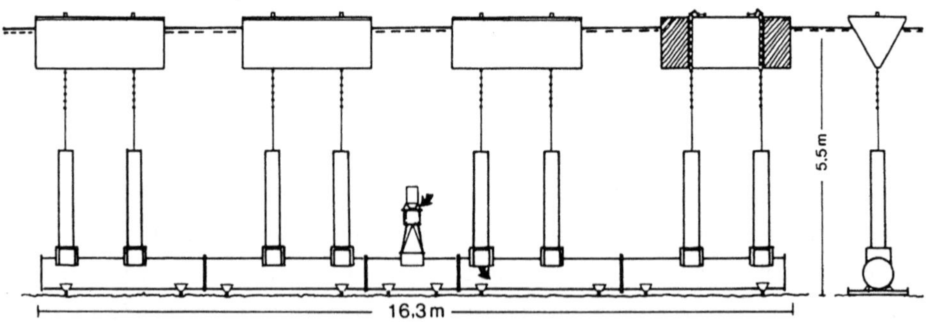

Fig.7. The 1kW Prototype in steel.

Combining Turbine and Wave Powered Pumps

The flow Q created by the piston pumps of the wave power converter can be written as:

$$Q = \alpha(\tfrac{\lambda}{b})\ \frac{H\,A_r}{T}\left(1 - (\beta(\tfrac{\lambda}{b}))^{-1}\ \frac{\Delta H\,A_r}{H\,A_w}\right) \tag{9}$$

The maximum power (2) is absorbed when:

$$Q = \tfrac{1}{2}\,\alpha(\tfrac{\lambda}{b})\ \frac{H\,A_r}{T} \tag{10}$$

$$\Delta H = \tfrac{1}{2}\,\beta(\tfrac{\lambda}{b})\ \frac{H\,A_w}{A_r} \tag{11}$$

For any size of wave power converter it is possible to find a suitable turbine size that will ensure this optimum in a large range of waves of constant steepness, $\frac{H}{\lambda}$ = constant.

The performance diagram, Fig.8, shows the the performance of the 145 mm diameter turbine of 1500 rpm mounted on the 1kW prototype. The performance of the wave powered piston pumps are calculated using (9) with values of $\alpha(\frac{\lambda}{b})$ and $\beta(\frac{\lambda}{b})$ taken from the least optimistic α-β Diagram, Fig.5b. Expected working points are found in each case at the point of intersection between turbine and pump performance curve.

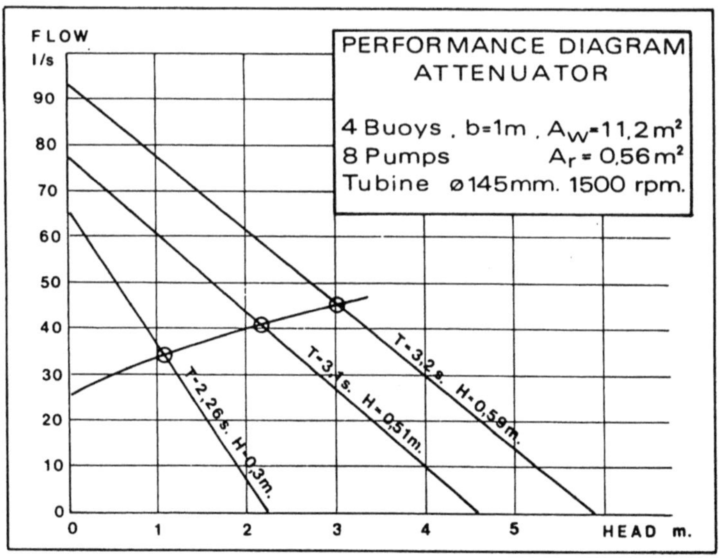

Fig.8. Performance diagram for 1kW prototype.

Test Results for the 1kW Prototype

Test results from the first test with the 1kW prototype are presented in Table I. The results are presented in terms of average head, ΔH, (measured with pressure cell), and the absorbed power calculated by use of both (2) and the turbine performance curve shown in Fig.8. The converted power is the average electrical power of 380 volt A.C. delivered to the grid from the 1kW rated generator running at 1500 rpm. The tests were conducted in regular waves of hight and period as given in the table. If compared to the small scale tests the performance of the prototype is between "the first-" and "the improved attenuator test", Fig.6.

TABLE I

WAVE PERIOD T sek.	2.26	2.40	2.52	2.65	2.76	2.89	3.01	3.10	3.20
WAVE HEIGHT H m.	.32	.32	.37	.41	.44	.46	.48	.52	.60
HEAD ΔH m.	1.39	1.20	1.54	1.90	2.00	2.15	2.28	2.65	3.13
ABSORBED watt.	490	412	559	736	784	862	928	1130	1412
CONVERTED watt.	62	31	121	261	335	395	443	613	794

Conclusions

A principle for utilizing ocean wave power has been proposed and the experimental procedure of investigating the concept has been described.

Test results for small scale experiments has been presented and the application of test results demonstrated when the choice of turbine has to be made.

The combination of wave powered single working piston pumps and submerged turbine and generator has been tested. The performance compares well to the predictions made on the basis of small scale experiments.

All the dominating forces involved in the KN principle of converting wave power are of inertial kind and the results can be presented in a non-dimensional form according to the model law of Froude. Changing the geometry, different results will occur and further research will reveal how much the performance of the wave power converter can be improved by introducing new geometrical forms and configurations.

List of Symbols

Q	Flow of water
ΔH	Head, measured in water columb
g	Gravitational acceleration
ρ	Density of water
P_{inc}	Incomming wave power
P_{abs}	Absorbed power
$a(i)$	Effective opening area of valve
i	Valve adjustment
ℓ	Total length of buoy
A_w	Total water plane area of buoys
b	Beam of buoys
A_r	Total cross-section area of pistons
L	Average piston stroke length
k_f	KN's ratio defined by (5) (damping ratio)
H	Incoming wave hight
T	Wave period
λ	Wave length
W	Capture width (length of wave crest containing the absorbed power)
$\alpha(\frac{\lambda}{b})$, $\beta(\frac{\lambda}{b})$	non-dimensional functions of $(\frac{\lambda}{b})$ related to the geometry of floats investigated

References

Nielsen, Kim: "Bølgeenergimaskiner", Instituttet for Skibs- og Havteknik, DTH, 1980.

Nielsen, Kim: "Bølgeenergimaskiner II", Instituttet for Skibs- og Havteknik, DTH, 1982.

Engelund, F.A. and Pedersen, Fl. Bo: "Hydraulik", Den Private Ingeniørfond og DTH, 1974.

Newman, J.N.: "The exciting forces on fixed bodies in Waves", Journal of Ship Research, Dec. 1962.

"Second International Symposium on Wave and Tidial Energy", Cambridge, England, organized by BHRA Fluid Engineering, Cranfield, Bedford, England, Sept. 1981.

Design Optimization of Axi-symmetric Tail Tube Buoys

T.J.T. Whittaker, F.A. McPeake.

The Queen's University of Belfast,
Dept. of Civil Engineering,
David Keir Building,
Stranmillis Rd.,
Belfast, N. Ireland.

Summary

This paper presents the results of an experimental test programme and theoretical analysis of the hydrodynamic design of wave powered axi-symmetric navigation buoys. A two degree of freedom system with power extraction damping the relative motion of the two heave modes is considered. Results are presented for a theoretical model which has been verified experimentally.

It is concluded that the power extraction efficiency and frequency bandwidth of the current design of wave powered navigation buoy can be considerably improved. Peak capture factor increases proportionally with the tube diameter for tube / buoy diameter ratios of up to 75%. Increased tube length widens the frequency bandwidth response. The level of applied damping has considerable effect on peak performance and frequency bandwidth response. With this two degree of freedom system, the levels of applied damping which produce maximum capture factor are approximately 20 times greater than the optimum values for the isolated water columns. Also the radiation damping associated with the two heave modes should be similar so that the damping applied by the power extraction system is matched at both natural response frequencies. The work has general application to the design of wave powered buoys with a central cylindrical water column.

Notation

B	radiation dampng.
Ba	applied damping; (turbine damping.)
Bab	optimum applied damping at water column resonance.
C	damping ratio; Ba / Bab.
D	non dimensional radiation damping.
F	wave excitation force.
k	buoyancy.
L	tail tube length from water line.
Ma	added mass, (dynamic.)
Ms	mass, (static.)
R	buoy hull radius.
r	tail tube radius.
x	displacement from equilibrium position.
ρ	water density.
ω	frequency in radians.
Suffix b.	refers to the buoy hull.
Suffix w.	refers to the water column.

Introduction.

During the past few years, a research team based in The Department of Civil Engineering, The Queen's University of Belfast, have been investigating the various stages in the power conversion chain of wave powered navigation buoys. This work has been conducted jointly with two of the lighthouse authorities in the British Isles, Trinity house and Irish Lights, and a local company Munster Simms Engineering Ltd., who manufacture Wells turbine generators. Although the main thrust of the work has been towards the design and testing of Wells turbine-generators for use on standard tail tube buoys, as described by Whittaker et al (1), it has been necessary to gain a fundamental understanding of the hydrodynamic interaction between the buoy and the incident wave field.

The use of waves to power navigation buoys is not a recent innovation as they were first introduced in the early nineteenth century. These early systems used the rolling and pitching motions of the hull to sound a bell. Around the turn of the century the oscillating water column was introduced. A long cylindrical tube open to the sea at the bottom was located on the vertical centre line of the buoy forming a water column and air chamber above. A typical wave powered buoy as used by The Irish Lights Lighthouse Authority, is shown in figure 1. Air forced into and out of the plenum chamber by the wave induced oscillation of the water column was used to sound a whistle. The modern wave powered buoy utilises this same principle only the whistle is often replaced by a turbine generator. Masuda (2) describes such a system with a conventional axial flow turbine requiring complicated ducts and non-return valves to rectify the oscillating air flow to obtain uni-directional rotation. Whittaker et al (3) describes the use of the Wells self-rectifying turbine which has greatly simplifed the power conversion system.

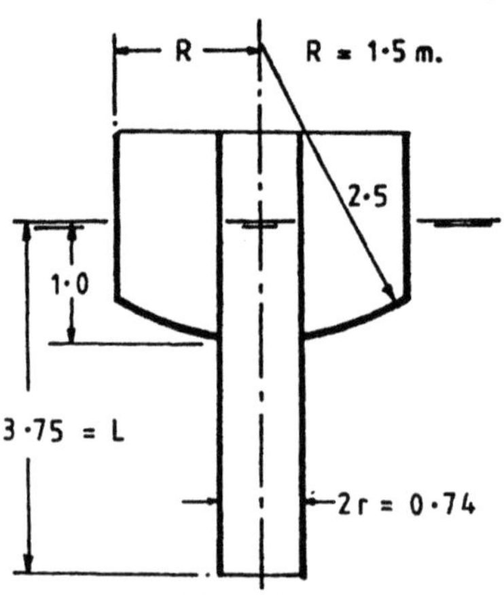

Fig. 1

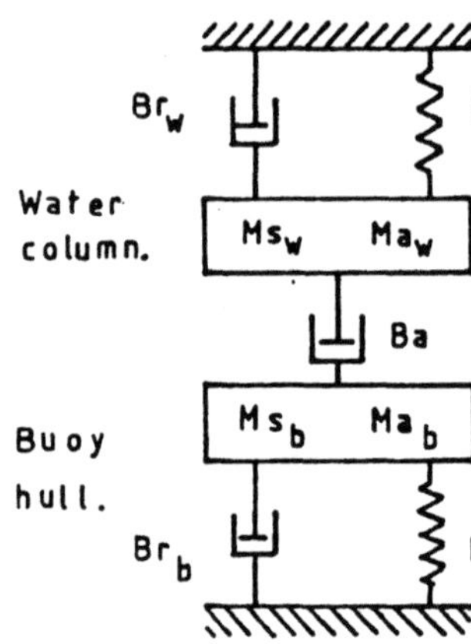

Fig. 2

Buoy hydrodynamics.

The buoy has a total of seven degrees of freedom; six to

describe the motion of the hull, plus the heave motion of the water column. The motion of the buoy hull in uni-directional waves can be described in terms of heave, pitch and surge. Calculation and model testing have shown that the natural pitch response frequency is outside the power extraction frequency bandwidth and surge motions are limited in this frequency range by a heavily damped mooring. Coupling between pitch and surge is minimal as a result of the mooring attachment on the bottom of the hull. Roll is not normally excited because of the bridal mooring which provides a two point attactment on the transverse axis of the hull. Air compliance in the plenum chamber was considered to be small and was not included in the mathematical model.

Therefore, the hydrodynamic model was limited to that of a two degree of freedom mass, spring, damper system, as shown in figure (2). One mode describes the heave motion of the buoy hull and the second is for the water column in the tail tube. The equation of motion for this model is presented in matrix form in equation 1.

$$\begin{bmatrix} Ms_b+Ma_b & Ma_{bw} \\ Ma_{wb} & Ms_w+Ma_w \end{bmatrix} \begin{bmatrix} -\omega^2 x_b \\ -\omega^2 x_w \end{bmatrix} + \begin{bmatrix} B_b+Ba & -Ba+B_{bw} \\ -Ba+B_{wb} & B_w+Ba \end{bmatrix} \begin{bmatrix} j\omega x_b \\ j\omega x_w \end{bmatrix} + \begin{bmatrix} k_b & 0 \\ 0 & k_w \end{bmatrix} \begin{bmatrix} x_b \\ x_w \end{bmatrix} = \begin{bmatrix} F_B \\ F_W \end{bmatrix} \quad \text{Equation 1.}$$

The heave modes are coupled in two ways. Hydrodynamic coupling exists as indicated by the off-diagonal terms in the added mass and radiation damping matrices. However, the dominant coupling results from the damping applied by the turbine ot the air flow from the plenum chamber which damps the relative motion of the buoy hull and water column.

Test Programme.

The primary objective of the research programme was to determine how the length and diameter of the tail tube and the magnitude of the turbine damping affected the pneumatic power output from this type of device. This was accomplished by testing a range of buoy / tail tube configuratons in the wave tank and measuring the pneumatic performance with a range of applied damping levels. In addition the added mass and damping coefficients were calculated using a 3 D. diffraction theory program and the performance calculated by solving the equations of motion using the 'frequency response method'.

The hull shape, shown in figure 1, was chosen for these experiments as it represented one of the standard hull forms used by the lighthouse authorities throughout the world. A 1/10 scale was chosen for the models to suit the characteristics of the wave tank. Initially the complete Irish Lights buoy with a 3.75m. long, 0.74m. diameter tail tube was modelled as a control experiment to establish the performance of a typical buoy currently in service. This work was also used to design a Wells turbine generator with matched damping characteristics which was tested on a fully instrumented buoy at sea as described by Whittaker et al (4).

Table 1 shows the full range of variables tested. For each tail tube geometric configuration, the applied damping is expressed in terms of damping ratio relative to the optimum applied damping for a fixed tail tube; optimum damping being defined as the value which produces peak pneumatic power output.

Tube diameters.	0.64	0.74	0.11	0.16	0.23			
Tube lengths.	0.2	0.3	0.4					
Damping ratios.	1	5	10	15	20	30	40	50

The above values refer to model scale.

Table 1.

Validation of the Mathematical Model.

The performance curves predicted by the theoretical model for the range of buoy configurations shown in table 1, were checked experimentally. A sample of the calculated performance curves is compared with those obtained from the wave tank experiments in figure 3.

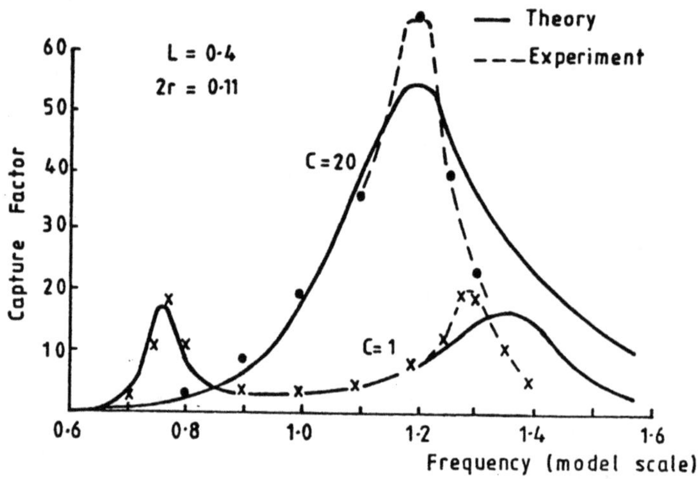

Fig. 3. COMPARISON OF THEORY – EXPERIMENT.

These curves relate capture factor to wave frequency at model scale for a range of applied damping. Capture factor is defined as the ratio of pneumatic power output to the wave power incident on the buoy diameter. The general trends of the curves are well modelled over most of the frequency range. However, some differences are evident at higher wave frequencies. This is a result of the experimental model being effectively in line array in the wave tank. As the wavelength is equal to the tank width at 1.33 Hz., array interaction enhances performance below this frequency and reduces performance beyond because of the phase of the radiated wave reflecting from the tank walls;

Whittaker (5). Array interaction was not included in the mathematical model as this phenomenon does not exist with isolated navigation buoys at sea. Therefore it was concluded that the calculated response was an accurate representation of the performance of a single buoy. The predictions of the mathematical model were also compared with results published by Evans (6) for a Norwegian buoy design. Excellent agreement was observed.

Performance of the Standard Buoy.

Before presenting the results of the parameter study it is necessary to discuss the performance of the standard Irish Lights buoy with a 0.74m. diameter 3.75m. long tail tube, as shown in figure 4.

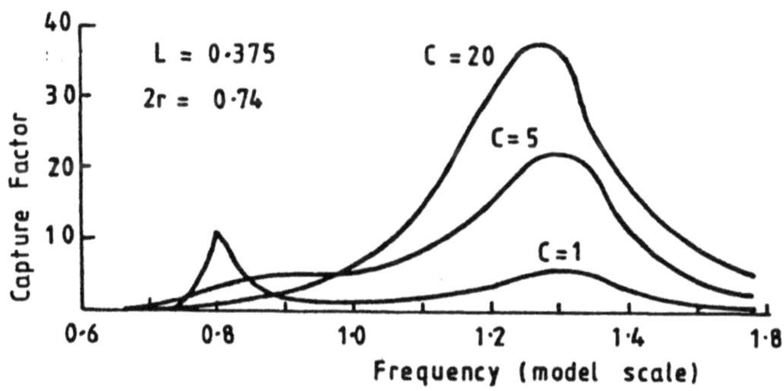

Fig. 4. IRISH LIGHTS WAVE POWERED BUOY.

Damping ratio significantly affects performance. A double peak performance curve identifying the heave resonant frequency of both the water column and buoy hull is only experienced with very light damping; a damping ratio of 1. However, the peak performance is only 30% of that obtained with a damping ratio of 20. At this higher damping level the water column resonance is completely suppressed resulting in a narrow frequency bandwidth response about the natural buoy heave frequency. The reason for this can be found by considering the magnitude of the radiation wave force components for both modes of motion, shown in figure 8. As there is a difference of nearly two orders of magnitude in the radiation damping, it is impossible to provide matched impedance for both modes with a single turbine design.

With the present buoy design the turbine damping is chosen to provide peak performance at the buoy heave frequency. This has the secondary effect of constraining the system to extract power only from low wave periods; typically 3 seconds at full scale. This period corresponds to a low spectral density in a typical sea spectrum even for an inshore location. It is therefore desirable to not only increase peak performance and bandwidth response but also to increase the response at larger wave periods and obtain a better matching between the transfer function for power extraction and the wave sprectrum.

Tail-tube Geometry and Performance.

The effect of increasing tail tube diameter is shown in figure 5. for a 0.4m. long tube and for a damping ratio of 1. This shows that capture factor increases proportionally with water column area for the range of tubes tested. It should be noted that the lowest curve is similar to that obtained for the Irish Lights buoy with unit damping ratio.

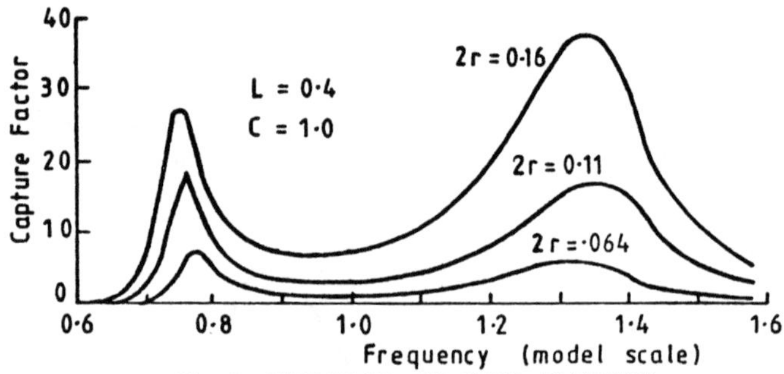

Fig. 5. VARIATION OF TUBE DIAMETER.

Figure 6. shows the effect of tail tube length on performance for unit damping ratio. Although the results for the intermediate diameter tube (0.11m.) are presented, the trend was similar for the other tubes. Lengthening the tail tube widened the response frequency band width as a consequence of the greater column mass lowering the stiffness / mass ratio and reducing the resonance frequency of the water column. The principal values of added mass and radiation damping varied insignificantly with tube length. However, a higher degree of hydrodynamic coupling was observed with the shorter columns, but this proved to be numerically insignificant when solving the equation of motion.

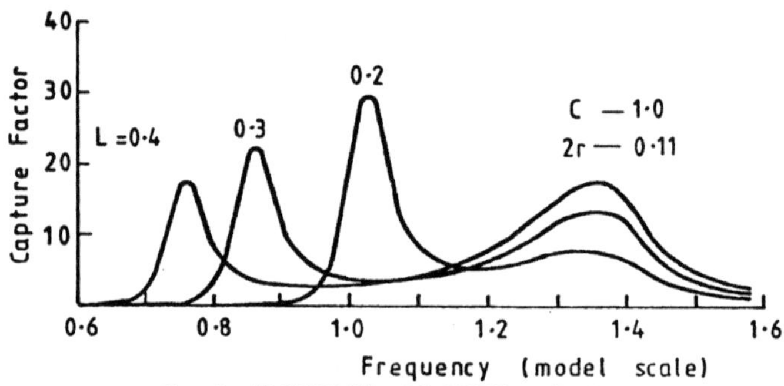

Fig. 6. VARIATION OF TUBE LENGTH.

As with the performance of the Irish Lights buoy damping proved to be the most significant variable irrespective of tube length or diameter. Figure 7. shows the performance of the 0.16m. diameter tube, with a range of damping levels. As with the standard buoy, increased damping ratio increased performance and a value was reached beyond which peak capture factor was reduced. However, a comparative examination of figures 4 and 7 reveals that with the larger diameter tube the peak capture factor occurs at a frequency between the resonance frequency of the buoy hull and water column. Examination of the performance curves for all the column diameters tested, revealed that with a damping ratio of 20 the frequency corresponding to peak performance shifted to a lower value progressively with increasing tube diameter. This reduction in frequency is advantageous as it corresponds to a longer wave length providing beter matching with a typical inshore sea spectrum.

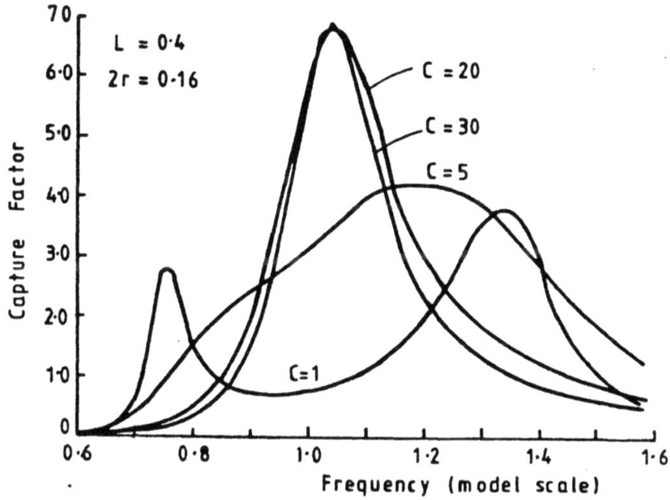

Fig. 7. VARIATION OF APPLIED DAMPING.

Optimum Buoy Design.

The standard navigation buoy represents one extreme in geometry when the water column cross sectional area is small in relation to that of the buoy hull. In this case the buoy heave mode predominates and the water column acts as an inertial reference for the power take-off. Wave excitation on the column is small compared to the pressure force in the plenum chamber. As the column diameter is increased, both the wave excitation force and the radiation force become greater until the water column motion predominates and the buoy hull becomes the inertial reference for the power take-off. Figure 8 shows the variation in radiation damping with increasing column diameter.

As the power take-off is not unique to each mode but acts as the coupling mechanism between the buoy and water column heave, impedance matching for both modes can only be achieved if the respective magnitudes of radiation damping are of similar magnitude. This can be achieved with the larger diameter tail tubes.

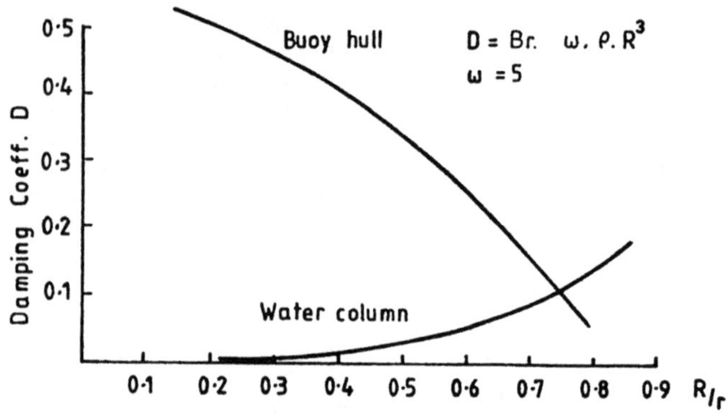

Fig. 8. VARIATION OF RADIATION DAMPING COEFF.

However, it is not possible to optimise the buoy design purely from hydrodynamic considerations as there are practical limitations to the length and diameter of the tail tube. Diameter must be restricted to leave sufficient plane water area in the hull to ensure floatation. As navigation buoys are heavily constructed, this limits the column diameter to approximately half that of the hull. Column length is restricted by the depth of water on site and the mooring configuration. Buoys can be moored in a water depth of ten metres and as the mooring must be attached to the bottom of the hull for stability, a bridal must be used. With large diameter columns the length is limited to about 4m. to prevent contact with the mooring chains. The connection between the bridal and the main mooring chain at the swivel must not rest on the sea bed to avoid twisting. Column length must also be limited to enable handling of the buoy on the deck of the ship.

Conclusions

From the first stage of an extensive research and development programme to determine the relationship between water column geometry and the pneumatic performance of axi-symmetric tail tube buoys for use as navigation aids, the following has been concluded:-

1. Pneumatic performance is proportional to column cross sectional area for tube / diameter ratios of up to 75%.

2. Increased tube length widens the frequency bandwidth response by reducing the natural column heave frequency. This is most significant with low damping ratios.

3. The level of applied damping is the most significant variable affecting performance. For small diameter tubes, as used in the current buoy designs, a change in damping ratio from 1 to 20 increses the peak performance by a factor of three and the buoy heave frequency predominates.

4. As the power extraction damps the relative motion of the two heave modes, matched impedance necessitates similar levels of radiation damping for both degrees of freedom.

5. The current wave powered navigation buoy designs are not efficient and respond only to very high wave frequencies. Doubling the tube diameter not only improves peak performance while reducing the response frequency, but also substanially widens the frequency bandwidth response when a damping ratio of 20 is used.

6. Although hydrodynamic performance improves with increased water column diameter and length, for the range of variables tested, full optimisation is not possible as there are practical limitations such as floatation, stability, handling, water depth and snaging between the tube and the bridal mooring.

Acknowledgements

Grateful thanks are accorded to the following:-

The Wolfson foundation for funding the research work.

Dr. E R Jeffreys for his considerable assistance with the mathematical modelling and calculation of the hydrodynamic coefficients.

The academic and technical staff of the University and the members of the wave power research team for their considerable efforts in the development of this project.

References

1. Whittaker, T.J.T., McPeake,F.A., Barr,A.G., The Development and Testing of a Wave Activated Navigation Buoy with a Wells Turbine. 3rd Int. Symp. on Offshore Mechanics and Engineering, New Orleans, LA, USA, Feb. 12-16, 1984, pp 267-273, Vol 2.

2. Masuda,Y. Wave Activated Generator. Int. Coll. on the Exploitation of the Oceans. (Trans.). Bordeaux, France 1971.

3. Whittaker,T.J.T., McIlhagger,D.S., Barr,A.G., Wells Turbines for Navigation Buoys. 3rd. Int. Conf. Energy for Rural and Island Communities. Inverness, Scotland Sept. 12-16, 1983, PP 289-297.

4. Whittaker,T.J.T., McIlhagger,D.S., Barr,A.G., THompson,A., Electrical Considerations with Wave Powered Navigation Aids Conf. Alternative Energy Systems: Elect. Integration and Utilisation. Coventry. U.K. Sept. 10-12.

5. Whittaker,T.J.T., Murray,M.A., Performance Characteristics of Hydropneumatic Buoys in Line Array. Proc. Int. Symp. on Hydrodynamics, Ocean Eng., Trondheim, 1981. pp 1047-1066.

6. Evans,D.V., McIver,P., The Hydrodynamical Characteristics of a Section of a Floating Sphere Making Vertical Oscillations. School of Mathematics University of Bristol. Report AM-82-09

Physical and Mathematical Modelling of a Point Absorber Wave Energy Conversion System with Nonlinear Damping

D. C. HICKS and C. M. PLEASS

College of Marine Studies
University of Delaware
Newark, Delaware 19716 USA

Summary

A point absorber type device, damped by a single-acting hydraulic cylinder, has been tested in the laboratory and modeled mathematically. The system's efficiency was found to be equal that of systems with double-acting pumps. Response and efficiency curves are presented from both the physical and mathematical models for the case of monochromatic wave forcing.

Introduction

A point absorber wave-energy system has been under development at the University of Delaware since 1976. The system, referred to as DELBUOY, is designed to desalinate seawater rather than produce electricity. The economic advantages of this end use have been discussed by Pleass [1,2]. The DELBUOY system comprises a light weight, shallow draft cylindrical buoy driving a submerged, single-acting positive displacement pump. The flow of pressurized seawater from the pump is rectified using check valves, with pressure surges damped by hydraulic accumulation. The large ratio of buoy to piston area enables the system to amplify the wave-induced dynamic pressure up to the 700-800 psi pressure required for the reverse osmosis desalination process. Designed primarily for trade wind wave conditions, the sea trials have been carried out at the University of Puerto Rico. Since deploying the first prototype in 1980, six units have been tested with the first fully operational unit producing a steady supply of fresh water in 1982.

In parallel with the engineering of the prototypes, laboratory and mathematical models were used to investigate the system's heave response and efficiency as a function of wave height, frequency, and relative damping. Due to the nonlinearities associated with the one-way pump force, linear theory cannot be used to describe the system's response. As a result the majority of the work to date on modeling wave energy conversion systems does not apply. The few exceptions are the work of Count [3,4], Parks and Tondl [5], Parks [6], and Tanaka [7].

Description of the System

The evident analogy to a mechanical oscillator can be extended, with some restrictions, to nonlinearly damped devices. The buoy, limited to a single degree of freedom, is damped by a positive displacement pump (Figure 1). Through the use of check valves and an accumulator, the pump exerts a constant resistive load, P, for $\dot{Z} > 0$ and zero load for $\dot{Z} < 0$. In addition, the pump and hydraulics introduce frictional and viscous damping forces. The equation of motion for the system can be written as,

$$(M + M_a)\,\ddot{Z} + B\,(\dot{Z} - w) + C_d(\dot{Z} - w)\,|\,\dot{Z} - w\,| + C_v\dot{Z}|\,\dot{Z}\,| + (R + P/2)\,\dot{Z}/|\dot{Z}| + SZ + P/2 = F_I(\omega,t) \tag{1}$$

where Z is the buoy position, with overdots denoting time derivatives and $F_I\ (\omega,\ t)$ is the wave force on the buoy.

Due to the nonlinear damping terms, equation (1) cannot be solved exactly. The alternatives are to generate response curves based on data from physical model tests, develop an approximate analytic solution or to numerically solve the equation. The limitations of the first approach are obvious. As for the second approach, the techniques described by Count [3] and Parks and Tondl [5] were shown to be effective for systems with a single nonlinear Coulomb damping term, but are limited to small resistive loads. The authors have applied the technique of equivalent linearization to a simplied form of equation (1) but found it unwieldly for use with the full nonlinear equation. Exercising the third option, Count [3,4] has applied an unspecified single step numerical method to systems with Coulomb damping, although his results have not been validated with actual data.

Whilst Coulomb damping is appropriate for the hinged and oscillating water column devices modeled by Count [3] and Parks and Tondl [5], flexibly tethered point absorbers like DELBUOY can only effectively extract energy from the upward motion of the buoy. In addition, the prototype tests showed the sliding friction and viscous loss terms to be significant. The goal of this study was to assess the effects of these nonlinear damping terms. The first phase entailed wave tank testing of a scale model DELBUOY system. Response curves for both monochromatic and random wave conditions were developed along with measurements of the wave excitation force, $F(\omega, t)$. In the second phase a time domain mathematical model was developed to solve equation (1). The standard fourth order Runge-Kutta method chosen for the model was found to be both stable and convergent with a time step of T/100. No special algorithm was required to handle the discontinuities resulting from the pump and friction terms. Once validated by hindcasting the physical model data, the mathematical model was used to predict system performance at full-scale. Due to space limitations only the results for monochromatic waves will be presented.

Scale Modeling of a Nonlinearly Damped Point Absorber

Using Froude modeling criteria a 1/10 scale DELBUOY system was constructed and tested in the 120' long x 8' wide x 5' deep random wave tank at the U.S. Naval Academy. The 10.5" diameter x 4" high cylindrical buoy was damped by a miniature hydraulic cylinder, complete with check valves and accumulator. The buoy displacement, external damping force, and incident waves were monitored electronically. To reduce the effects from reflected waves, only data from the first five wave cycles were used for the analyses. With a buoy mass of 14 lbs. and 8 lbs. of counterweight the system's natural period and equilibrium draft were 1.0 sec and 2", respectively. The system was tested at several operating pressures in a range of frequencies. The wave heights and periods, chosen by scaling the wave conditions at the DELBUOY deployment site, were slightly nonlinear.

Wave Forces on the Buoy

Using the partitioning suggested by Newman [8] the wave force on a floating body can be separated into three terms,

$$F(\omega,t) = F_F(\omega,t) + F_D(\omega,t) + F_R(\omega,t) \tag{2}$$

If the incident wave and body motions are of small amplitude the first two terms, representing the Froude-Krylov and wave diffraction forces, can be evaluated with the body held fixed. The last term is the radiation force induced by the body motion in the absence of the incident wave. The sum of the terms $F_I(\omega, t) = F_F(\omega,t) + F_D(\omega,t)$ was measured by holding the buoy fixed at its equilibrium position and monitoring the force exerted by incident regular waves. The non-dimensional wave force amplitude $F_o/S\eta_o$, and phase lag Θ, are plotted in Figure 2. Included for comparison are the theoretical curves predicted by Black [9] for a half immersed sphere and by Garrett [10] for a vertical cylinder in shallow water. The wave force is seen to decrease rapidly in the shorter waves as a result of the increased wave diffraction force $F_D(\omega,t)$. In the following sections this decrease in the wave force will be seen to result in a reduction in the optimum operating pressure as well as a drop in the rate of energy extraction at the higher frequencies.

Newman [11] has shown the linear wave making or radiation force coefficient to be related to the wave excitation force as

$$B = \frac{F_o^2 \omega^3}{2 g^3 \eta_o^2} \tag{3}$$

By inserting the experimentally determined values for F_o into equation (3) the linear damping coefficients for the model were calculated. From equation (3) and Figure 2 it can be seen that the energy lost to outgoing waves, generated by the buoy's oscillations, increases for the shorter waves.

These values for F_o and B were used as input for the mathematical model. The remaining model parameters M, C_v, R, P and S were measured directly for the scale model; C_D was chosen from data in the literature; M_a was calculated using the expression given by McCormick [12]; and the vertical water particle velocities, w, were calculated using linear water wave theory.

Heave Response in Regular Waves

The gain values calculated from the physical model data are presented in Figure 3. The relative damping is defined as $(P + R)/F_o$, where the sliding friction in the model has been treated as additional pump force. The dashed lines correspond to the steady state solution for a linearly damped harmonic oscillator [13], with the experimentally determined values for F_o used as the driving force and the critical linear damping defined as $B_{cr} = 2\ (M + M_a)\,\omega_o$. With the average power output from the system proportional to $Z_o P\ [ka]^{\frac{1}{2}}$, it can be seen from Figure 3 that the rate of power production drops off rapidly with increasing ka.

The response curves for the linearly and nonlinearly damped systems are similar in shape. Like the critically damped linear system, the stiff pump forces suppress system resonance. The two sets of curves tend to diverge at the higher and lower frequencies. This results from the linear damping force tending to zero with $ka \rightarrow 0$ and to ∞ as $ka \rightarrow \infty$ while the pump force is constant for all frequencies. It should be noted that the system can operate at relative damping levels of up to 1.40 as compared to the limiting value of 1.0 predicted by Count [3,4] and Parks and Tondl [5] for systems with two-way resistive damping.

Validation of the Mathematical Model

The time domain mathematical model was exercised to simulate the physical model experiments. In Figure 4 the gain values predicted by the mathematical model are presented with actual data for six wave frequencies. The relative damping has been redefined as $P + R/S\eta_o$, to clearly illustrate the relationship between decreasing F_o with increasing ka, and the system power output. The model's stability, rate of convergence and sensitivity to variations in the input parameters were checked and found to be satisfactory.

Although Figure 4 shows a good correlation with respect to gain, it does not convey the mathematical model's accuracy in simulating the nonlinearities in the system's response. In Figure 5, time series are presented for both models, at two levels of relative damping. From the overplots it can be seen that the mathematical model simulates the buoy set-down resulting from the one-way pump force; skewness in the heave response resulting from the nonlinear damping; square wave and superimposed velocity squared external damping force; as well as the proper displacement amplitudes and phase lags. It should be noted that it is the buoy set-down that enables the system to function at damping ratios greater than unity. The high frequency signals in the wave and force records result from the harmonics generated by the nonlinear buoy displacements. These high frequency waves, as well as for the memory effects on the instantaneous wave force resulting from the waves generated by the buoy motion (Newman [8]), were considered to be of second order and were not incorporated into the model.

Extrapolation to Full-Scale

Once proven as an effective hindcasting tool the mathematical model was used to predict system efficiencies at full-scale. Sliding friction and viscous losses are proportionately smaller in the full-scale prototype and the coefficients C_v and R were therefore reduced by 70% and 50%, respectively. To assess the effects of the single acting pump on efficiency, relative to double acting systems [3,4,5], the model was run with both one-way and Coulomb type resistive loading. This was accomplished by modifying the pump force terms in equation (1); the $[P/2\ \dot{z}/|\dot{z}| + P/2]$ term representing a one-way pump was replaced by $P\dot{z}/|\dot{z}|$ for a two-way resistive load. The results for three frequencies are presented in Figure 6 as capture width versus relative damping, P/F_o. The shape of the curves for Coulomb damping are in agreement with those predicted in the literature [3,4,5] for terminator type devices, although omitting the frictional force from the relative damping term used to define the abscissa, causes the curves to fall to zero short of P/F = 1.0. The additional damping terms in the present model and the rapid drop off in F_o versus ka for three dimensional systems results in a maximum capture width, 1/2a, of approximately one-third the value given by Count [3,4] and Parks and Tondl [5] for two dimensional devices.

As expected the capture widths, 1/2a, for the present system are several times lower than the theoretical value of $\lambda/2\pi$ predicted for linearly damped, resonant point absorbers. The nonlinear damping suppresses resonance and limits the buoy to gain factors of approximately one or less. The efficiency of the one-way pump is seen to equal that of the two-way system and to exhibit a broader response curve with respect to the relative damping. This result is consistent with the concept of equivalent work or damping.

Conclusions

It has been shown that a mathematical model based on standard numerical integration methods is capable of describing the response of point absorber systems with multiple nonlinear damping terms. The accuracy of the mathematical model has been validated with experimental data. It was found that flexibly tethered point absorber systems pay no price in reduced efficiency for their one-way pump damping and in fact exhibit a broader response curve. This increased tolerance to variation is wave height is a major practical advantage.

Acknowledgements

The authors would like to thank the University of Delaware Sea Grant College Program and the U.S. Agency for International Development for their financial support, as well as the Departmento de Ciencias Marinas of the University of Puerto Rico and the U.S. Naval Academy for the generous use of their facilities.

References

1. Pleass, C. M. The Use of Wavepowered Systems for Desalination: A New Opportunity. Proc. Int. Symp. on Wave and Tidal Energy, Canterbury, England, Sept. 1978.

2. Pleass, C. M. Seawater Desalting Using Wavepowered Reverse Osmosis Systems. ACS Int. Symp. on the Chem. and Econ. of Ocean Resources, Hawaii, April 1979.

3. Count, B. M. The Theoretical Analysis of Wave Power Devices with Non-Linear Mechanical Conditioning. CEGB Laboratory Note R/M/N1008.

4. Count, B. M. On the Hydrodynamic Characteristics of Wave Energy Absorption. Proc. of the 2nd Int. Symp. on Wave Energy Utilization, Norwegian Inst. of Tech., Tronheim, Norway, June 1982.

5. Parks, P. C. and Tòndl, A. Non-linear Oscillations in Wave Power Machines. Proc. 8th Int. Conf. on Non-Linear Oscillations, Prague, 1978.

6. Parks, P. C. Wedges, Plates and Waves - Some Simple Mathematical Models of Wave Power Machines. In: Power From Sea Waves, B. Count (ed.), Academic Press, London, 1980.

7. Tanaka, H. Sea Trial of a Heaving Body Wave Power Absorber. Proc. of the 2nd Int. Symp. on Wave Energy Utilization, Norwegian Inst. of Tech., Tronheim, Norway, June 1982.

8. Newman, J. N. Marine Hydrodynamics. M.I.T. Press, 1977.

9. Black, J. L. Wave Forces on Vertical Axisymmetric Bodies. J. Fluid Mech. 67(2) 319-376, 1975.

10. Garrett, C. J. R. Wave Forces on a Circular Dock. J. Fluid Mech. 67(1) 129-139, 1971.

11. Newman, J. N. The Exciting Forces on Fixed Bodies in Waves. J. Ship. Res. 6(4) 10-17.

12. McCormick, M. E. Analysis of Optimal Ocean Wave Energy Conversion. J. Waterways, Port and Coastal and Ocean Eng. 109(2) 180-198, 1983.

13. Halliday, D. and Resnick, R. Physics, John Wiley and Sons, 1978.

Nomenclature

A_b = πa^2 - buoy water plane area
a - buoy radius
B - linear damping, or wave making coefficient
C_d = $1/2\ \rho A_b C_D$ - total buoy drag coefficient
C_D - drag coefficient
C_v - viscous loss coefficient for the hydraulic system
F_o - wave force amplitude
g - gravitional acceleration
ka = $2\pi a/\lambda$ - nondimensional buoy radius
$k_o a$ - resonant wavelength for the system
1/2a - capture width, defined as the power output from the system divided by the wave energy flux per buoy width
M - buoy mass
M_a - added mass
P - pump force
R - sliding friction
S = $\rho g \pi a^2$ - buoyant spring constant for the buoy
w - wave induced water particle velocity
Z - buoy displacement from its equilibrium position
ρ - fluid density
ω - angular wave frequency
ω_o - natural frequency
λ - wave length
η_o - wave amplitude
Θ - phase lag between the incident wave and the force exerted on the buoy

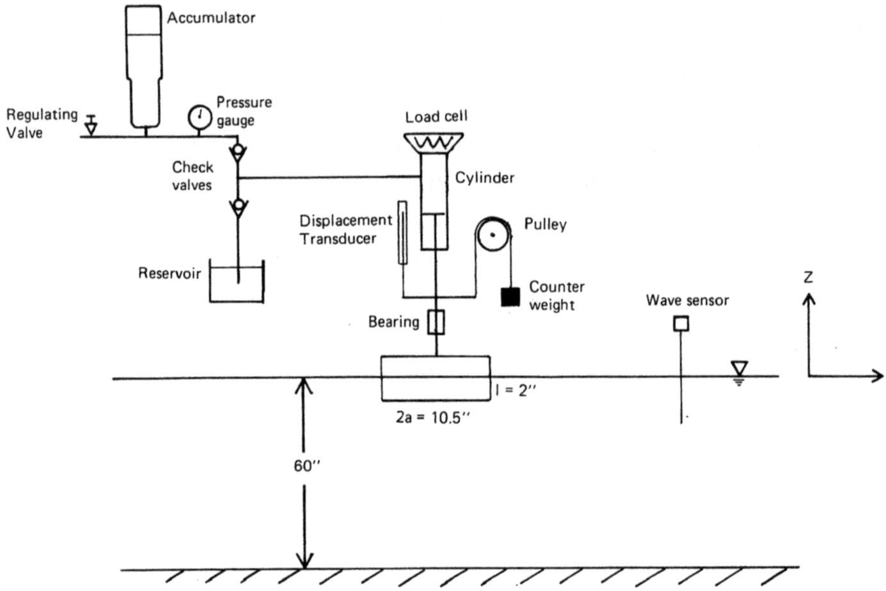

Figure 1. Lay-out of the nonlinearly damped point absorber system modeled in the present study.

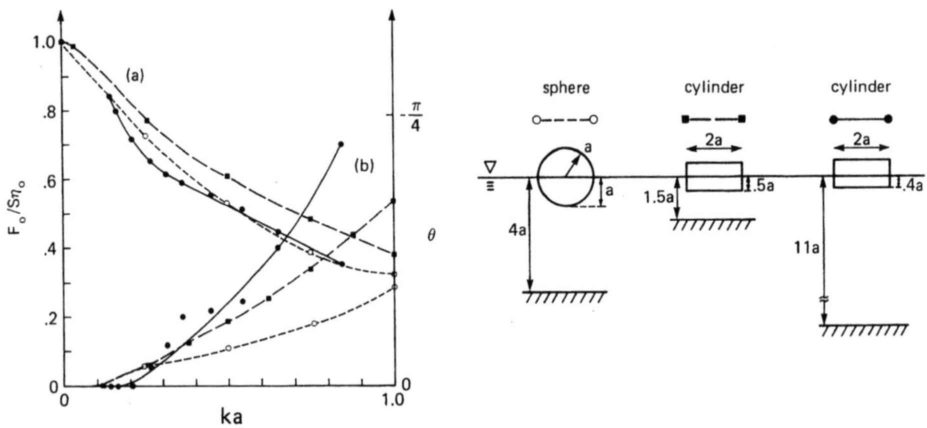

Figure 2. The nondimensional wave force amplitude (a) and phase lag (b) as a function of ka for a truncated cylinder, (●——●) present study, (■— —■) Garrett [10]); and a half immersed sphere, (○----○) Black [9].

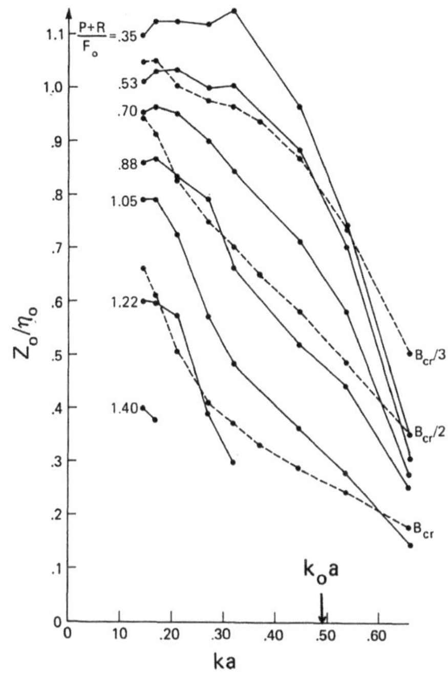

Figure 3. The system gain versus ka for a point absorber with one-way resistive loading (•——•), and for a linearly damped harmonic oscillator (•------•).

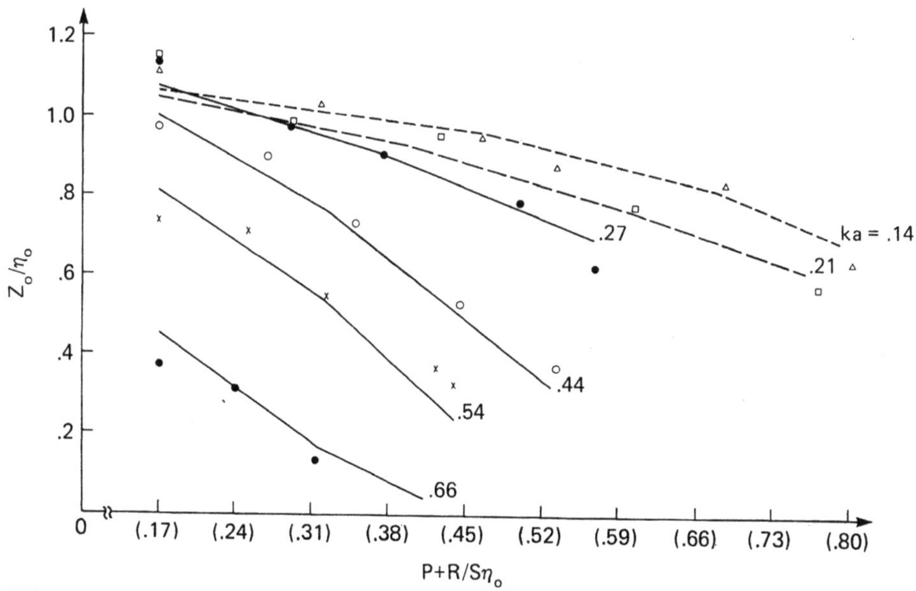

Figure 4. The system gain versus relative damping load for a point absorber with a single-acting hydraulic cylinder. The solid line represents output from the mathematical model with symbols showing data from the physical model.

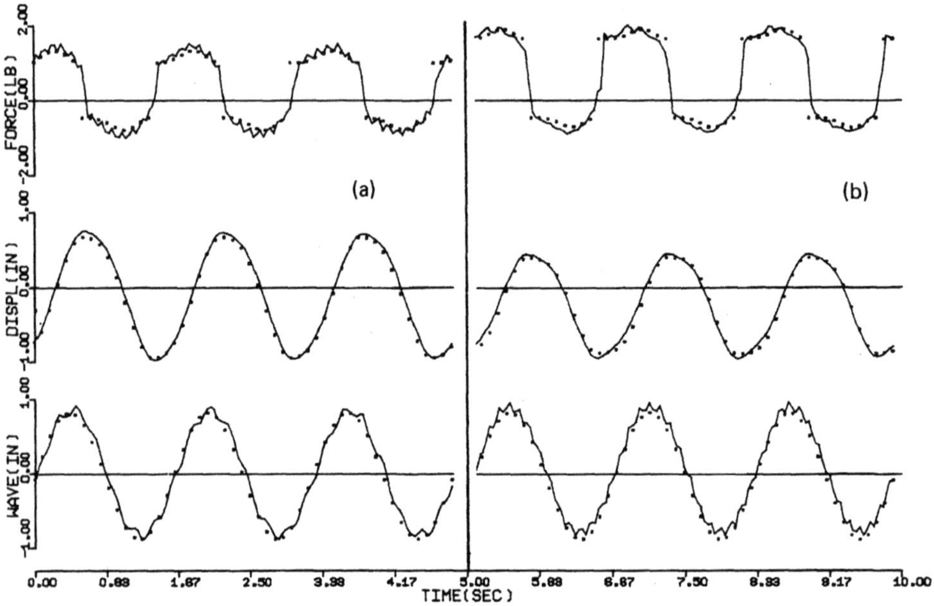

Figure 5. Wave, buoy displacement and damping force records for a point absorber driving a single-acting pump; P + R/F_o = .55 (a) and .85 (b); (———) physical model, (••••••••) mathematical model; ka = .21.

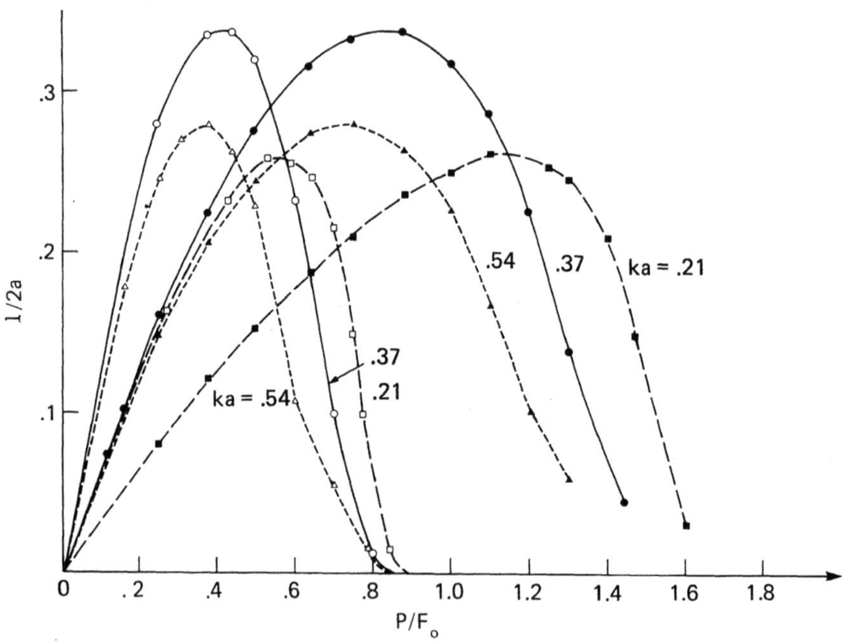

Figure 6. Capture width versus relative damping load for a point absorber driving a hydraulic cylinder. The data is from the mathematical model with solid symbols representing a single-acting pump and open symbols a two-way pump; R/Sη_o = .08 and C_v a/M = 1.4.

Progress in China's Developmental Research on Wave Energy

Li Guixiang, Guo Jialiang

Institute of Marine Scientific and Technological Information, NBO, Tianjin, the People's Republic of China

Summary

This paper describes the research in China on offshore wave energy resources and makes an estimate of the deposits of resources. It also gives an account of the history and status of the developmental research.

With the progress of ocean development, great progress has been made in the study of wave energy utilization in China in recent years. More than ten institutions under the Chinese Academy of Sciences, the Ministry of Communications, the ministry of Education, the Ministry of Water Resources and Electric Power and the National Bureau of Oceanography have undertaken the evaluation of wave energy resources and the study of wave-activated electricity generation one after another, and some encouraging achievements have been obtained in wave energy development technology.

A. Evalution of Wave Energy Resources

Close to the broad Pacific Ocean, China borders on the Bohai-sea, the Yellow Sea, the East China Sea and the South China Sea on the east and south, and the total area of the four seas is over 4.7 million square kilometers. The mainland coastline of China from the mouth of the Yalu River in the north to the mouth of the Beicang River in the south extends to 18000 kilometers, which strides across 40 degrees of latitude. China is situted in the east Asia monsoon climatic zones. The bias north wind prevails in winter and the bias south wind in Summer. The wave along the coast is mainly wind wave caused by wind force with bias north wave prevailing in winter and the bias south wave in Summer, and secondly the swell from the Pacific Ocean. The mean wave height is bigger in winter and the maximum wave height observed from January to February is 11.5 meters. The wave height in Summer is smaller. But exceptionally big wind waves with a maximum height of 9.5 meters may rise from June to September due to typhoon. The wave along the coast of the Bohai Sea and the Yellow Sea is the smallest among the wave along the coast of China. The mean wave height is only below 0.6 meters and the annual mean wave period is about 5 seconds. The wave height along the coast of the East China Sea is the biggest along the Coast of China. The mean wave height comes to 0.9-1.2 meters and the annual mean period of wave is 6-7 seconds. The wave height along the coast of the South China Sea takes the second place. The mean wave height is 0.6-0.9 meters and the mean period is about 6 seconds. In view of the above, preliminary evaluation and analysis of the surface wave energy and coastal wave energy resources of the Bohai Sea, Yellow Sea, East China Sea and South China Sea have been made by chinese oceanographers.

1. Evaluation Method

The wave energy depends on the wave conditions. That is to say, the wave energy is related to the height and period of wave. The formula for calculating total wave energy at the width of unit wave front is

$$E = \frac{1}{16} \rho g h_s^2 \qquad (1)$$

The formula for calculating the mean wave power at the unit wave front width is

$$P = \frac{\rho g}{64} H_s^2 \overline{T} \qquad (2)$$

where ρ is the density of sea water; H_s is the effective height; $\overline{T}$ is the mean period of wave. The wave energy resources in the coastal waters of China and its adjacent waters can be calculated with formulas (1) and (2).

2. Calculation of Coastal Wave Energy Resources

The coastal wave energy reserves are calculated with wave data observed by oceanographic stations along the coast of China. Rough calculation indicates that the theoretical reserves of wave energy along the coast of China are rich, with total power up to 70 million kW. But the density of wave energy on the unit coastline is small and the mean density is only 3kW/m. The density of wave energy along the coast of Zhejiang Province and Fujian Province is the biggest, being up to 6kW/m.

The evaluation of wave energy resources along the coast of China is given in Table 1. It is seen from the table 1 that the wave energy is dispersed mainly over the coast of the East China Sea and South China Sea, each occupying 46%. The reserves of wave energy along the coast of the Bohai Sea and Yellow Sea is only 8% of the total wave energy reserves of the whole country. The reserves of wave energy along the coast of Zhejiang and Fujian Provinces are the biggest, being over 11 million kW and over 20 million kW respectively.

Table 1 Calculation of Wave Energy Resources along the Coast of China

Sea Area	Mean Wave Height $H_{1/10}$	Mean Period $\overline{T}$(s)	Density of Wave Energy	Theoretic Power(kW)
Bohai Sea	0.7 m	2.7	0.74kW/m	1.511×10^6
Yellow China Sea	0.7 m	3.5	1.05 "	4.267×10^6
East China Sea	1.3 m	5.2	4.49 "	32.521×10^6
South Sea	1.1 m	3.8	2.70 "	32.321×10^6
Total				70.620×10^6

3. Calculation of Wave Energy Resources in the Adjacent Seas

The reserves of wave energy in the Bohai Sea, Yellow Sea, East China Sea and South China Sea have been calculated quantitatively with "Marine Survey Data" published by the National Bureau of Oceanography and "Ship Report Data" published by the National Bureau of Meteorology.

The reserves of wave energy in the adjacent sea of China are rather rich. The total wave energy is about 806.3×10^{13} Jouls and the total power is about 574000 million kW through calculating the theoretical reserves (see Table 2). If 1% of the total power is used, the energy for development is about 5740 million kW. And if the total energy is converted into electric power, 5×10^6-50×10^6 million kilowatt-hours can be generated each year.

Table 2 Calculation of Wave Energy Reserves in the Four Seas

Sea Area	Energy ($\times10^{13}$ Jouls)	Power ($\times10^{13}$ W)
Bohai Sea	12.9	1.1
Yellow Sea	60.1	4.7
East China Sea	185.5	13.3
South China Sea	574.8	38.3
Total	806.3	57.4

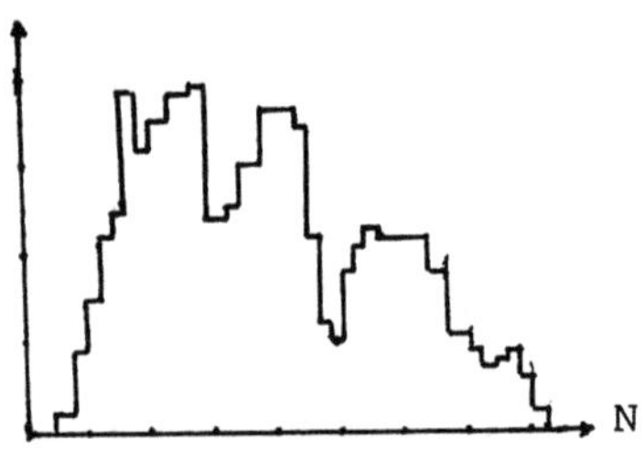

Fig.1. Latitudinal Distribution of Wave Energy

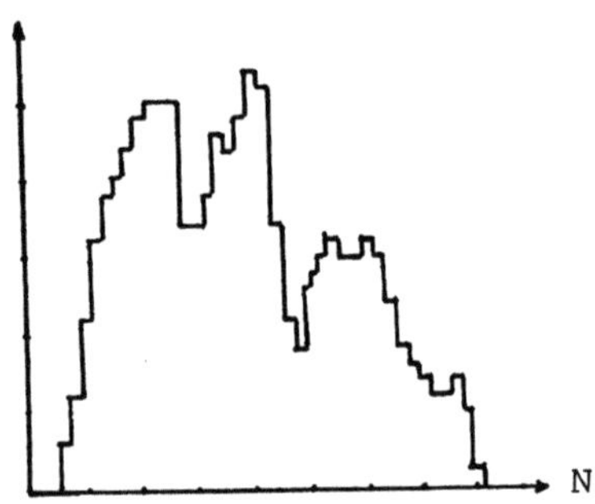

Fig.2. Latitudinal Distribution of Wave Power

The rough latitudinal distributions of the total wave energy and power in the adjacent sea of China are given in Figs. 1 and 2. Their distributions are identical on the whole. It is seen from the two figures that the three fifths of the wave energy are concentrated on the South China Sea.

The rough longitudinal distributions of the total wave energy and power in the adjacent seas of China are given in Figs. 3

and 4. The energy is concentrated mainly between 110°-119°E, and the South China Sea is just right in this range. The longtudinal distribution of wave energy is identical bassically with the latitudinal distribution of wave energy.

Based on the above calculation and analysis, we have found that the wave energy resources of China have the following features:
(1). The geographic distribution is uneven. As fas as the sea areas are concerned, the wave energy is small in the northen sea area and big in the southen sea area. The wave energy is concentrated mainly in the East China Sea and South Sea. With respcet to locations, the wave energy along the coast of islands is richer than that along the coast of the mainland. The wave energy along the coasts of islands off Fujian Province and Zhejiang Province, east Guangdong Province and the Xisha islands is the richest along the coast of China.

(2). Theseasonal variation is conspicuous. The wave energy resources in China, especialy the wave energy along the coast of China, are significantly affected by monsoons, generally higher in Autumn and Winter, and lower in Spring and Summer. The variation of wave energy near islands is more obvious than that near the mainland.

(3). The long coast of China, except the longer shallow coast of the Bohai Sea and Yellow Sea, is bassically the erosion coast, suitable for building wave power stations.

(4).The sea area rich in wave energy is also an important area for developing industry and fisheries which is short of conventional energy sources and urgently needs electric power. there are the favourable factors for developing wave energy.

B. Studies on Electricity Generation by Wave Energy

Through primary calculation of the wave energy resources, we have found that there are favourable conditions for generating power by wave energy along the coasts of many islands. So during the transition period towards a variety of energy sources, it will have immeasurable effects on the national defence building, islands development and ocean engineering to actively undertake the applied studies on wave power generation. The study of wave power generation in China began as early as mid-70s. During the early days, only one institution in Shanghai studied the wave-activated generating sets. Since the beginning of 1980s, the study of electricity generation by wave energy has been developing rapidly. More than ten institutions under the Ministry of Communication, the Ministry of Water Resources and Electric Power, the Ministry of Education and the Chinese Academy of Sciences have begun to study electricity generation by wave energy. So far, great progress has been made in the principles test, device development and conversion technique.

The 10-watt wave-activated valve turbine generating sets on navigation buoys passed the technical appraisement in Shanghai in 1982, have been produced in batches and used extensively. This kind of generating sets are the results of the joint work of the Institute of Design of Shaghai Waterway Bureau and Shanghai Jiaotong University. The set can generate electricity under the conditions of gentle breeze and wave height of 0.2 meters. Its maximum power amounts to 60 W and the electric energy production a day can be used for about three days for a navigation buoy.

The 10-watt wave-activated valveless turbine generating sets on navigation buoys passed the technical appraisement in Guangzhou in March, 1985, which is developed by the Guangzhou Institute of Energy Resources, the Chinese Academy of Sciences. Two sample generating sets of this kind were deployed in the South China Sea in 1984. The primary test on the sea indicated that the electric energy production by a generating set a day is 6.9 times the power consumption by a navigation buoy a day. After having gone through two tests of typhoon,the set runs well.

Besides, Tianjin University, the South China Institute of Technology, Shandong College of Oceanology and the Institute of Hydraulics of the Chinese Academy of Water Resources and Electric

Power are undertaking studies on wave energy development. During the past over ten years, China has obtained some achievements in the study of electricity by wave energy and come to the stage of building special wave energy devices for experiment and undertaking theoretical analysis and calculation model tests with wave pools for ship tests and harbour engineering from the early stage of imitation design and primitive direct determination at sea. At present, not only 10-watt wave -activated generating sets on navigation buoys have been developed, but also the medium sized wave-activated generating sets and choosing sites for building wave power stations are under way. In the last ten years, institutions and scientific menbers undertaking studies on wave energy development have increased gradually. China now has more than ten institutions and about one hundred scientist workers at this work.

C. Prospects of Studies on Wave Energy Development in China

There are over 5000 islands in China's Seas. Conventional energy resources are in short supply there, whire affects the economic construction and improvement of standards of both the material and cultural life of the people on the islands. As wave energy is one of the important natural energy resources of the island, development of wave energy resources is important for developing China's marine economics. This has been paid great attention to by the country. At present, China is strengthening the planning and actively undertaking studies on the development of wave energy. In the days to come, the major work will be to consolidate and perfect the achivements in the mini-wave-activated generating set and undertake the earlier stage study on the medium-sized wave-activated generaing set. It is estimated that by 1990 the 100-watt wave-activated generting set can be produced in batches and the 1000-watt wave-activated generating set will become practical to meet the power needs of the inhabitants on islands.

By 2000, the goal of produing the 10-watt and 100-watt wave-activated generating sets commercially will be achieved, and the 1000-watt wave-activated generating sets will be produced in small batches, with a total capacity of several thousands of kW.

References

1. Wang Chuankun: Primary analysis of ocean energy sources of China. The Ocean Engineering, 2 (1984) 58-67.
2. Ma Huaishu; Yu Qingwu: The preliminary estimate for the potential surface wave energy resources in the adjacent sea areas of China. Marine Science Bulletin, 3 (1983).
3. Yearbook of Ocean Technology (1982). Edited by Institute of Marine Scientific and Technological Information, NBO.PRC, published by China Ocean Press, 1983.
4. Sun Xiangping et al.: The outline of marine hydrology and meteorology along the coast of China. Published by China Sciences Press, 1981.

Wave Climate and the Wave Power Resource

DENIS MOLLISON

Department of Actuarial Mathematics and Statistics,
Heriot-Watt University, Riccarton, Edinburgh EH14 4AS, Scotland.

Summary

Wave climate statistics are reviewed from the wave power engineer's point of view. It is emphasised that there is no fixed answer to the question "What do we need to know about the wave power resource ?" but rather a feedback loop between our knowledge and the development of any particular type of wave power device.

The mean power available averages over 40 MW/km along the best oceanic coasts, such as those of western Europe; of which at least 50% is potentially economically extractable. The total resource worldwide is roughly equal to present world electricity consumption. (Electricity is not the only possible use; direct use of wave power may also be attractive, for instance for desalination.)

Current interest is more in small scale devices, perhaps 10 to 30 metres wide, which enjoy a number of advantages, including the 'point absorber effect', whereby they have an effective capture width considerably greater than their physical width. Mean outputs of 1 to 2 MW should be feasible, and even devices with outputs as low as 100 kW may be economically attractive.

The outstanding data requirement is for better information on directional spectra, for specific analyses of device performance and survival rather than for power output estimates.

Contents

1 Introduction

That there is energy stored in the oscillations of the sea surface is obvious to anyone who has experienced an ocean storm. Wave climate statistics, especially of extremes, have been of interest for a long time to ship and harbour designers, and more recently to the oil industry. That wave energy travels, and that it forms a substantial renewable energy resource, has only become widely appreciated over the last dozen years, though patents for wave power devices go back to the eighteenth century.

My intention here is to answer the question - What is the wave climate like? - from this new point of view, emphasising those aspects of the resource of most interest to the designers of wave power devices, and indicating aspects where more work is needed. The answer is necessarily somewhat open. On the one hand it is important to describe the wave climate in as simple terms as possible so as to clarify the designer's task. On the other, the development of designs, and new design ideas, often lead to new demands on our knowledge of the wave climate, and on our ability to summarise aspects which are newly seen to be important.

Some of the basic aspects of the resource were recognised early as of importance; for instance, its character, as mechanical oscillations of a medium, its characteristic frequency range (.05 to .2 Hz) and scale, and its variability, both from wave to wave and over longer time periods. One cannot consider the resource in isolation from the type of device proposed for its exploitation; a distinction which is increasingly recognised as important is that between the large and small scale resources.

The recent development of wave power studies owes its origin to the panic over energy supplies which followed the Yom Kippur war of 1973. Understandably therefore, the early emphasis was on establishing and trying to exploit the large scale resource, researching and designing installations that could make a significant contribution to a large population's electricity demands.

In the oil glut of the 1980s governmental attitudes, especially in the UK, have changed to indifference or hostility towards large scale use of renewable energy sources. Emphasis has therefore shifted to where perhaps it should have started: to the design of small scale devices, both to meet local demands, and as prototypes for the time when interest in renewable energy sources revives.

The distinction between large and small scale devices is not just political and social. A device small compared with the wavelength of sea waves can capture power on a front significantly greater than its physical width, and this provides a clear scientific reason for distinguishing between large and small scale resources in this review.

Before trying to characterise these resources, I shall first describe how wave power arises, its general characteristics, how the resource is estimated, and how these estimates can be used in the design and evaluation of devices.

2 General character of the resource

Considered in terms of the earth's energy balance, wave power is a minor byproduct of wind power, which is in turn a minor byproduct of solar power. The atmospheric circulation aids in redistributing the sun's energy input (see e.g. Battan [3]), and loses a very small proportion of this energy through drag on the sea surface, thus creating gravity surface waves. The net power input from wind to waves is seldom as much as 1 watt per square metre, compared with the mean solar radiation of 350 watts/m^2.

However, once energy has been put into gravity surface waves, in deep water it dissipates slowly (hardly at all for low waves of more than 9 seconds period) and meanwhile travels, typically at 500 to 1000 kilometres a day. Thus the power input over an area of ocean can be harvested along its boundary, with power densities higher than those of the original solar input: as we shall see, the net resource crossing a line facing prevailing oceanic winds can average 40 to 50 MW/km (or kW/m; it seems appropriate to use MW/km for the large scale resource, kW/m for density at a point or for the small scale resource).

This advantage of concentrating the earth's solar energy input is to some extent shared by wind power, as is its form as mechanical energy rather than heat. The overall wind power flux is much greater, but is spread over a height of several kilometres, whereas the wave power flux is largely contained in the top 10 to 20 metres of the sea, so that it can achieve power densities about ten times as large. Also, wave power enjoys a theoretical advantage: with wind power the medium travels, and there is a limit to the possible efficiency of extraction of $^{16}/_{27}$, ≈ 57% (see e.g. Lipman *et al* [24]); with wave power, the medium only oscillates, and 100% efficiency of extraction is theoretically possible (and can very nearly be attained in experiments).

Both wave and wind power have the practical advantage of being in general terms concentrated when and where they are wanted: in the temperate zones, and in winter rather than summer, since this is where the solar energy gradient with latitude is greatest. The wave power resource is thus concentrated around the oceanic countries, particularly those with prevailing onshore winds, such as western Europe, the (western) United States and Australia. Large scale installations could make a substantial contribution to those countries' energy requirements. Small scale installations are an exciting prospect for a number of oceanic communities presently disadvantaged by expensive power supplies. The direct use of wave power for desalination is also a possibility for some of these communities.

The main disadvantage of wave power, which it shares with wind, is its variability. There is great variation in power levels, with the passage of each wave, from day to day, and even from month to month (though, as already noted, the general pattern of seasonal variation is favourable). This poses problems for the wave statistician as well as for the engineer. Long data runs are needed to provide reliable wave climate estimates, and there are considerable problems of correlation in choosing representative samples for tank tests, and in estimating the return period of extreme wave heights. The accuracy of climatic estimates can be improved by the use of background *wind* climate information.

The waves generated by a particular wind show a directional spread about the wind's principal direction. We need to distinguish between the gross wave power level, which may average nearly 100 kW/m in mid ocean, and the net resource crossing a line which will only be a proportion (the *directionality coefficient*) of that figure. In mid ocean the directionality coefficient will typically average only around 50% even for the best (fixed) direction; near coasts the combination of sheltering and refraction (which bends waves towards the slope direction in shoaling water) leads to directionality coefficients close to 100%.

While accurate measurement of wave heights, and of the frequency domain, can be carried out routinely by a variety of techniques, notably wave buoys, the directional domain is more problematic. Directional measurements are expensive, and subject to theoretical limitations on their resolution. Although there has been substantial progress on a variety of techniques in recent years, for wave climate estimation we still have to rely on syntheses and wind-wave models, calibrated by routine non-directional measurements and using formulae for directional spread based on the best measurements that have been made.

In evaluating the small scale resource, we need to take account of the available advantages of geographically special sites, of refraction 'hot spots', and above all of the point absorber effect: namely that devices can potentially capture power from an additional 'capture width' of *wavelength*/π (or *wavelength*/2π for an axisymmetric device) - for a small device this can multiply the potential resource by a factor of five or more.

3 Wave spectra

Deep water waves can, to a good approximation, be regarded as composed of the linear superposition of a large number of simple components. Each of these basic components is a sinusoidal wave train, with period T say (typically 5 to 20 seconds), which appears to travel at its *phase velocity* $U = gT/2\pi$. The phase velocity is typically similar to the velocity of the wind generating the waves (see next Section). The wavelength is $L = UT = gT^2/2\pi$.

In fact the water particles do not (to first order) travel, but only oscillate in circular orbits, whose amplitude falls off exponentially with depth ($a_d = a\exp(-kd)$, where k is the wave number, $= 2\pi/L$).

The energy of the wave train, per unit area, is $E = \rho g H^2$, where ρ is the density of sea water, and H is the root mean square wave height : $H^2 = \frac{1}{2}a^2$. This energy is made up from equal components of kinetic energy and potential energy.

The power flux P, per unit of wave front, can be evaluated from the simple observation that, by moving in circular orbits at constant speed, water particles are transferring potential energy in the apparent direction of the waves. This shows that P is equal to the potential energy per unit area, $\frac{1}{2}E$, multiplied by the phase velocity U, i.e. $P = cH^2T$, where the constant $c = \rho g^2/4\pi$, $\approx$ 7.87 kW/m^{3}sec.

This calculation only applies to a wave train in steady state. More generally, the total energy E can be shown to travel at the *group velocity*, which in deep water is equal to half the phase velocity. (Drop a stone into a pond, and note how the group travels more slowly than the individual ripples, which first grow and then decay as they pass through the centre of the group.) Thus $P = \frac{1}{2}EU = cH^2T$, as before.

For the overall sea state, the movements, energy and power of the individual sinusoidal components add up to give a Gaussian random process, whose spectrum (strictly spectral density) $e(f,\theta)$ describes how the total energy E is distributed over frequencies f and directions θ. Thus

$$E = \iint e(f,\theta)\, df\, d\theta = \rho g H_{rms}^2 .$$

Here H_{rms}^2 is the variance of the surface elevation. A common alternative is to work in terms of the *significant wave height* H_S : this was originally defined as the height of the highest one-third of waves (measured trough to crest between successive zero crossings), which $\approx 4.0\, H_{rms}$ for a narrow spectrum, but is now normally used simply as a synonym for $4H_{rms}$.

It is often convenient to work in terms of the spectral distribution $s(f,\theta) = e(f,\theta)/E$, whose overall integral equals unity. The gross power flux is

$$P = \iint \tfrac{1}{2}U\, e(f,\theta)\, df\, d\theta = E \iint \tfrac{1}{2}U\, s(f,\theta)\, df\, d\theta = cH_{rms}^2 T_e ,$$

where $c = \rho g^2/4\pi$ as before, and the *energy period* T_e (Mollison *et al* [34]) is the average value of T with respect to the spectral distribution, i.e.

$$T_e = \iint T\, s(f,\theta)\, df\, d\theta .$$

Since $T = 1/f$, we may write $T_e = m_{-1}$, where m_n denotes the nth moment of the spectral distribution,

$$m_n = \iint f^n s(f,\theta)\, df\, d\theta .$$

(Note: it is common to work with $S(f,\theta) = H_{rms}^2 s(f,\theta)$ instead of $s(f,\theta)$; this leads to formulae $H_{rms}^2 = m_0$ and $T_e = m_{-1}/m_0$.) It is convenient to define the mean phase velocity $U_e = gT_e/2\pi$, and the corresponding wavelength $L_e = gT_e^2/2\pi$; the ratio H_{rms}/L_e is then a measure of the steepness of the sea state.

The net power flux P_θ in direction θ is found by multiplying the energy of each component by its group velocity resolved in direction θ, i.e.

$$P_\theta = cH_{\rm rms}^2 \int \int_{\theta-\pi/2}^{\theta+\pi/2} T\ s(f,\phi)\ \cos(\phi-\theta)\ df\ d\phi\ .$$

Note the limits of integration: waves from opposite directions pass through each other, they do not cancel out. For either an individual sea state, or a wave climate average of sea states, we can define the *directionality coefficient* $d_\theta = P_\theta/P$.

Various measures of spectral width have been proposed. The most convenient from a theoretical point of view (see Longuet-Higgins [25]) are defined in terms of higher moments of the spectrum, including at least m_2. For wave power purposes it is best to have a parameter which is dependent much more on the more powerful low frequency end of the spectrum; one such is σ_T, the relative standard deviation of the period T (Mollison [31]), i.e.

$$\sigma_T = \sqrt{m_{-2} - m_{-1}^2}\,/m_{-1} = \sqrt{(m_{-2}/m_{-1}^2) - 1}\ .$$

(See Section 8 for a use of m_{-3} !)

4 Wave growth, travel and decay

The slow processes of wind and wave interactions that generate surface waves are highly nonlinear, and still not fully understood (see e.g. Leblond & Mysak [23]).

Briefly, energy and momentum are transferred from the turbulent atmospheric boundary layer to the sea through air pressure and shear stress. Energy is input mostly to the high frequency end of the wave spectrum, and subsequently transferred to lower frequencies through nonlinear wave-wave interactions (Hasselmann [17, 18]). These interactions conserve the overall energy and momentum, and there is also a transfer to higher frequencies, but this is balanced by dissipation through breaking (see Figure 4.1).

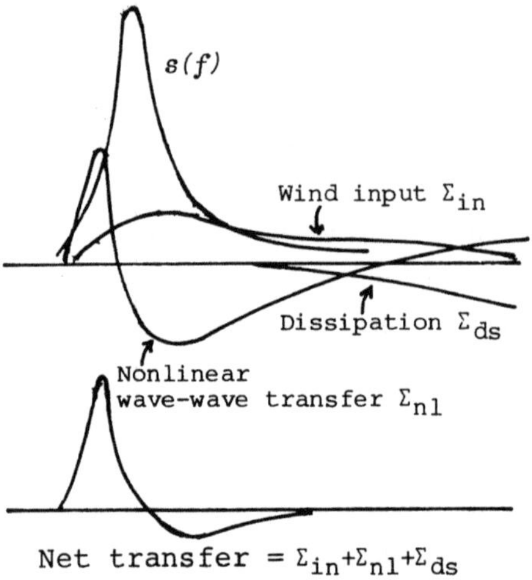

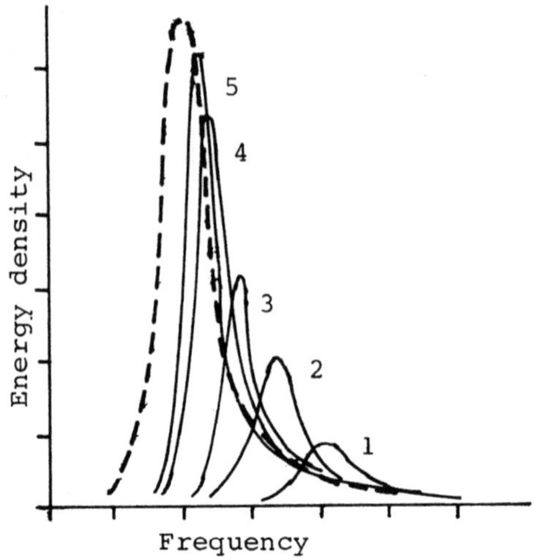

Figure 4.1
Energy transfers in the spectrum of a growing sea (after Hassselmann *et al* [19]).

Figure 4.2
Growth of wave spectrum with fetch (after Hasselmann *et al* [19]), with tentative fully-developed spectrum (dashed curve).

The growth of waves from flat sea to near full development was studied in detail in the JONSWAP project (Hasselmann *et al* [19], Ewing [13]), confirming this general theory of wave growth. They found that wave energy density E grew approximately linearly with fetch x; thus

$$H_{rms}^2 \approx 1{\cdot}6 \times 10^{-7}\ U_{10}^2\ x/g$$

where U_{10} denotes the windspeed measured at 10 metres above the sea surface. Note that this relation, which was observed to hold for fetches up to 10^3 - 10^4 times U_{10}^2/g, does not mean that the rate of net energy input is constant, as is sometimes assumed in discussing this result. The rate of net energy input equals the rate of change of power, and thus will be proportional to T_e, which is increasing. Figure 4.2 shows the kind of growth of wave spectra observed during the JONSWAP experiments. Hasselmann *et al*'s formula fitted to the JONSWAP data gives T_e proportional to $x^{\cdot 33}$, but this, like the energy growth formula, can only hold for growing seas.

The JONSWAP experiments unfortunately tell us little about the limits to wave growth. For this we must turn to earlier work, such as the classic papers of Pierson & Moskowitz [35, 38, 37]. Following the scaling law approach of

Kitaigorodskii, and on the basis of data for apparently fully developed seas in winds of up to 40 knots ($\approx$ 20 m/sec), they proposed a wave spectrum which can be written as

$$\int_0^f S(y)\, dy = H_{rms}^2 \exp[-.675(fT_e)^{-4}]$$

where T_e = $\cdot 625\ U_{19\cdot 5}$, H_{rms} = $\cdot 0053\ U_{19\cdot 5}^2$. This spectrum has a high frequency tail $\sim f^{-5}$ as originally proposed by Phillips [36].

Its spectral width σ_T, $\approx$ ·28, and its steepness, $H_{rms}/L_e \approx 1/115$, are independent of the windspeed. Its mean phase velocity is almost exactly the same as the windspeed, $U_e = \cdot 975 U_{19\cdot 5}$; this seems a suggestive coincidence, but there is no direct theoretical explanation for it.

As one would expect from dimensional considerations, the power in such a fully developed sea is proportional to the fifth power of windspeed U^5, and the fetch required to generate it is proportional to U^2, and can thus be expressed as a multiple of the wavelength. If we assume that the JONSWAP energy growth formula applies up to the stage where the Pierson-Moskowitz value of H_{rms} is attained, we find that this fetch is approximately $3000L_e$. For instance, for $U_{19\cdot 5}$ = 20 m/sec, we obtain H_{rms} = 2·12 m, T_e = 12·5 secs, and the required fetch is about 700 km (the power density attained is 440kW/m, so in this example the mean rate of energy input during growth is about 0·6 watts/m^2).

Fully developed seas associated with windspeeds greater than about 20 m/sec will be rare because of the fetch required. In practice, one of the most favourable conditions is where a strong depression moves so as to stay on top of some of the waves it is creating, which will occur if the depression moves at approximately their mean group velocity (= $\frac{1}{2}U_e$). In the North Atlantic, depressions have an average eastward velocity of around 8 m/sec (Bartholomew & Herbertson [2]), and so the effective fetch of westerly seas is often significantly increased.

Nevertheless, most occurrences of extreme power levels and steepness are associated with growing seas, for which the narrower JONSWAP shape of spectrum is more appropriate ($\sigma_T \approx \cdot 2$).

The directional spread of wind seas is less accurately known than their frequency spectrum. Mitsuyasu [30] fitted a spreading function of $\cos^{2r}(\theta/2)$

form,

$$s(f,\theta) = s(f)\ c_{\theta}\ \cos^{2r}((\theta-\theta_0)/2)$$

where θ_0 is the wind direction, and the frequency dependent parameter r reaches a maximum of 15·85 at the peak frequency f_p; that is, the spectrum is most narrowly directional around the peak frequency. Note that this formula allows for energy travelling in all directions, unlike the alternative $\cos^r(\theta)$ $(\theta_0-\pi/2 < \theta < \theta_0 + \theta/2)$ form of spreading function which is commonly fitted to swell, appropriately enough as swell normally has a very narrow directional spread.

In the open ocean, waves decay only very slowly once their steepness is reduced below the point at which breaking losses are significant. Snodgrass *et al* [47] observed swells which had crossed the whole Pacific Ocean. Numerical models (e.g. Golding [15]) tend to assume that no losses occur. Certainly losses from viscosity are minimal: a wave of 9 seconds period would need to go three times round the world to lose even 10% of its amplitude (Crapper [10]): but losses to other kinds of ocean wave, such as internal gravity waves, are possible (see e.g. LeBlond & Mysak [23]).

As waves enter shoaling water, their direction is altered by refraction so that their principal direction is brought nearer the perpendicular to the depth contours, while their directional spread tends to be narrowed. Note that it is the net power crossing a line that is conserved, so that as the directional coefficient increases the gross power level will decrease even if there is no actual energy loss.

Over uneven bathymetry, refraction will cause focussing over a convex bottom (defocussing over a concave one) leading to 'hot spots' of above average power: Mehlum [29] gives examples of sites whose mean power is more than doubled by this effect.

It is well known that waves lose the last of their energy through breaking in shallow water. Energy losses in intermediate depths (20 to 100 m) are mainly due to bottom friction (Bagnold [1], Vitale [50], Hasselmann *et al* [19]). Mollison [33] estimated the bottom friction coefficient required to account for observed wave climate losses over the rough but gently shoaling bottom off South Uist. The estimated loss rate was (e.g.) around 1% per km in 35 m depth, and 70% of the energy was lost between 100 and 15 m depth (see also Section 7). Backscattering and reflection may also account for significant losses in appropriate circumstances (Shemdin *et al* [46]).

5 Measurement, estimation and simulation of waves

Wave buoys at present afford the most reliable and accurate wave height measurements, and hence estimates of gross power levels and frequency spectra. Calibration errors are normally only a few per cent, less than the absolute error in estimating spectral parameters from a perfect wave record of typical length (around 100 wave periods) (see e.g. Tucker *et al* [49]).

While good quality data runs over considerable periods exist for wave buoys (e.g. an over 90% data return during 1976-78 for the Institute of Oceanographic Sciences 42 m depth buoy off South Uist, NW Scotland), wave power variability from year to year is such that it is desirable to use longer term wind information when estimating the long term average power or selecting a subsample representative of the wave climate. Two methods, stratified sampling (as used by Crabb [8]) and a 'wind input' method are compared by Mollison [31]. Mollison [31] also estimated that even with 13 years of exact wave measurements, the estimate of the long term average power would only be known within 95% confidence limits of ± 6%; however this variability is unduly influenced by rare extreme storms, so that estimates of mean output power will be considerably more reliable.

Directional buoys have been available for some time (Longuet-Higgins *et al* [26]), but have only recently been deployed on a routine basis. Alternative methods include arrays of small buoys (Mehlum [28]), and radar backscattering (Shearman [45]), while satellite techniques offer great promise for the future (Tucker [48]). All these methods are to an extent indirect, since assumptions about the form of the directional spectrum are required for their analysis. Present estimates of directional wave climate mostly rely on syntheses and numerical hindcast methods.

Crabb [8] selected a representative set of 399 wave records for a site in 42 metres depth off South Uist and synthesised their directional spectra, using local wind speed and direction information for the wind sea, and meteorological chart analysis for swell. (See Mollison [31, 33] for comments on the method of selection and effect of finite depth respectively.)

The estimates presented here (Section 7) are largely based on the numerical hindcasting method developed by the UK Meteorological Office (Golding [15]), which has been in routine use for the North Atlantic since 1978. Wave climate estimates from this model for South Uist are in close agreement with Crabb's synthesis. Also a comparison with the NORSWAM hindcasting model, developed to analyse North Sea storms, showed reasonable agreement of the two models with each other and with measured wave data (Ephraums *et al* [11]).

Where estimates of wave climate are required for a new area, it will probably be best (or at least most economical) to use numerical hindcasts if available, calibrated by direct wave buoy measurements; and otherwise to use directional measurements or syntheses. Where, as will usually be the case, several possible sites are to be covered by a single measurement site, it is much preferable to measure (or hindcast) waves at a deep water site with good exposure, from which the climate at inshore and sheltered sites can be estimated using numerical wave models; whereas one cannot use measurements at an inshore or sheltered site to estimate waves anywhere else.

When it comes to reproduction of directional wave climate samples in test tanks, satisfactory methods exist using either discrete spectra (Jeffrey *et al* [22], Salter [42]) or filtered white noise (Bryden [4]). An advantage of discrete spectra is that these can be measured and thus verified with an accuracy that is not possible for more realistic continuous spectra. Two errors to avoid when using discrete spectra are: using more than one component at any frequency, which leads to non-ergodic seas (see, e.g. Jefferys [21]); and not allowing for the inherent variability in (e.g.) energy between samples from the same sea state (see Tucker *et al* [49], but note that the right variability *can* be achieved with a small number of components if their individual energies are chosen correctly, i.e. chi-squared distributed with the appropriate number of degrees of freedom).

6 Wave climate estimates in relation to device design

As described by Mollison [31] in reviewing the prediction of device performance, the detail in which we require knowledge of wave climate advances hand in hand with the development of a device design. At each stage it is important for understanding to present analyses clearly, showing the assumptions behind any conclusions on performance and estimating the errors involved.

Usually the first requirement is for the frequency distribution of the wave climate. One helpful presentation, suggested in this context by Stephen Salter, is as a 'stretchy' axis (Mollison [31]), in which the distance between points on the frequency axis is stretched to represent the proportion of power available in that frequency interval. If efficiency curves are plotted against such an axis (see e.g. Figure 8.1), the area under the curve shows the overall average efficiency in that wave climate; thus providing a good indication as to whether the frequency range(s) where the device is most efficient are well matched to the wave climate. If this presentation is to be a realistic guide to device optimisation, we must discount frequencies most associated with high power levels at which the device will be at full output, for instance by setting a height limit on the data used for calculating the stretchy axis.

Tests in wave spectra, rather than single frequency waves, are essential for the optimisation of device overload and power take-off limits (for more discussion see Mollison [31]). Except for devices with energy storage, such as Ducks and Tapchan (Salter [43], Mehlum [28]), these limits will begin to affect performance while the device is on average still far below full load.

When it comes to selecting a set of spectra for device tests, our aim should be to allow us (1) to estimate output in the overall wave climate adequately (say to within ± 5%), and (2) to evaluate correctly whether particular changes in device design or parameters constitute an improvement relative to the particular wave climate (which may require considerably greater accuracy in comparing pairs of tests).

A sample of 'real seas', however chosen, is not the best answer to this problem. We shall still need to estimate performance in other real seas, both from the same and other wave climates. We may also want to consider the performance of our device scaled up or down in size. And we will want to set error bounds on our output estimates for a particular wave climate. All these considerations argue for using a parametric description of seas, and for testing in a set of seas chosen for its coverage of a range in parameter space including the wave climates we are likely to be interested in (at the appropriate scale or scales). We should develop an interpolation formula for estimating performance in seas not tested in; the role of modelled real seas comes in evaluating whether our interpolation formula is sufficiently accurate.

The choice of an economical set of sea states for tests is a problem of multivariate optimal design, but one whose most important aspect is the choice of an appropriately economical parametric description of sea states. At worst, a description allowing five parameters (H_{rms}, T_e, σ_T, direction θ, and directional spread) each for the wind sea and any swell components should suffice. But for most devices one would expect some much smaller set of parameters to explain most of the variation in output; for instance, for a device with broad frequency response and approximate $\cos(\phi-\theta)$ directional characteristic, the two parameters P_θ and T_e might provide adequate accuracy (here θ is the direction the device faces, ϕ the direction of a component of the wave spectrum).

In any case, it is important to choose test values covering the outer limits of the range of parameter values of interest; just as it is easier to predict the past than the future, so interpolation is to be preferred to extrapolation.

For survival tests, parametric seas are again appropriate; especially as the honest engineer will be searching for the particular circumstances which test her device most severely. This will not necessarily be an extreme 100 year wave, and may be something which might well not be included in a standard test set, such as steep seas within a particular frequency band, or seas in a direction which excites resonance in a spine.

7. The large scale resource

No accurate estimates are available for the world wave power resource; Tornqvist, cited in Glendenning [14], gave worldwide estimates based on visual wave observations (Hogben & Lumb [20]), which are reproduced in Figure 7.1. (For some estimates of the U.S. resource see Hagerman [16].) Arthur, cited in Pond & Pickard [40], estimated the mean power incident on the world's coasts at 2000 GW. From these figures, the world's potentially exploitable wave power resource may be described as being of roughly the same magnitude as present world electricity consumption, i.e. around 1000 GW. Electricity is not the only possible use: direct use of wave power may also be attractive, for instance for desalination (Salter [44]); or for chemical production, for instance of ammonia or hydrogen, where floating factories might be able to harvest the mid-ocean resource.

In mid-ocean the gross power can average nearly 100 kW/m, with substantial contributions from all points of the compass. Nearer to coasts, the onshore components will be broadly similar, but the offshore components of course much reduced (Crabb [7]). Estimates of gross wave power averages for the NE Atlantic

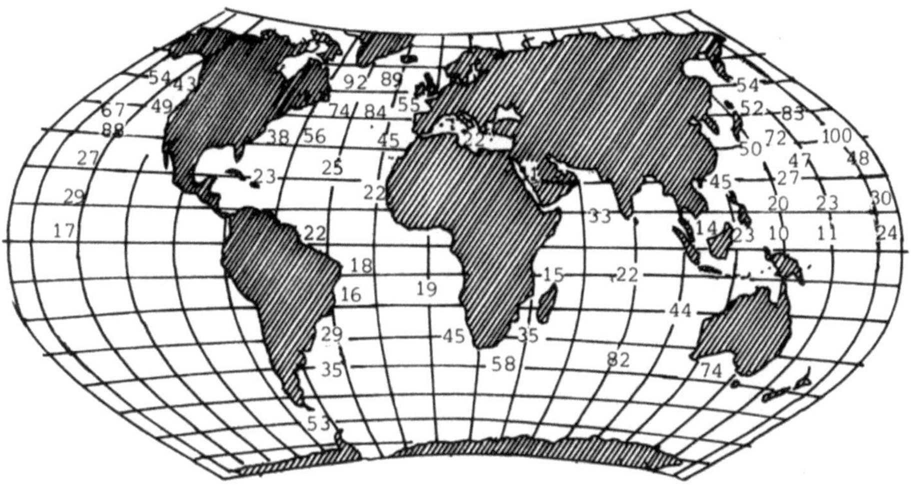

Figure 7.1
Estimates of worldwide mean gross wave power levels in kW/m, from Tornqvist's calculations based on visual wave observations.

are shown in Figure 7.2, with 'wave roses' showing the directional distribution.

In the course of the UK wave energy programme, much attention has been given to various sites off South Uist where measurements have been made by the Institute of Oceanographic Sciences (Crabb [9]), and particularly to a site in intermediate depth (42m) for which Crabb [8] produced a synthesised set of directional spectra (see Section 6). However, as pointed out by Mollison [33], this directional synthesis did not allow for changes of direction due to refraction, and so as it stands is more appropriate for the deeper (100m) site. For this deep water site, there is good agreement concerning the net resource between the adjusted IOS synthesis (46 MW/km estimated long term average, Mollison [33]) and the Met Office hindcasts (45 MW/km average for 1978-81). The latter's higher estimate of gross power (67 as against 59 kW/m) is accounted for by its not taking full account of the site's being sheltered from the east, and including 8 kW/m from that sector.

Off South Uist the sea bottom is mostly very rough, but shelves at a low mean angle (the 100 m site is about 30 km offshore); the significant power loss at intermediate and shallow water sites is thus not surprising (the net power level

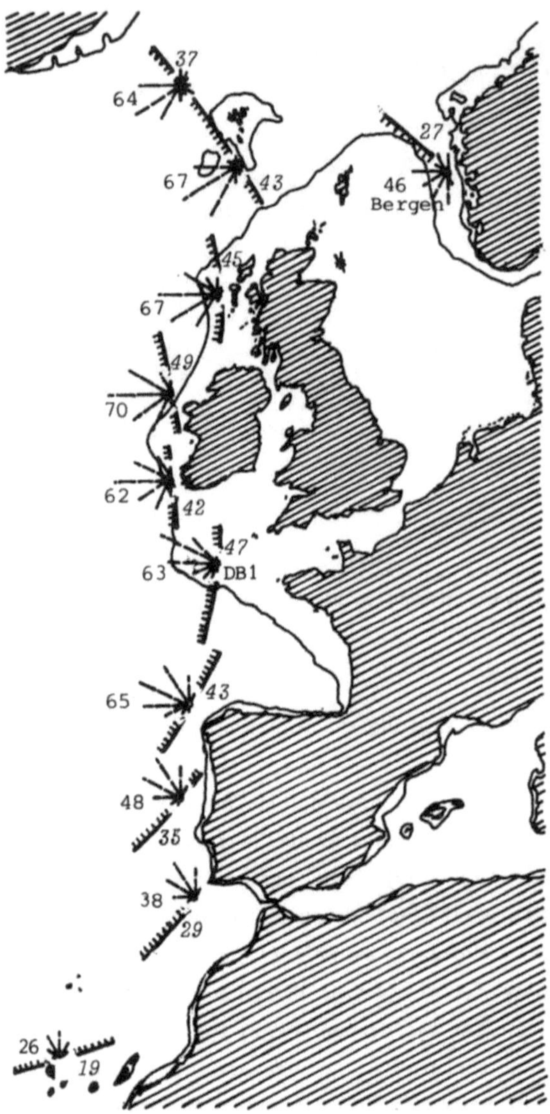

Figure 7.2
Wave power estimates for western Europe based on the UK Met Office wind-wave hindcast model (Norwegian Met Inst model for Bergen). Wave roses show mean power from each 30^0 sector ($22 \cdot 5^0$ for DB1), with marks at 5 kW/m intervals. Gross power levels are shown (in kW/m); and also net power P_θ (numbers in *italics*) crossing lines (⊥⊥⊥⊥⊥⊥⊥) whose direction θ maximises P_θ for the particular site.

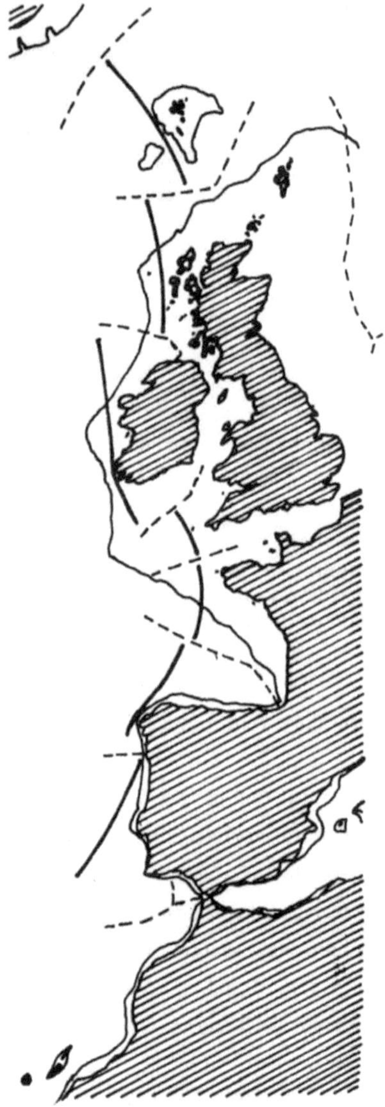

Figure 7.3
Tentative lines for the maximal exploitation of Europe's Atlantic wave power resource, with international sea bed boundaries (– – – –).

(In both figures, the 200m depth contour is shown.)

reduces to 40 MW/km in 42m depth, 30 in 25m, 15 in 15m, Mollison [33]). Even at the 100m site, a few per cent has already been lost in crossing nearly 100 km of continental shelf (depth ≈ 125m). Along most of Europe's west coasts, deep water can be found somewhat closer to shore.

Lines in directions crossed by maximum mean power are also shown in Figure 7.2, together with the value of the maximum, which for coastal sites is typically about 75% of the gross power level. Naturally, these lines face towards the areas of the North Atlantic whence most of the power originates, rather as though they were ripples caused by large stones dropped in the middle of a pond.

Off Scotland and Ireland, these lines run roughly parallel to the coast. Portugal is less favoured, with a net resource of up to nearly 40 MW/km, but with the best direction to face (at a deep water site) being angled at about 45^0 to the coast. The resource component perpendicular to the coast is thus considerably lower, ranging from about 30 MW/km at the Spanish border to 20 MW/km off C. de S. Vicente; and following refraction in intermediate depths, one would expect the gross power level at an inshore site to be only slightly higher than these figures, *before* allowing for any energy losses. These estimates are thus in reasonable agreement with Portuguese estimates of mean gross power from inshore wave buoy measurements, of 27 kW/m in 30 m depth off Oporto (Pontes & Perdigao [41]), and of 26 kW/m off C. da Roca and 16.5 kW/m off Sines (Pires and Pessanha [39]).

Potential output power is best discussed in terms of an actual device. The obvious choice here is the Salter Duck, whose design has been aimed, more than any other, at extracting as much of the available resource as possible. Of course, this is not to deny that other present devices might be more economic, or future ones more productive.

The estimated mean output for Ducks at a deep water site off South Uist is 19 MW/km (Salter [43]); the Met Office model and IOS's data agree for this site to the nearest MW/km. (This is for a device with a peak power rating of 52 MW/km.) The best coastal Atlantic sites would appear to be off western Ireland, with estimated output about 10% higher than off South Uist (Mollison [32]).

While flat calms can occur even in winter, the general pattern of wave power is favourable, with the mean output in the six winter months (October to March) being about 50% higher, nearly 30 MW/km off the Hebrides. (Thus the load factor averages over 55% for the half of the year when demand is greatest.)

All these output figures are for electricity landed (in the case of South Uist, including transmission to central Scotland). That the mean output is less than half the 45 MW/km available in the sea can be explained by three factors of roughly equal importance: the contribution from high power levels which it is not economical to extract, the efficiency of extraction from the sea (about 75% in small waves), and the efficiency of power conversion and transmission (about 80% on average); an allowance has also been made for losses through non-availability, about 5% with economically optimal maintenance effort (Salter [43]).

While clearly there is scope for improvement on current designs, it is difficult to imagine an economic device which extracted more than about two-thirds of the mean resource at such a site. Thus it seems reasonable to summarise the large scale resource of the NE Atlantic, from Iceland to North Portugal, as having a net resource of 40-50 MW/km, of which 20-30 MW/km is potentially economically extractable.

It is not clear how this oceanic resource should be shared out. Stephen Salter has suggested that wave energy should belong to the country on whose shores it would otherwise have been lost. In terms of the law of the sea, wave power is clearly a resource of the high seas rather than of the sea bed (Chapman [5]), but it is not a 'living resource' to which the UN Convention on fishing might apply.

Figure 7.3 shows a rough allocation of Europe's Atlantic wave power resource based on the median line boundaries used for sharing sea bed resources (Chapman [5]). Using a conservative mean output estimate of 15-20 MW/km, the total potential is around 50 GW. Of this, the UK and Ireland each have about 12 GW, while the Faeroes (Denmark), France, Spain and Portugal each have between 5 and 8 GW. The UK also has the advantage of being the natural customer for the excess resource of its less populous neighbours (in particular, the 'Cape Wrath spur', stretching 800 kilometres from Scotland to Iceland).

This discussion of sharing the resource of course looks well into the future, beyond the technical, economic and political problems of exploiting fully such a large and variable source of power. The way towards that future seems presently to lie via the development of the small scale resource.

8 The small scale resource

As mentioned in the introduction, there have been political and commercial reasons for the recent shift of emphasis towards the small scale wave power resource. But at least as important has been the realisation that small scale devices have a number of advantages to balance against the economies of scale of larger installations.

Firstly, advantage can be taken of geographically special sites, for instance where deep water comes close to the shore. More specifically, parts of the shore itself may be conveniently shaped for adaptation to wave power extraction, such as natural blowholes, or clifftop reservoir sites (Mehlum [28]).

Secondly, refraction of waves over variable bottom topography creates local concentrations of above average power ('hot spots'). Whether onshore or in intermediate depths, device sites can be chosen to take advantage of such concentrations, which even averaged over the variety of wave climate conditions can amount to a doubling of the available resource (Mehlum [29]).

Both the above may operate over a scale of kilometres, and thus apply to quite large installations (tens of MW). The third advantage of small scale devices is the point absorber effect, which operates on a scale small compared with the wavelength (tens of metres, output up to one or two MW).

Evans, Newman and Budal (Evans [12]) proved that a device can absorb power from a 'capture width' wider than its own physical width, by up to L/π for a device of general shape ($L/2\pi$ for an axisymmetric device). Experiments, including those on the full scale Kvaerner Brug prototype (Malmo & Reitan [27]), confirm this.

For large scale devices the implication is that the resource can be fully exploited by a line with gaps, provided these are short compared with a wavelength; for instance, Duck units of 37 metres on a 45 metre spine extract very nearly as much power as 45 metre units.

For small scale devices, it is interesting to calculate the maximum potential gain for a single unit. If the device is of width W, its potential capture width in waves of period T is $(W + L(T)/\pi)$, where the wavelength $L(T) = gT^2/2\pi$. Thus the potential resource in a mixed sea is

$$P^* = \int (W + L/\pi)\ kT\ s(f)\ df = pW + (cgH_{rms}^2/2\pi^2) \int T^3 s(f) df$$

where $s(f)$ denotes the wave spectral distribution (with allowance for directionality if appropriate), and p is the large scale resource in kW/m. Hence

$$P^* = p(W + W^*),$$

where W^* represents the potential mean additional capture width. W^* can be expressed as $\mu_{-3}L_e/\pi$, where L_e is the wavelength corresponding to the mean wave climate period, i.e. $gT_e^2/2\pi$, and μ_{-3} is a standardised third moment of the spectrum defined by

$$\mu_{-3} = \text{"E}\left[(T/T_e)^3\right]\text{"} = m_{-3}/m_{-1}^3 .$$

For instance, for a Pierson-Moskowitz spectrum $\mu_{-3} = \Gamma(1\tfrac{3}{4})/\Gamma(1\tfrac{1}{4})^3 = 1{\cdot}234$ (note that μ_{-3} is always $\geqslant 1$).

For the average wave spectrum at the deep water South Uist site W^* is approximately 70 metres, and the additional average resource is therefore over 3 MW. Similar levels may be expected at other good oceanic sites.

An example of how the point absorber effect increases the available resource is shown in Figures 8.1 and 8.2. The first shows the performance of a 29 metres wide solo Duck, from an early test at 1:100 scale in unidirectional seas, compared with efficiency curves from narrow tank experiments simulating a close-packed spine based Duck. The horizontal 'stretchy' axis represents the frequency distribution of the large scale resource at the 42 metres depth site off South Uist (with a 1 metre cut-off on H_{rms}). The upper curve shows the theoretical limit on 'efficiency' of $100(1+L/(\pi W))$%.

In Figure 8.2 the same performance curve is redrawn as a percentage of this upper limit, and the horizontal axis correspondingly re-stretched to represent the frequency distribution of the resource available to the isolated device. Note the 'red shift' in the wave spectrum, i.e. towards long periods, due to the proportionality of the added capture width to T^2. While this design promises a full-scale output of approximately 1 MW, there is clearly still scope for improvement, especially at these longer periods; and this possibility is being actively pursued.

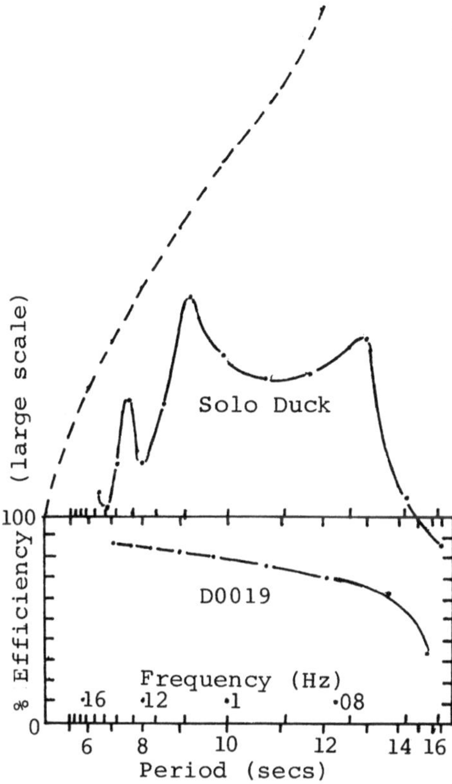

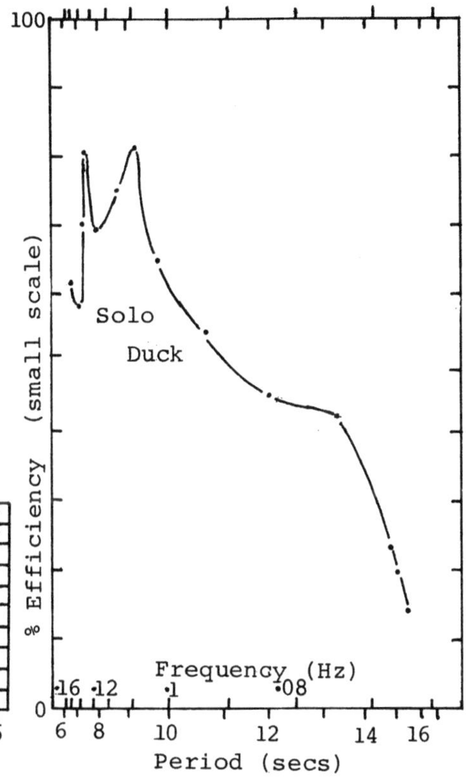

Figure 8.1
Dependence of output on frequency for early solo Duck model @ 1:100 scale, compared with efficiency curve for a spine based Duck (D0019, Dec 1979). The horizontal axis represents the distribution of the large scale resource at the 42 m depth South Uist site (with cut-off at H_{rms} = 1m). The dashed curve shows the theoretical output limit $100(1+L/(\pi W))\%$ for the solo device.

Figure 8.2
Output curve for solo Duck as in Figure 8.1, replotted as a percentage of the theoretical output limit, against a horizontal axis representing the distribution of the small scale resource at the same site. (Thus in both figures the area under the curve is the same, and represents the overall mean efficiency in this wave climate.)

For the present, we can look forward to the exciting demonstration projects of small scale devices in Norway. In an (inshore) wave climate averaging only 15 kW/m, the Kvaerner Brug prototype (Malmo & Reitan [27]) is estimated (based on a year's tests on-site) to produce 180 kW from a front of only 10 metres, i.e. with a final output 'efficiency' averaging over 100% thanks to the point absorber effect. The Tapchan (Mehlum [28]) is on a similar scale, but works by gathering waves over a wide front rather than the point absorber effect; and has the advantage of incorporating storage.

The Norwegians' achievement is considerable, in building a first generation of devices which are already economic or nearly so, in a relatively modest wave climate. This promises well for the exploitation of the best oceanic sites, such as our own Atlantic resource.

8 References

1. Bagnold, R.A.: Motion of waves in shallow water: interaction between waves and sand bottoms. Proc. R. Soc. Lond. A 187 (1946) 1-15.

2. Bartholomew, J.G.; Herbertson, A.J.: Bartholomew's Physical Atlas, Vol. 3: Atlas of Meteorology. Bartholomew, Edinburgh, 1899.

3. Battan, L.J.: Fundamentals of Meteorology. Prentice-Hall, Englewood Cliffs, N.J., 1979.

4. Bryden, Ian: Long floating cylinders in three-dimensional random seas. Ph.D. Thesis, Univ. of Edinburgh, 1983.

5. Chapman, W.M.: The law of the sea and public policy. *In* Ocean Engineering: goals, environment, technology (ed: Brahtz, J.F.), Wiley, New York, 1968, 112-156.

6. Count, Brian: Power from Sea Waves. Academic Press, London, 1980.

7. Crabb, John: Wave power levels to the west of the Hebrides. UK Dept. of Energy, WESC (78) DA 64b, 1978.

8. Crabb, John: Synthesis of a directional wave climate. *In* Count [6], 41-74.

9. Crabb, John: Assessment of wave power available at key UK sites. Inst. of Oceanogr. Sci. Report no. 186, 1984.

10. Crapper, G.D.: Introduction to Water Waves. Ellis Horwood, Chichester, 1984.

11. Ephraums, J.J.; Ewing, J.A.; Golding, B.W.; Worthington, B.A.: Comparisons of the Met Office and NORSWAM wave models with measured wave data collected during March 1980. UK Dept. of Energy, WESC (81) DA 130, 1981.

12. Evans, D.V.: Some theoretical aspects of three dimensional wave-energy absorbers. Proc. Symp. Ocean Wave Energy Utilization. Gothenburg, 1979, 77-113.

13. Ewing, J.A.: Some results from the joint North Sea wave project of interest to engineers. Marine Vehicles, 1974, 41-46.

14. Glendenning, I.: Wave energy. CEGB (Marchwood) Report, 1976(?).

15. Golding, Brian: Computer calculations of waves from wind fields. *In* Count [6], 115-134.

16. Hagerman, G.M.: Oceanographic design criteria and site selection for ocean wave energy conversion. *(This volume)*

17. Hasselmann, K.: On the nonlinear energy transfer in a gravity-wave spectrum. 1. General theory. J. Fluid Mech. 12 (1962) 481-500.

18. Hasselmann, K.: On the nonlinear energy transfer in a gravity-wave spectrum. 2. Conservation laws, wave-particle correspondence, irreversibility. J. Fluid Mech. 15 (1963) 273-281.

19. Hasselmann, K.; Barnett, T.; Bouws, E.; Carlson, H.; Cartwright, D.E.; Enke, K.; Ewing, J.A; Glenapp, H.; Hasselmann, D.E.; Kruseman, P.; Meerburg, A.; Muller, P.; Olbers, D.J.; Richter, K.; Sell, W.; Walden, H.: Measurements of wind-wave growth and swell decay during the Joint North Sea Wave Project (JONSWAP). Dtsch. Hydrogr. Z. A8, 12 (1973) 1-95.

20. Hogben, N.; Lumb, F.E.: Ocean Wave Statistics. H.M.S.O., London, 1967.

21. Jefferys, E.R.: Directional seas should be ergodic. *(Submitted for publication)*

22. Jeffrey, D.C.; Keller, G.J.; Mollison, D.; Richmond, D.J.E.; Salter, S.H.; Taylor, J.R.M.; Young, I.A.: Edinburgh Wave Power Project Fourth Year Report, Vol. 3, 1978.

23. LeBlond, P.H.; Mysak, L.A.: Waves in the Ocean. Elsevier, Amsterdam, 1978.

24. Lipman, N.H.; Musgrove, P.J.; Pontin, G.W.W. (eds.): Wind Energy in the Eighties. Peregrinus, Stevenage, 1982.

25. Longuet-Higgins, M.S.: Statistical properties of wave groups in a random sea state. Phil. Trans. R. Soc. Lond. A 312 (1984) 219-250.

26. Longuet-Higgins, M.S.; Cartwright, D.E.; Smith, N.D.: Observations of the directional spectrum of sea waves using the motions of a floating buoy. *In* Ocean Wave Spectra. Natl. Acad. Sci., 1963, 111-136.

27. Malmo, O.; Reitan, A.: Development of the Kvaerner multiresonant OWC. *(This volume)*

28. Mehlum, Even: Tapered channel wave power plants. *(This volume)*

29. Mehlum, Even: Wavetrack. Norwave AS, Forskningsvn 1, Blindern, Oslo 3, Norway *(No date)*.

30. Mitsuyasu, H.: Observations of the directional spectrum of ocean waves using a cloverleaf buoy. J. Phys. Oceanogr. 5 (1975) 750-760.

31. Mollison, Denis: The prediction of device performance. *In* Count [6], 135-172.

32. Mollison, Denis: Ireland's Wave Power Resource. Natl. Board for Sci. and Tech., Dublin, 1982.

33. Mollison, Denis: Wave energy losses in intermediate depths. Appl. Ocean Res. 5 (1983) 234-237.

34. Mollison, Denis; Buneman, O.P.; Salter, S.H.: Wave power availability in the NE Atlantic. Nature 263 (1976) 223-226.

35. Moskowitz, L.: Estimates of the power spectrums for fully developed seas for wind speeds of 20 to 40 knots. J. Geophys. Res. 69 (1964) 5161-5180.

36. Phillips, O.M.: The equilibrium range in the spectrum of wind-generated ocean waves. J. Fluid Mech. 4 (1958) 426-434.

37. Pierson, W.J.: The interpretation of wave spectrums in terms of the wind profile instead of the wind measured at a constant height. J. Geophys. Res. 69 (1964) 5191-5203.

38. Pierson, W.J.: Moskowitz, L.: A proposed spectral form for fully developed wind seas based on the similarity theory of S.A. Kitaigorodskii. J. Geophys. Res. 69 (1964) 5181-5190.

39. Pires, H.N.O.; Pessanha, L.E.V.: Wave power climate of Portugal. (*This volume*)

40. Pond, S.; Pickard, G.L.: Introductory Dynamic Oceanography. Pergamon, Oxford, 1978.

41. Pontes, M.T.; Perdigao, J.N.A.: Wave power resource in Portugal. (*This volume*)

42. Salter, S.H.: Physical modelling of directional seas. Proc. Symp. on Description and Modelling of Directional Seas, Copenhagen, 1984, 81-89.

43. Salter, S.H.: Progress on Edinburgh Ducks. *(This volume)*

44. Salter, S.H.: Wave power desalination. Proc. 4th Conf. on Energy for Rural and Island Communities, Inverness. Pergamon. *(To appear)*

45. Shearman, E.D.R.: Radio science and oceanography. Radio Science 18 (1983) 299-320.

46. Shemdin, O.; Hasselmann, K.; Hsiao, S.V.; Herterich, K.: Nonlinear and linear bottom interaction effects in shallow water. *In* Turbulent Fluxes Through the Sea Surface, Wave Dynamics and Prediction (Favre, A. and Hasselmann, K., eds.), Plenum Press, New York, 1978, 347-372.

47. Snodgrass, F.E.; Groves, G.W.; Hasselman, K.F.; Miller, G.R.; Munk, W.H.; Powers, W.M.: Propagation of ocean swell across the Pacific. Phil. Trans. R. Soc. Lond. A 259 (1966) 431-497.

48. Tucker, M.J: Observation of ocean waves (*with discussion*). *In* The Study of the Ocean and the Land Surface from Satellites (Houghton, J.S., Cook, Alan H., and Charnock, H., eds.), Royal Society, London, 1983, 129-138.

49. Tucker, M,J; Challenor, P.G; Carter, D.J.T: Numerical simulation of a random sea: a common error and its effect upon wave group statistics. Appl. Ocean Res. 6 (1984) 118-122.

50. Vitale, P: Sand bed friction factors for oscillatory flows. Proc. ASCE 105 (WW3) (1979) 229-245.

Wave Power Climate of Portugal

H.N. OLIVEIRA PIRES

L.E. VASCONCELOS PESSANHA

Instituto Nacional de Meteorologia e Geofísica - Portugal

Summary

The wave power available at the portuguese coast is evaluated by using three chronological series of wave records representing the northern and southern parts of the West coast and the South coast of Portugal. Analytical models for the annual mean spectra and probability distributions for wave power are derived. Statistics for extreme wave heights are also included.

1 - Wave Power

Consider $\eta(t)$, the height of the sea surface varying with time due to wave motion at a fixed place. There is no satisfactory deterministic approximation to $\eta(t)$, therefore it must be faced as a stochastic process.

If, for convenience, we chose the mean sea level as the reference plan we have

$$\mathrm{Var}\,[\eta(t)] = \langle \eta^2(t) \rangle \qquad L^2 \qquad (1)$$

It is easily seen that the variance is a measure of the mean potencial wave energy per unit area W_p.

$$W_p = \frac{\rho}{2}\, g \, \langle \eta^2(t) \rangle \qquad M\,T^{-2} \qquad (2)$$

with ρ the water density and g the gravitational acceleration.

$\eta(t)$ may be considered as gaussian, and either the autocovariance function $c(\tau)$, or its Fourier transform, the variance spectrum $P(f)$, define the statistical properties of the process.

$$c(\tau) = \langle \eta(t)\,.\,\eta(t+\tau) \rangle \qquad (3)$$

$$P(f) = \mathrm{TF}\,[c(\tau)] \qquad (4)$$

and

$$c(0) = \int_{-\infty}^{+\infty} P(f)\,.\,df = m_0 = \mathrm{var}\,[\eta(t)] \qquad (5)$$

being

$$m_n = \int_{-\infty}^{+\infty} P(f)\,.\,f^n\, df \qquad (6)$$

From linear wave theory it is known that there is equipartition of the total wave energy W between potential W_p and kinetic energy W_c and that the energy travels in deep water at a group velocity $u(f) = g/(4\pi f)$.

$$W/2 = W_p = W_c \tag{7}$$

So the mean omnidirectional power per unit width can be approximated to first order by

$$P = \frac{\rho g^2}{4\pi} \int_{-\infty}^{+\infty} \frac{P(f)}{|f|} df \qquad M\,L\,T^{-3} \tag{8}$$

and it is therefore possible to calculate the wave energy flux in a sea state knowing the corresponding variance spectrum. Assuming, at this time scale, stationarity and ergodicity, it is possible to estimate de variance spectrum of the sea state from a limited record of $\eta(t)$. At a different time scale, the sea state changes with time and so does the corresponding wave power as the weather conditions change.

From chronological series of wave power measurements (variance spectra) at a certain place, it is possible to estimate a probability distribution for the wave power density at that place. It is also possible to calculate the annual mean variance spectrum and the annual mean wave power density.

2 - Wave height and wave period

The significant wave height is defined as the average wave height of the 1/3 highest waves in a wave record, each wave (zero down crossing wave) being defined as the segment of the record between sucessive zero down crossings. The reason for the use of the significant height is mainly in the fact that the significant height has roughly the same value as the characteristic height observed visually.

It is also in use the zero crossing period T_z defined as the mean time interval between sucessive zero down crossings.

It is known, that in a stationary sea state the wave heights fit a Raleigh distribution with parameter $8 . m_0$. This statement is valid for trough to crest heights in the case of narrow spectra, and in any case for zero crossing waves.

The significant height H_s is:

$$H_s = 4.00 \sqrt{m_0} \tag{9}$$

A pèriod, here called equivalent period T_e , can be computed from (8) in such a way that if the wave power should come from a spectrum with the same variance, infinitesimal bandwidth, and located at that period, the resulting power would be the same as with the actual spectrum.

The relation between T_e and T_z depends on the spectral shape.

3 - Analysis of the wave records

The chronological series of H_s and T_z obtained from all the wave records taken near the Portuguese coast four times a day from 1974 to 1980 (near 20000 records) were subjected to statistical analysis. For Sines and Faro the results are published [1], [2].

This work was done mainly for coastal engineering purposes, and so only the records of more than 4 metres of significant height were subjected to spectral analysis by Tuckey method.

For the present paper the spectral analysis was carried by Welch method [3], much faster. Even so, the number of records processed in this way must necessarily be kept small. The calculated spectra were then corrected by using the transfer function of the waverecorder. For most spectra the correction is neglegible.

The power density is then computed from this spectra using (8).

4 - Serial correlation in the chronological series of significant height, variance and zero crossing period. Bartels number.

The analysis of the chronological series of significant height in Sines and Faro [1], [2], showed that the series present a seasonal variation, and a diurnal variation corresponding to wind diurnal variation near the coast due to the sea-land effects. The analysis also showed strong serial correlation. The autocorrelation functions of lag 1 have values close to 0.9. Even seven days apart the correlation is still close to 0.25. Apart the seasonal and diurnal variations the time series may be well represented by an autoregressive first order model.

In a weakly stationary sequence of random independent observations with mean μ and variance σ^2 the mean of n observations has mean μ and variance σ^2/n. If the observations are dependent with autocorrelation coeficients c_j the variance of the mean of n observations is

$$\frac{\sigma^2}{n} \left[1 + \frac{2}{n} \sum_{j=1}^{n-1} (n-j)\, c_j \right] \tag{10}$$

Bartels number of equivalent independent observations (Bartels 1943) is then defined as

$$K_n = n \Big/ \left[1 + \frac{2}{n} \sum_{j=1}^{n-1} (n-j)\, c_j \right] \tag{11}$$

K_n/n expresses the "useful" fraction of independent observations [4] .
If the sequence is auto-regressive of first order, then $c_j = c_1^j$ and

$$K_n / n \longrightarrow (1-c_1) / (1+c_1) \qquad (12)$$

In the case of a sequence of significant wave heights observed every 6 hours, with $c_1 = 0.9$, the fraction $K_n/n = 0.05$, meaning that only the observations spaced five days or more are "useful" for estimating the mean significant height.

Similar results hold for the chronological series of variance and zero crossing period, but in this cases, the lag 1 correlation being slightly smaller (0.88 and 0.86 respectively), the observations four days apart may be considered independent.

Using the corresponding values of K_n the standard deviation of the calculated mean values of H_s and T_z can be estimated. As for large n a gaussian distribution can be assumed, these values limit the corresponding 68% confidence intervals (standard error). Twice and three times these values limit respectively the 95% and 99% confidence intervals.

The results for Sines, Roca and Faro series with respectively 7456, 5319 and 5249 observations are as follows:

Mean significant height (m) and mean zero crossing period (s):

Sines	1.57 ± 0.04	6.90 ± 0.08
Roca	2.01 ± 0.07	7.96 ± 0.11
Faro	0.93 ± 0.04	4.77 ± 0.07

As can be seen the statistical error is quite small (5% for the significant height and 1% for the zero crossing period). It is easily seen that relative errors smaller than 10% in the wave height and smaller than 2% in zero crossing period can be achieved by using just 60 independent observations for calculating these mean values. That is not the case for wave power density as we shall see.

5 - Estimation of the annual mean wave power density from small random samples of independent wave records

From the series of wave records of Sines, Roca and Faro were selected three random samples of 60 independent (more than four days apart) wave records, uniformely distributed over the months of the year and over the hours of the day to avoid bias due to seasonal and diurnal variations. These records where processed and the corresponding wave spectra and power densities were computed.

The corresponding annual mean power densities can be estimated as the mean values of each sample, with standard errors equal to the corresponding sample standard deviation divided by $\sqrt{60}$.

Annual mean wave power densities (kW/m) :

Sines :	15.8	$\pm$ 3.1
Roca :	24.8	$\pm$ 5.5
Faro :	3.2	$\pm$ 0.5

The relative standard errors are respectively 20, 22 and 16% .

To reduce the relative standard error to 10% the size of each sample should increase to near 300. That could not be done.

6 - Annual mean spectra

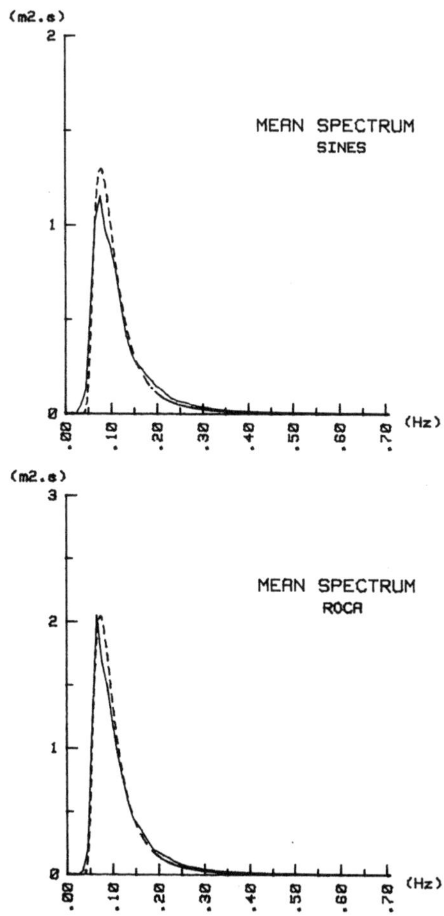

Fig 1

An annual mean spectrum $P_a(f)$ can be obtained from each sample of 60 spectra. The annual mean wave power is

$$P_a = \frac{\rho g^2}{2\pi} \int_0^{+\infty} \frac{P_a(f)}{f} \, df \qquad (13)$$

The physical meaning of these annual mean spectra is that if we have some device that extracts power from the waves, with some known transfer function $T_d(f)$, the annual mean power obtained with that device is

$$P_d = \frac{\rho g^2}{2\pi} \int_0^{+\infty} \frac{P_a(f)}{f} |T_d(f)|^2 \, df$$

If the transfer function of the capturing device is not constant, for instance if the device adapts itself to the wave conditions, the annual mean spectra do not provide enough data to compute the annual mean captured power. The annual mean spectra obtained for Sines, Roca and Faro are shown in Figs 1 and 2.

The mean spectra are broader than the spectra of individual sea states, so the analytical models generally used for wave spectra are not suitable to the mean wave spectra. An analytical

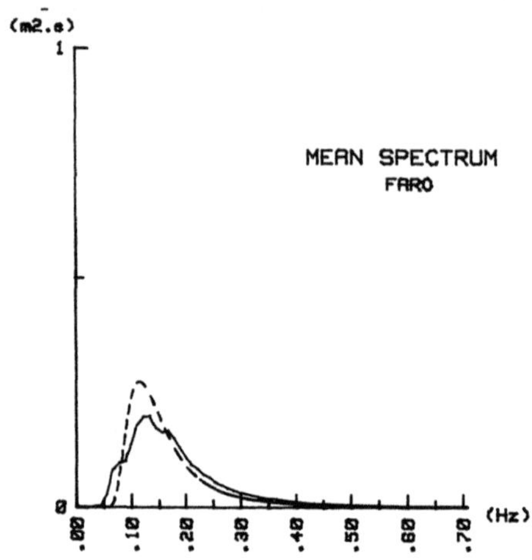

Fig 2

model of the form

$$P_a(f) = a\ f^{-c}.\ e^{-b\ f^{-(c-1)}} \qquad (14)$$

with $c = 4$ was found to fit quite well the observed annual mean spectra for Sines, Roca and Faro - Figs 1 and 2.

For these spectra the first moments are found to be:

$$m_{oa} = a\ /\ [\,b(c-1)\,] \qquad (15)$$

$$m_{2a} = \left(\frac{ab}{c-1}\right)^{(3-c)/(c-1)} \Gamma\,[(c-3)/(c-1)] \qquad (16)$$

with Γ representing the gama function

$$\Gamma(x) = \int_0^\infty u^{(x-1)}.\quad e^{-u}\quad du$$

And the annual mean power density can be expressed as:

$$P_a = K \left(\frac{ab}{c-1}\right)^{-c\ /\ (c-1)} .\ \Gamma\,[c\ /\ (c-1)] \qquad (17)$$

The frequence of maximum spectral density is

$$f_{max} = [\,c/\,[\,b(c-1)\,]\,]^{-1/(c-1)} = 1\ /\ T_{max} \qquad (18)$$

And,with $c = 4$, a and b may be computed using the expressions (15)and (17), that is, the analytical annual mean spectrum can be defined from the annual mean wave variance m_{oa} and the annual mean power P_a. As the annual mean zero crossing period is different from the period corresponding to the annual mean spectrum, the annual mean zero crossing period should not be used for computing a and b.

The results for the analytical models for the annual mean spectra for Sines, Roca and Faro can be summarized as follows:

	Mean wave variance (m^2)	Mean wave power P_a (kW/m)	Parameter a ($m^2\ s^{-3}$)	Parameter b (s^{-3})	Peak period (s)
Sines	0.1937	15.8	3.7 E(-4)	6.3 E(-4)	12.8
Roca	0.2889	24.8	4.7 E(-4)	5.4 E(-4)	13.5
Faro	0.0602	3.2	4.0 E(-4)	2.2 E(-3)	8.5

7 - Wave power estimation from wave variance and zero crossing period

If the equivalent period T_e is known the wave power density can be computed as $P = K m_o . T_e$ with $K = (\rho g^2) / (4\pi)$

Although the relation between T_e and T_z depends on the shape of the spectra, some correlation may be expected, and so we may expect that $P' = K m_o . T_z$ will be correlated with the power density P. Besides, as the relative variance of m_o is much larger than the relative variance of T_z, we may expect this correlation to be strong.

The results are:

Sines:	$P = 1.33 P' - 0.16$	correlation	0.997
Roca:	$P = 1.36 P' - 1.21$	correlation	0.996
Faro:	$P = 1.26 P' + 0.08$	correlation	0.995

These linear equations provide a way of estimating the power densities just by knowing the wave variances and the zero crossing periods.

The standard residual errors of these estimates are of 1.8, 3.9 and 0.4 kW/m respectively.

With this method it is quite easy to obtain large series of power densities, providing a way of extending the relatively small samples used in [5] and [6]

Even if in each power density estimate, the relative error is important, the annual mean power density can be estimated with less than 10% relative standard error using series of independent wave records (more than four days apart) obtained from the three original series of wave records.

Annual mean wave power densities (kW/m):

Sines	16.5 ± 1.1	(643 records)
Roca	26.3 ± 2.1	(360 records)
Faro	4.8 ± 0.4	(365 records)

8 - Improved estimation of the annual mean spectra

Adopting the annual mean power densities obtained in [7] and the corresponding annual mean wave variances, the estimates of the annual mean spectra obtained in 6 can be improved.

If we assume the same spectral shapes obtained in [6] the analytical models of the spectra can be derived by using these values for computing a and b. The results can be summarized as follows:

Analytical models for annual mean spectra:

	Mean wave variance (m^2)	Mean wave power P_a (kW/m)	Parameter a ($m^2\ s^{-3}$)	Parameter b (s^{-3})	Peak period (s)
Sines	0.2047	16.5	4.0 E(-4)	6.5 E(-4)	12.7
Roca	0.3072	26.3	5.0 E(-4)	5.5 E(-4)	13.5
Faro	0.0839	4.8	4.7 E(-4)	1.9 E(-3)	9.0

9 - Probability distributions of wave power density

The series of wave power densities obtained from independent wave records referred to in [7] provide data for estimating probability distribution functions for wave power density.

It is found that the series fit Weibull distributions with shape parameter smaller than one.

$$F(x;s,q) = 1 - \exp\left[-(x/s)^q\right]$$

$$f(x;s,q) = (1/s)(x/s)^{(q-1)} \cdot \exp\left[-(x/s)^q\right]$$

The scale parameter s and the shape parameter q estimated by the method of the moments are:

	Scale parameter	Shape parameter
Sines	10.2	0.57
Roca	19.3	0.65
Faro	3.0	0.57

These results (Fig 3) are accepted by the qui square test at the 5 % significance level.

10 - Extreme wave conditions

Considerable efforts have been made in order to obtain reliable extreme wave statistics for Sines needed for coastal engineering purposes.

In [5] , the series of Sines wave records were analysed to obtain extreme significant wave height statistics. Two methods were used. The former method consists in fitting a Gumbel distribution to the annual maxima of significant wave heights. In the latter method a compound Poisson-Exponential distribution was used to model the occurrence of significant wave height peaks over a threshold (POT) of 4 metres.

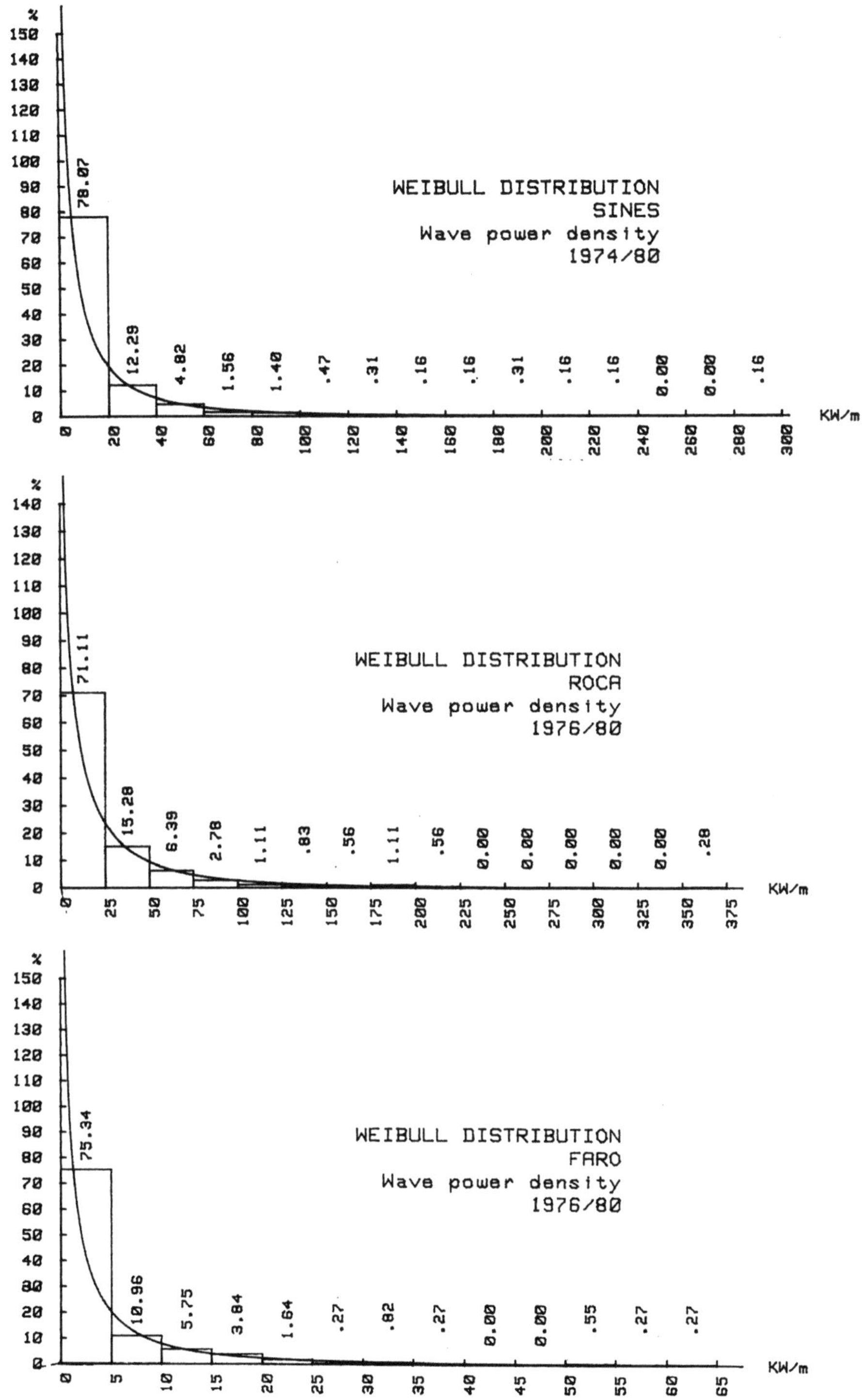
%
WEIBULL DISTRIBUTION
SINES
Wave power density
1974/80
78.07
12.29
4.82
1.56
1.40
.47
.31
.16
.16
.31
.16
.16
0.00
0.00
.16
KW/m
WEIBULL DISTRIBUTION
ROCA
Wave power density
1976/80
71.11
15.28
6.39
2.78
1.11
.83
.56
1.11
.56
0.00
0.00
0.00
0.00
0.00
.28
KW/m
WEIBULL DISTRIBUTION
FARO
Wave power density
1976/80
75.34
10.96
5.75
3.84
1.64
.27
.82
.27
0.00
0.00
.55
.27
.27
KW/m

Fig 3

In [6], D.T. Resio derived extreme wave statistics from hindcast results at a grid point near Sines. The results can be summarized as follows:

Return periods of significant wave height (m) at Sines

Return period (year)	Gumbel	POT	Hindcast
5	7.2 ± 0.8	7.3 ± 0.5	9.1
10	8.1 ± 1.0	8.0 ± 0.6	10.0
25	9.2 ± 1.3	8.8 ± 0.7	11.2
50	10.0 ± 1.5	9.5 ± 0.8	12.0
100	10.8 ± 1.8	10.1 ± 0.9	12.7

Design significant wave height (m) for a useful life of 50 years at Sines

Risk (%)	Gumbel	POT	Hindcast
1	15.4 ± 3.1	13.8 ± 1.4	
5	13.5 ± 2.5	12.3 ± 1.2	16.0
10	12.6 ± 2.3	11.6 ± 1.1	
15	12.1 ± 2.2	11.2 ± 1.1	14.0
25	11.5 ± 2.0	10.6 ± 1.0	

From the Raleigh distribution, and assuming that the maximum significant height persists for 3 to 6 hours the mode for the maximum wave height is found to be close to twice the corresponding significant height.

For Roca there is no statistical analysis of extreme waves, but only the results of fitting a lognormal distribution to the serie of significant height, which can be compared to the same results for Sines.

It is found that for large values the significant heights with the same probability are greater at Roca than at Sines by a factor of 1.3. This factor and the results of the POT method for Sines were used to estimate return periods and design waves for Roca.

This aproach was used because the chronological serie of records at Roca with many periods without observations does not provide enough data neither to fit a Gumbel distribution nor to use the POT method.

In [2] a Gumbel distribution was fitted to the annual maxima (1977 to 1980) of significant height at Faro.

The results for both locations are:

Return period of significant height (m)			Design significant height (m) for a useful life of 50 years		
Return period (year)	Roca	Faro	Risk (%)	Roca	Faro
5	9.5	4.8	1	18.0	8.7
10	10.4	5.2	5	16.0	7.8
25	11.4	5.7	10	15.1	7.4
50	12.4	6.1	15	14.6	7.1
100	13.1	6.5	25	13.8	6.8

11 - Conclusion. Comments

We can assume in first approximation that the wave climate derived from the Roca wave records represents the wave climate of the northern part of the Portuguese west coast, that Sines represents the southern part of the west coast, and Faro the south coast, and say that for these zones the annual mean available wave power is of the order of respectively 26, 16 and 5 kW/m with mean spectra close to those presented. For the same zones design significant wave heights of the order to 16, 13 and 8 m and design maximum wave heights of twice these values may be assumed. More data are required for more detail.

Important questions refer to sheltering from the coast, bottom effects,wave direction and wave multidirectionality. The correspondent need for series of directional wave spectra will become evident.

References

1 Pires, H. N. Oliveira and Pessanha, L.E.Vasconcelos - Elementos sobre o clima de agitação marítima na região de Sines
INMG 1982

2 Pessanha, L.E.Vasconcelos and Pires, H.N.Oliveira - Elementos sobre o clima de agitação marítima na costa sul do Algarve
INMG 1981

3 Welch, Peter D. - The use of fast Fourier transform for the estimation of power spectra: a method based on time averaging over short modified periodograms
IEEE Trans. Audio and Electroacoust. 1967

4 Oliveira, J. Tiago - A estatística na meteorologia
Procedings - II Colóquio de estatística e investigação operacional 1981

5 Pires, H.N.Oliveira and Pessanha, L.E.Vasconcelos - Estima da distribuição de probabilidade dos valores extremos utilizando séries climatológicas curtas

6 Resio, D.T. - Wave environment at Sines, Portugal
Water Ways Experiment Station. U.S.Army Corps of Engineers 1979

Oceanographic Design Criteria and Site Selection for Ocean Wave Energy Conversion

GEORGE M. HAGERMAN, JR.

Gibbs & Cox, Inc.
1235 Jefferson Davis Hwy.
Arlington, VA 22202
U.S.A.

Summary

This paper describes an approach to site selection that considers not only the magnitude of the wave energy resource, but also the limitations on its utilization by seastates too calm or severe for sustained conversion operations. A figure of merit is introduced that compares this operational resource with the most extreme wave that a platform would have to survive during its service life. This approach is illustrated in evaluating the ocean wave energy conversion potential of 36 areas off the coast of the United States and in selecting an optimum water depth for siting an ocean wave energy conversion platform.

Introduction

Ocean wave energy conversion potential has usually been characterized solely by the magnitude of the wave energy resource. Although some resource studies (13,21) have considered device efficiency limitations, they have not considered that under certain conditions, a device would not be operational at all. Some seastates contain an insufficient number of sufficiently large waves to sustain conversion. Other seastates threaten the safety of personnel or would damage operating equipment. An additional factor to consider is the severity of the most extreme seastate expected during a platform's service life. This is often characterized by the 100-year maximum wave height and can have a major influence on system acquisition cost. Dividing the mean annual wave energy flux, excluding non-operational seastates, by the 100-year maximum wave height results in a figure of merit that more accurately reflects the economic potential of a site than the magnitude of the resource alone.

The findings of this study and opinions expressed herein are those of the author and do not necessarily reflect the views of Gibbs & Cox, Inc.

Evaluation of Mean Annual Wave Energy Flux

The data base used to evaluate the wave energy resource in a given area was the Summary of Synoptic Meteorological Observations (SSMO; (18)). Areas covered by this study are shown in Figure 1. Quayle and Changery (16,17) have adjusted SSMO wave height statistics to agree with those hindcast by the U.S. Army for sites in the western North Atlantic Ocean (5). Estimates of wave energy flux were based on these adjusted wave heights, which correspond to the significant wave height of combined sea and swell. Each significant wave height was represented by a family of Ochi-Hubble spectra (14). The wave energy flux associated with each spectrum was computed analytically, based on application of linear wave theory to a random sea (6).

For a given SSMO area, each month's wave height frequency distribution was used to estimate the mean monthly wave energy flux and the results for all months were averaged to obtain the mean annual flux. A similar estimate was made including only operational seastates. These were defined as any seastate having a significant wave height greater than 1 m and a maximum wave height (in 1000 waves) less than 12.5 m.

Evaluation of 100-Year Maximum Wave Height

Both hurricanes (tropical cyclones) and winter storms (extratropical cyclones) can generate potentially destructive waves that a wave energy conversion system must be designed to survive. In Hawaii and along the northeast and middle Atlantic coast, the joint probability distribution for both storm wave populations was used to determine the 100-year maximum wave height. Along the southeast Atlantic coast and in the Gulf of Mexico, hurricane wave statistics alone were used. Along the Pacific and Alaskan coasts, winter storm wave statistics alone were used.

Hurricane wave heights were based on published hindcasts for the Standard Project Hurricane (SPH) along the Atlantic and Gulf of Mexico coasts (2) and in Hawaii (3). Winter storm wave heights were based on extrapolation of extreme wave height observations in the SSMO (for method, see (4)). SPH and SSMO wave statistics were used because they could be consistently applied to all SSMO areas. Where site-specific hindcasts were available (1,7,9,11,22; see Figure 1 for locations), revised figures of merit were developed for comparison.

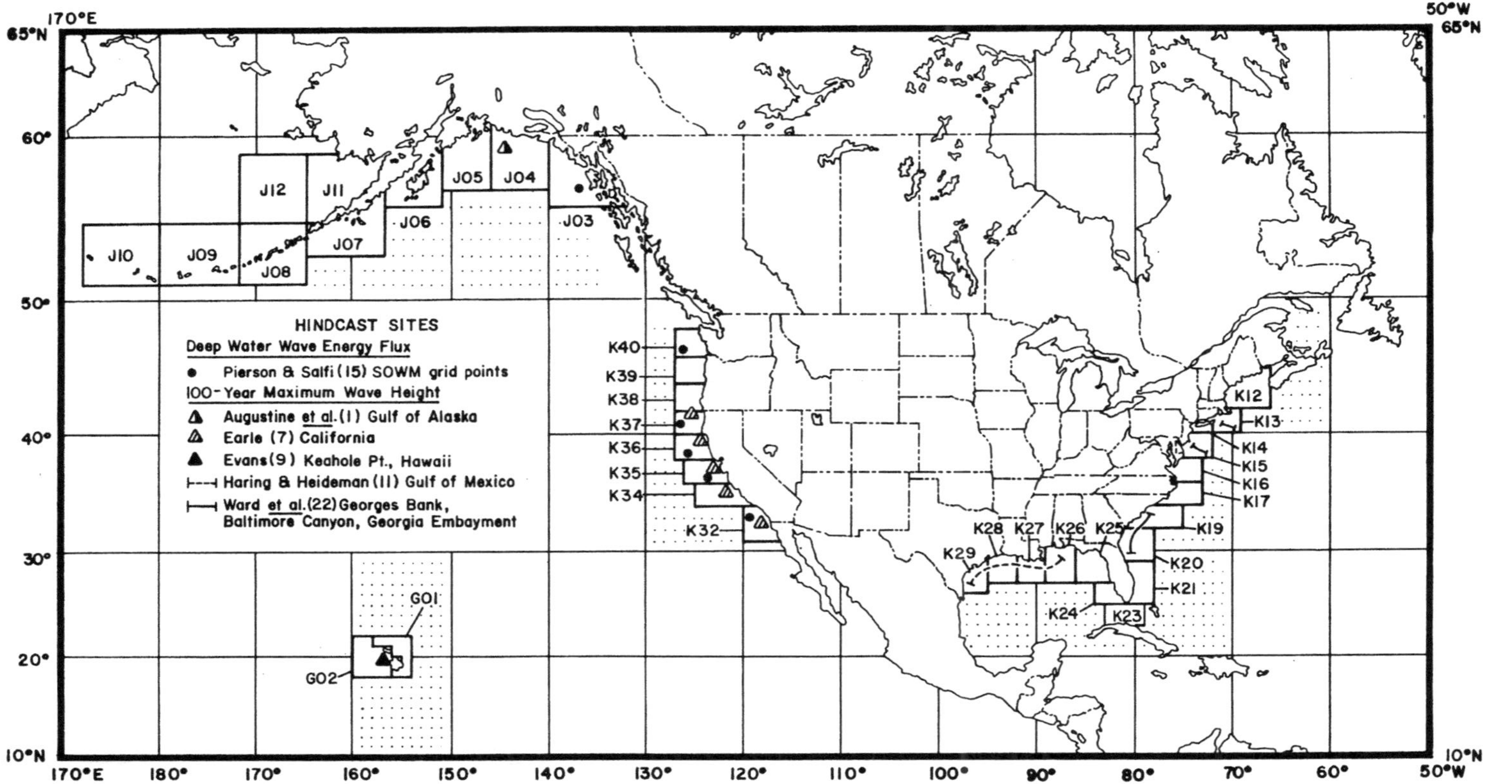

Fig. 1. SSMO areas covered by this study. SSMO data in Alaska and Hawaii cover the years 1963-1970; those off the U.S. west coast, 1949-1973; those off the U.S. east coast and in the Gulf of Mexico, 1949-1970. Pierson and Salfi's Spectral Ocean Wave Model hindcast data were for nine months in 1974 and three months in 1975.

Selection of a Deep Water Coastal Area

A figure of merit (FOM) for each SSMO area was developed by dividing the mean annual wave energy flux in operational seastates by the 100-year maximum wave height. These are plotted in Figure 2. Clearly, the U.S. coastal waters with the greatest potential are off California, Washington, Oregon and Alaska.

Month-to-month variations in wave energy flux for six areas along the Pacific coast and in the Gulf of Alaska are plotted in Figure 3. In addition to the range of fluxes predicted by the family of Ochi-Hubble spectra, wave energy fluxes estimated from Spectral Ocean Wave Model (SOWM) hindcasts (15) are also shown. The high mean monthly fluxes in more northerly areas during the winter are due to a few individual storms each month that have peak wave energy fluxes of more than 200 kW/m (13,15). Wave energy from these storms may not be recoverable in hazardous seas near storm centers (for example, see (10)). More distant areas experience less severe wave conditions (8) and elevated levels of wave energy that may therefore be recoverable.

Another feature of Figure 3 is the higher wave energy flux further south during the summer. This trend is not as clear in the SOWM data, since these represent individual years rather than the several years' average of the SSMO. Furthermore, for the lower summer wave heights, SOWM hindcasts in the north Pacific are generally lower than U.S. Army hindcasts at all latitudes and measured buoy data south of 50°N (19).

Selection of an Optimum Water Depth

Onshore-offshore variation in FOM was evaluated for SSMO area K36 (Point Arena, California). To simplify the analysis, resource waves were represented by a Pierson-Moskowitz spectrum and cosine-squared spreading function. For each deep water significant wave height, transformed spectra were computed at water depths ranging from 10 m to 200 m, using the methods outlined in (12) for plane offshore bathymetry. This was done for three mean angles of deep water wave approach: 0°, 30°, and 60°.

The site-specific hindcast 100-year maximum wave height for this area was transformed using linear wave theory to estimate shoaling effects. A 0° angle of wave approach was used to represent the worst case (no wave height reduction by refraction).

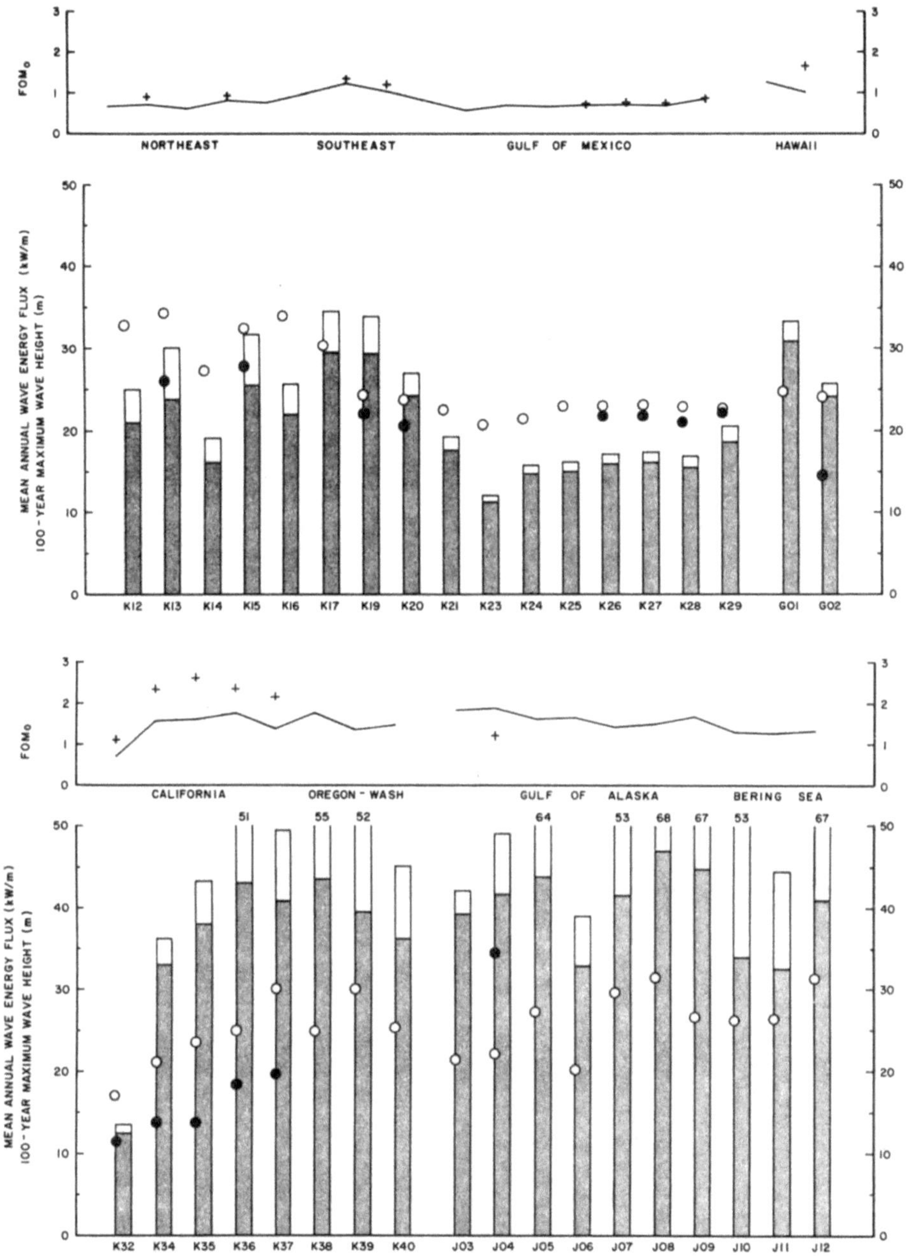

Fig. 2. Development of deep water figure of merit (FOM_0). Bars are mean annual wave energy fluxes; open bars include all seastates, shaded bars include operational seastates only. Circles are 100-year maximum wave heights; open circles are based on SPH and SSMO wave heights, closed circles are from site-specific hindcasts. Crosses (upper plot) are revised figures of merit developed from site-specific hindcast 100-year wave heights.

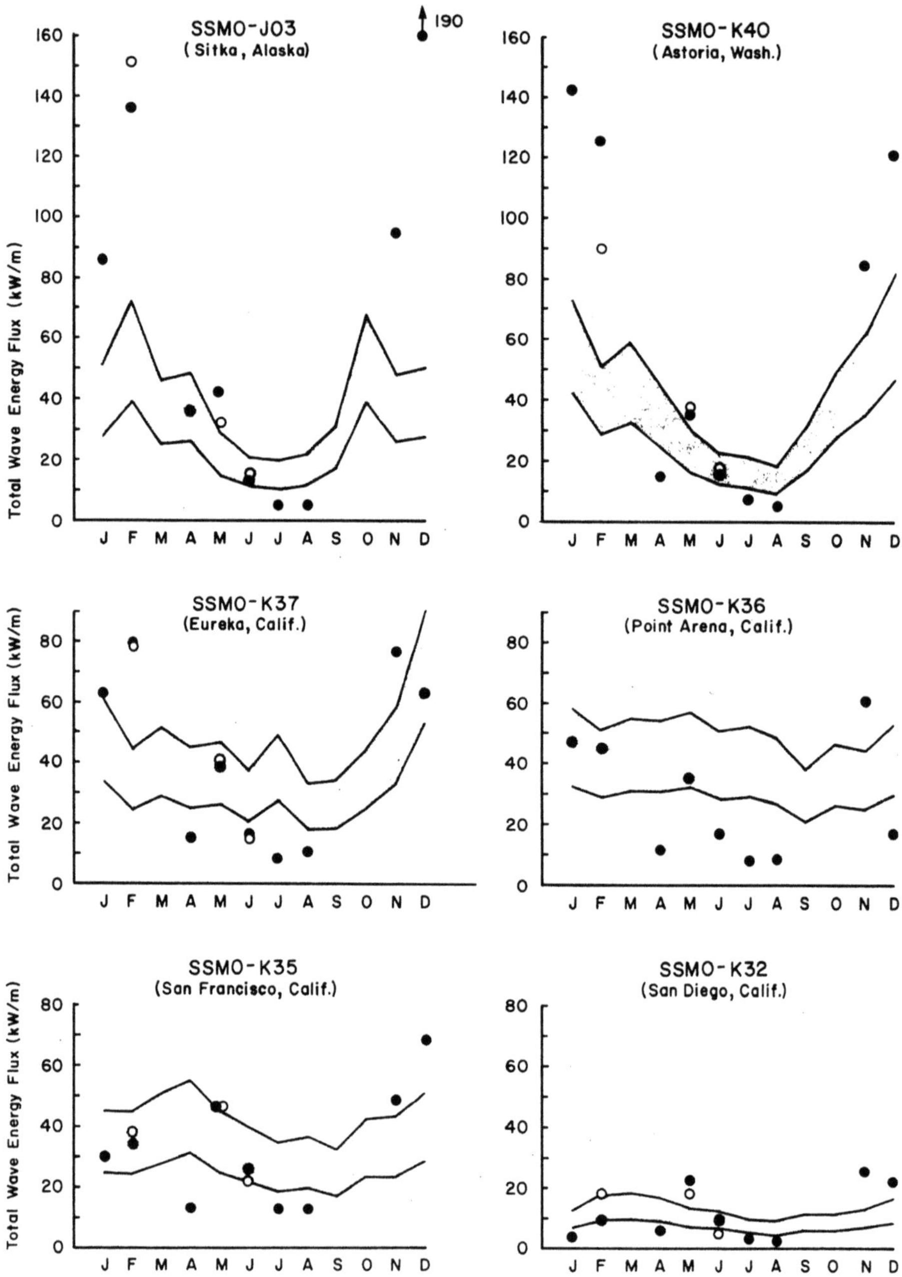

Fig. 3. Month-to-month variation in wave energy flux off the U.S. west coast and in the Gulf of Alaska. Shaded areas cover the range of fluxes estimated using SSMO data and Ochi-Hubble spectra. Circles are fluxes estimated by Pierson and Salfi (15) from SOWM hindcasts; dark circles = 1974, open circles = 1975.

Breaking-limited wave heights were computed using Miche's formula (20). Three wave periods were checked: 15, 19, and 24 seconds. The upper limit of wave heights associated with this range was used to compute the FOM at each water depth.

Results are plotted in Figure 4A. For all angles of deep water wave approach, FOM increases slightly in water depths between 80 and 180 m, due to the reduction in 100-year maximum wave height by shoaling. Further shoaling causes the 100-year wave to peak up, resulting in a substantial decrease in FOM from 80 m to the point of breaking. The FOM then rises dramatically as extreme wave heights become depth-limited.

The increase in FOM shoreward of the 100-year wave breakpoint does not necessarily indicate a more economical location for a wave energy conversion platform. Although the breaking wave is lower than its deep water counterpart, it imposes slamming loads on plate structure and snap loads on mooring components. Furthermore, several waves during the 100-year storm will be large enough to break in these water depths, and the platform may be in a veritable surf zone. Finally, storms of less than 100-year return period can generate breaking waves in these depths (although fewer per storm), and this must be factored into the loading history of the platform.

The stepwise increase in FOM in depths of 100 m and 36 m (shaded area in Figure 4A) is explained as follows. Consider a seastate in deep water that is too severe for safe platform operations. As the most extreme wave heights are reduced to within operational limits by refraction, the substantial amount of remaining wave energy becomes available for recovery. It may be economical to take advantage of this effect by designing a wave energy conversion system that does not have to operate in such extreme waves. To test this hypothesis, the same refraction and shoaling analysis described above was carried out for a system designed to operate in seastates having a maximum wave height of 9.5 m, as opposed to 12.5 m. The results are plotted in Figure 4B. Comparing the two plots for a 60^{0} angle of deep water wave approach, in 40 m water depth, it is evident that the system in Figure 4B could potentially recover 90-95% of the wave energy available to the system in Figure 4A, yet have power system components designed to operate at only 75% of the limiting wave

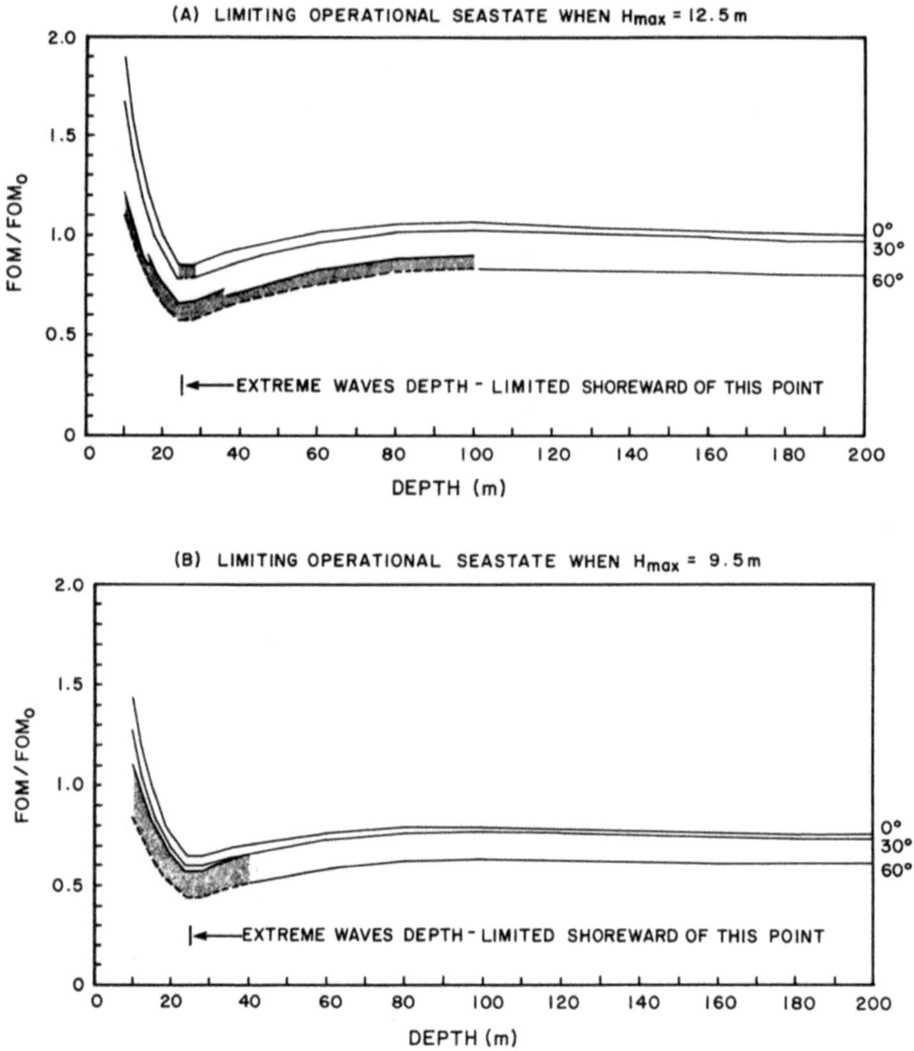

Fig. 4. Onshore-offshore variation in figure of merit (FOM) relative to its deep water value (FOM_0) for 0^0, 30^0, and 60^0 mean angles of deep water approach for resource waves. Shaded areas represent the increase in FOM/FOM_0 due to seastates that are non-operational in deep water, but whose extreme waves are reduced by wave refraction to the point where conversion operations would be possible in shallow water.

height. Presumably such components would cost less to build, and decreased wear and fatigue over the platform's service life would reduce costs and down-time associated with periodic maintenance and repair or replacement.

Conclusions

Deep water areas off California show the greatest economic potential for ocean wave energy conversion in coastal waters of the United States. Although areas off Oregon, Washington, and Alaska have a greater mean annual wave energy flux, a substantial portion of this flux is in seastates too severe for safe platform operations. Moreover, site-specific hindcasts of 100-year maximum wave heights indicate that a higher design penalty would have to be paid to utilize the resource in these areas.

In selecting an onshore-offshore location within a given coastal area, serious consideration should be given to fairly shallow sites. Where the 100-year maximum wave height becomes depth-limited, the platform may be designed to survive a much lower wave, although it is a breaking wave and occurs more frequently than once during the 100-year storm. Furthermore, in shallow water, power train components may be designed to operate at lower wave heights and still capture most of the wave energy available to more rugged systems. This is because refraction reduces extreme wave heights to within operational limits before substantially affecting wave energy flux.

References

1. Augustine, F.E.; Maxwell, F.D. and Lazanoff, S.M.: Extreme wave heights in the Gulf of Alaska. Proc. 10th Ann. Offshore Technol. Conf., 1978, pp. 1551-1562.
2. Bretschneider, C.L.: Revisions to hurricane design wave practices. Proc. 13th Conf. Coast. Eng., 1972, pp. 167-195.
3. Bretschneider, C.L.: Hurricane design winds and waves and current criteria for potential OTEC sites. Univ. Hawaii Look Lab. Tech. Rep. 79-45 (1979).
4. Bretschneider, C.L. and Rocheleau, R.E.: An evaluation of extreme wave climate at Keahole Point, Hawaii. Proc. 16th Conf. Coast. Eng., 1978, pp. 152-173.
5. Corson, W.D.; Resio, D.T.; Brooks, R.M.; Ebersole,B.A.; Jensen, R.E.; Ragsdale, D.S. and Tracy, B.A.: Atlantic hindcast deepwater significant wave information. U.S. Army Eng. Waterways Exp. Sta., WIS Rep. 2 (1981).
6. Det Norske Veritas: Estimates of wave power at the coast of Norway. Veritas Rep. 77-570 (1977).

7. Earle, M.D.: Wave conditions for the California coast and continental shelf. Report prepared for Environ. Data Info. Svc., Natl. Oceanic Atmos. Admin. (1977).

8. Earle, M.D.; Bush, K.A. and Hamilton, G.D.: High-height long-period ocean waves generated by a severe storm in the northeast Pacific Ocean during February 1983. J. Phys. Oceanogr. 14 (1984) 1286-1299.

9. Evans, D.J.: personal communication (1985).

10. Hamilton, G.D.: Severe buoy wave conditions. Mariners Weather Log 26 (1982) 122-126.

11. Haring, R.E. and Heideman, J.C.: Gulf of Mexico rare wave return periods. Proc. 10th Ann. Offshore Technol. Conf., 1978, pp. 1537-1550.

12. Le Mehaute, B. and Wang, J.D.: Wave spectrum changes on sloped beach. J. Waterway, etc. Div., Proc. Am. Soc. Civ. Eng. 108 (1982) 33-47.

13. Mogridge, G.R.; Funke, E.R.; Baird, W.F. and Mansard, E.P.D.: Analysis and description of wave energy resources. Second Intl. Symp. Wave Energy Utilization, 1982, pp. 59-79.

14. Ochi, M.K. and Hubble, E.N.: Six-parameter wave spectra. Proc. 15th Conf. Coast Eng., 1976, pp. 301-328.

15. Pierson, W.J. and Salfi, R.E.: The temporal and spatial variability of power from ocean waves along the west coast of North America. Wave and Salinity Gradient Workshop Proc., U.S. Energy Res. Dev. Admin. Rep. C00-2946-1 (1976).

16. Quayle, R.G. and Changery, M.J.: Estimates of coastal deep-water wave energy potential for the world. Oceans '81 Conf. Record, 1981, pp. 903-907.

17. Quayle, R.G. and Changery, M.J.: Preliminary height and period adjustments for visual wave data. Mariners Weather Log 26 (1982) 2-3.

18. U.S. Navy: Summary of Synoptic Meteorological Observations (SSMO). Selected series: Hawaiian and selected north Pacific island coastal areas, Vol. 1 (1971); Alaskan and British Columbian coastal marine areas, Vols. 11-14 (1970); North American coastal marine areas - revised, Vols. 2-4 (1975), Vols. 5-6 (1976).

19. U.S. Navy: U.S. Navy hindcast spectral ocean wave model climatic atlas: north Pacific Ocean. NAVAIR 50-1C-539 (1985).

20. Tayfun, M.A.: Breaking-limited wave heights. J. Waterways etc. Div., Proc. Am. Soc. Civ. Eng. 107 (1981) 59-69.

21. Triton Systems, Inc.: Wave power estimation from SSMO data. Report prepared for Sea Energy Corp., New Orleans, LA (1983).

22. Ward, E.G.; Evans, D.J. and Pompa, J.A.: Extreme wave heights along the Atlantic coast of the United States. Proc. 9th Ann. Offshore Technol. Conf., 1977, pp. 315-324.

Accuracy of the Estimate of Extreme Wave Conditions

L. BERTOTTI (1) and L. CAVALERI (2)

(1) A.S.C.O., Via Beato Pellegrino 14/20, Padua, Italy
(2) I.D.G.M.-C.N.R., San Polo 1364, 30125 Venice, Italy

Summary

The hindcast procedure to obtain long term estimates of extreme events is a known technique. This paper focuses its attention on the statistical aspect of the procedure, discussing the possible results and their accuracy.

Introduction

When long term estimates of extreme events are required and reliable data are not at hand, it is standard practice to make use of the hindcast procedure. This consists of hindcasting, usually by mathematical modeling, some suitable chosen heavy storms of the past. These provide an artificial set of data to serve as input to the statistical section.

We recently carried out an extensive hindcast study on the Tyrrhenian Sea, west of Italy. The aim of this paper is a presentation of the statistical part of the work (section 2) and a discussion of the accuracy of the final results (section 3). For any detail of the meteorological analysis and the mathematical modeling of the study the interested reader is referred to the extensive report on the subject "STONE-TIRRENO"(1984)[1].

Results of the Hindcast Procedure

The area of interest is shown in Fig. 1. A climatological analysis of the area showed that the storms affecting it can be classified into 4 different weather types, named A,B,C,D. Two separate (subjective and objective) procedures led to the choice of 71 storms, chosen according to the overall distribution. The storms were hindcasted by wind and wave modeling, and for each grid point the maximum of each storm was recorded. A comparison with experimental data indicated a r.m.s. error

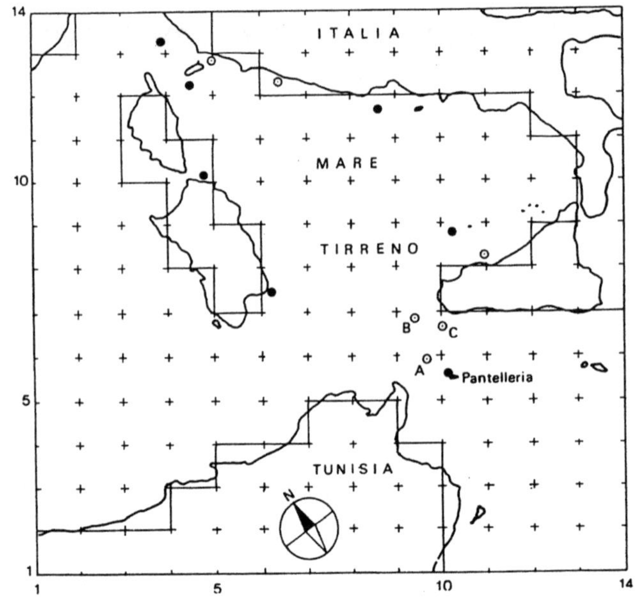

Fig. 1. Grid covering the Tyrrhenian Sea. The grid step is 70 km.

for the maxima of 0.4 m. The 71 maxima, subdivided into 4 weather types, were the input for the extremes statistics.

Extremes Statistics

The aim of the extremes statistics is to provide estimates of the probability that a given value will not be exceeded within a certain period of time. We tried 7 different extremal distributions (FT-1, logistic, exponential, normal, lognormal, Weibull, Pareto) checking their applicability (to each grid point and for each weather class). The comparison between the correlation coefficient of each fit and the associated critical value led to the choice of the FT-1 (or Gumbel) for the overall statistics with the expression

$$F(x)=\exp(-\exp(-ax + b)) \qquad (1)$$

Given (1), the return time of an event x is defined as $T(x)=1/(1-F(x))$. The no-exceedance probability of x within t years is given with good approximation [2] by

$$\bar{P}(x)=\exp(-t/T(x) \qquad (2)$$

Expression (2) is immediately generalized to n classes of events as the product of the probability of each class,

$$P^{-}(x) = \prod_{1}^{n} {}_i P_i^{-}(x) = \exp(-t \sum_{1}^{n} {}_i (1/T_i(x))) \qquad (3)$$

The results for a grid point are summarized in Fig. 2. This provides statistics for each single weather class and for the whole set of extremes according respectively to (2) and (3). The upper vectorial diagram provides information on value and direction of the single maxima.

Discussion of Accuracy of Extremal Estimates

The accuracy of the results of a hindcast procedure depends on several factors. Assuming the meteorological inputs to be correct, the accuracy of the models (wind + wave) depends on their theoretical bases, on the kind of storm, on the grid step and the numerical methods used for the computation.

While a discussion of the first two points is out of the scope of this paper, the latter two points are of course a trade-off between accuracy and computational effort. We mentioned before that a comparison with experimental data showed the r.m.s. error for the wave maxima to be close to 0.4 meter.

A crucial point of the whole procedure is the uncertainty associated with the choice of the storms. Notwithstanding any care put in it, there is certainly a good amount of subjectivity and randomness present in this choice and a different series of events would have led to different final results. Their statistical significance can be explored by the Jackknife technique [3],[4]. This consists in fitting the extremal distribution to each one of the N possible data set of (N-1) points each, obtained from the available extremes neglecting in turn the i-th point of the set. This provides N new estimates $\hat{\Theta}_i$ (i=1,N) of the parameter or quantity of interest, whose average $\hat{\Theta}_.$, when compared with the original estimate $\hat{\Theta}$, is a good indicator of its statistical significance.

The difference

$$\Delta\hat{\Theta} = \hat{\Theta}_. - \hat{\Theta} \qquad (4)$$

is in itself an estimate of the bias of the results.

This procedure provides also an estimate of the variance associated with the sampling of the events. The variance can be defined [5] as

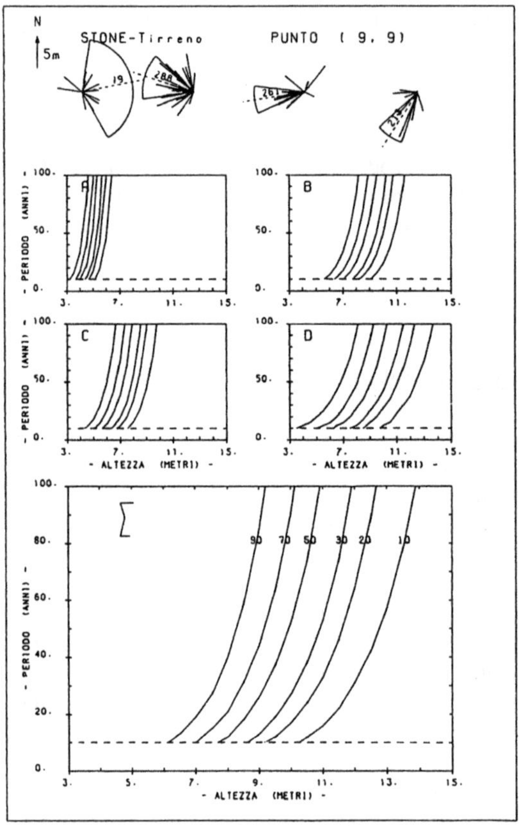

Fig. 2. Results of the extremal analysis for one of the grid points in Fig. 1. The curves indicate the encounter probability for given period and wave height. Smaller diagrams refer to the single weather classes, the lower one to the overall statistics. The upper vector arrays show modulus and direction of the single maxima.

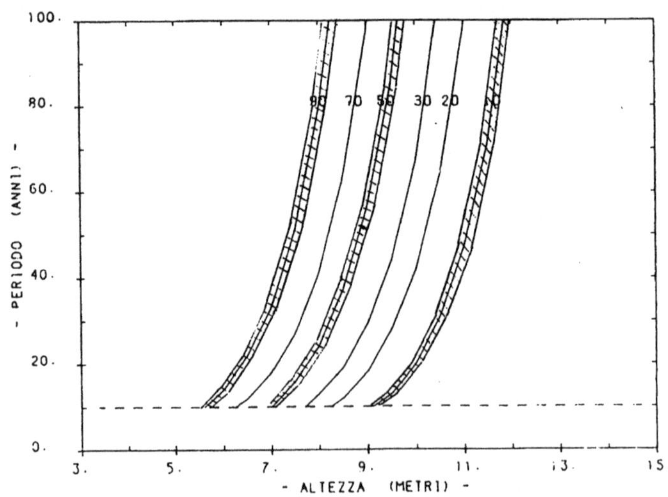

Fig. 3. Statistical significance of the diagrams of Fig. 2 because of the storms choice. The shadowed areas include one standard deviation.

$$\sigma_{\hat{\Theta}} = 1/(N-1) \cdot \sum_{1}^{N}{}_i\ (\hat{\Theta}_i - \hat{\Theta}_.)^2 \tag{5}$$

from which the standard deviation is

$$\mathrm{sd}\ \hat{\Theta} = (\sigma_{\hat{\Theta}})^{\frac{1}{2}} \tag{6}$$

The results of the Jackknife analysis for the 90, 50, 10% encounter probability diagram for some point of Fig. 2 (storm class B) are shown in Fig. 3. The bias is quite limited and the bias-corrected diagram is practically undistinguishable from the original one. The shadowed areas include one standard deviation of the order of 0.3 meter. If this was true for all the grid points, the result would be excellent. Anyhow we cannot assume this to be true for the whole grid. The Jackknife is particularly suitable for spotting in a set of extremes the value "out of place", i.e. the extreme that, associated with a much longer time, has happened to appear within the analyzed period. In this case, if we fit the extremal distribution only to the other N-1 data, the distribution will go to much lower values providing results strongly different from the original ones, with large values of $\Delta\hat{\Theta}$ and sd $\hat{\Theta}$. This is not a rare case as, when analyzing 20 years of data, we must expect the 100 year wave to appear in one case out of 5!

We have then to consider the statistical consequences of the 0.4 m r.m.s. error in the model output. The comparison with the experimental data has excluded any bias in the mathematical models and we can consequently assume with good certainty no bias in the results too. Nevertheless the error will cause a variance in the results. This can be evaluated by considering the analytical relationship between the quantities of interest, in our case the elapsed time t, the no-encounter probability P^- and the x-event.

Using eq.(2), and defining the encounter probability $P^+ = 1-P^-$, it is easy to obtain

$$t = t(x) = \ln(1-P^+)/\left[n_y(F(x)-1)\right] \tag{7}$$

with n_y the average yearly number of events. Knowing the variance of x (in our case 0.4 meter), the consequent variance of t is given [6] by

$$\sigma_t = \partial t(x)/\partial x \cdot \sigma_x \tag{8}$$

From eqs. (1), (7) and (8) we obtain

$$\sigma_t = -(1/n_y) \cdot \ln(1-P^+)F(x)a(ax+b)e^{-(ax+b)}\ \sigma x/(F(x)-1)^2 \tag{9}$$

For the same example used for Fig. 3 expression (9) indicates a similar variance for the lower encounter probability (P^+= 10 and 50%), rapidly growing for the higher P^+ values as a consequence of the term $\ln(1-P^+)$.

Formulas (5) and (9) refer to a single class of storms. The extension of (9) to the overall probability case is obtained considering the frequency of appearance p of each single weather class. This leads to the useful result

$$\sigma_t \leq \sum_{1}^{n}{}_i \, p_i \cdot \sigma_{t,i}$$

with $\sigma_{t,i}$ the variance associated with the i-th class of storms.

The very interesting conclusion of this analysis is that the statistical uncertainty present in the hindcast procedure is the major source of error in the final results of Fig. 2. To reduce it within the allowable limits of our aim we proceed in two different directions: a) from one side we can improve the models, trying to decrease the r.m.s. error in their final output; b) from the other we can increase the number of storms (i.e. of maxima), reducing the statistical uncertainty associated with it. Before making any effort in this sense it would be worthwhile to make a careful analysis of the effort required by each of the two possibilities and of their probable results.

References

1. The STONE Group, STONE-Tirreno, Statistica delle Onde Estreme. Consiglio Nazionale delle Ricerche, Padova, 1984, 86 pp.

2. Gumbel, E.J., Statistics of Extremes. Columbia University Press, New York, 1958, 375 pp.

3. Quenouille, M.: Approximate tests of correlation in time series. J. Roy. Statist. Soc. Serie B, 11, 1949, 18-84.

4. Efron, B., The Jackknife, the bootstrap and other resampling plans. Reg. Conf. Ser. in Appl. Math., Philadelphia, PA, 1982, 93 pp.

5. Blackman, R.B. and J.W. Tukey, The Measurement of Power Spectra from the Point of Communications Engineering. Dover, New York, 1958.

6. Borgman, L.E.: Feasibility for quantitative assesment of available margins inherent in flood protection of nuclear plants - Final report. A report to the Wave Dynamics Division, U.S. Army Waterways Exp.. Station, Vicksburg, MI, 1983.

On-line Identification and Prediction of Waves

J.L. Forsberg

Technocean AB
Box 9006, 400 70 Gothenburg, Sweden

Summary

One algorithm of a maximum likelihood estimation of an ARMA model for wind waves is described. The time variation of the parameters is taken care of. Examples from on-line identification of the model parameters is shown.

1. Introduction

Today sophisticated signal processing is available to engineers thanks to the break through in the field of system identification and to the access of cheap and powerful microprocessors.

On-line identification and prediction of waves could be useful in the control of wave energy convertors. The parametric description of wave records would save data in transmission lines and facilitate the simulation of wave trains compared to ordinary FFT methods.

2. Discrete time series modelling

It is known, Åström [1], Koopmans [2], Goodwin & Payne [3], that a sampled stationary process y(t) can be approximated by a discrete autoregressive-moving average ARMA(m,n) model on the innovations form

$$y(t) = -a_1 \cdot y(t-T) - a_2 \cdot y(t-2T) - \cdots - a_m \cdot y(t-mT) + \lambda(e(t) + c_1 \cdot e(t-T) + \cdots + c_n \cdot e(t-nT)) \quad (1)$$

where

a_j autoregressive (AR)-coefficient
c_j moving average (MA)-coefficient

e(t) uncorrelated white noise, innovation, Gaussian N(0,1) with variance 1 and zero mean

T sampling interval

λ gain factor

A more simple form of (1) is

$$A(q^{-1})\cdot y(t) = \lambda C(q^{-1})\cdot e(t) \tag{2}$$

with polynomials $A(q^{-1})$ and $C(q^{-1})$ in the back shift operator q^{-1}

$$\begin{aligned} A(q^{-1}) &= 1+a_1q^{-1}+\dots+a_mq^{-m} \\ C(q^{-1}) &= 1+c_1q^{-1}+\dots+c_nq^{-n} \\ q^{-1}y(t) &= y(t-T) \end{aligned} \tag{3}$$

Equation (2) can be recognized as a description of a digital filter with the output y(t) and the innovation e(t) as input. This is easily realized in a computer

$$y(t) = \lambda\frac{C(q^{-1})}{A(q^{-1})}\,e(t) \tag{4}$$

The polynomials A and C give the poles and zeros respectively of the digital filter.

Replacing q^{-1} by $e^{-i\omega T}$ in the filter transfer function, the squared modulus gives the continous spectral estimate $S(\omega)$ of y(t)

$$S(\omega) = \lambda^2\,\frac{|C(e^{-i\omega T})|^2}{|A(e^{-i\omega T})|^2}\cdot 1 \tag{5}$$

According to Jones [4] spectral estimates out of ARMA-parameters are more sensitive to sharp peaks and dips in the real spectrum of y(t) than common fast fourier transforms (FFT). Overvik & Houmb [5] and Johnsen [6] have done comprehensive off-line studies of ARMA(m,1)-models of records from Norwegian waveriders.

3. Identification of the ARMA-parameters

Maximum likelihood estimation (MLE) methods are extensively used in system identification. A modified recursive aproximate MLE method with very nice convergence properties has been suggested by Ljung et al [7]. They have shown that the algorithm converge with probability 1 to a local minimum, which is the best one can hope for.

The algorithm can be written

$$\hat{C}_{t-T}(q^{-1})\cdot z(t) = \phi(t)$$

$$K(t) = \frac{P(t-T)\cdot z(t)}{1+z^*(t)\cdot P(t-T)\cdot z(t)}$$

$$\varepsilon(t) = y(t) - \hat{\theta}^*(t-T)\cdot\phi(t) \qquad (6)$$

$$\hat{\theta}(t) = \hat{\theta}(t-T) + K(t)\cdot\varepsilon(t)$$

$$P(t) = P(t-T)\cdot\left(I- \frac{z(t)\cdot z^*(t)\cdot P(t-T)}{1-z^*(t)\cdot P(t-T)\cdot z\ (t)}\right)$$

where * denotes the transpose of a vector or matrix and

$$\hat{\theta}(t) = (a_1,\cdots, a_m, c_1,\cdots, c_n)^*$$

is the update estimate of the parameter vector.

$$\phi(t) = (-y(t-T),\cdots, -y(t-mT), \varepsilon(t-T),\cdots, \varepsilon(t-nT))^*$$

saves old measured data and estimates ε of the innovations e. The polynomial $\hat{C}_{t-1}$ is recognized as

$$\hat{C}_{t-T}(q^{-1}) = 1+c_1q^{-1}+\cdots+c_nq^{-n}$$

K(t) is a gain matrix, P(t) a covariance matrix and I the unity matrix.

4. One step-ahead predictor

The best one step-ahead prediction of y(t+T) at the time t is

$$\hat{y}(t+T) = \hat{\theta}(t)\cdot\phi(t+T) \tag{7}$$

where $\phi(t+T)$ is the update of $\phi(t)$ now including y(t) and $\varepsilon(t)$.

5. Tracking time-varying ARMA parameters

In real time identification it will be impossible to accurately track fast varying parameters.

Fortunately the sea-state vary slowly enough to be well followed by the identification algorithm used.

By introducing a forgetting factor g in the loss function

$$J_N(\theta) = \sum_{k=1}^{N} g^{N-k}\cdot\varepsilon^2(k) \tag{8}$$

earlier prediction errors will get decreasing influence on the crieterion function. The forgetting factor is here supposed to be constant but could vary in time.

Define a memory time constant T_m so that the prediction error $\varepsilon(t)$ older than T_m are wheighted less than e^{-1} i.e. 36%. If the forgetting factor is close to 1, this yields

$$g \approx 1-\frac{T}{T_m} \tag{9}$$

In Swedish waters 6-8 minutes long wave records are convenient, using a sampling interval of 0,5 s. It is resonable to consider the wave climate stationary during 15 minutes at the site near to Vinga island outside Göteborg on the Swedish west coast.

A memory time constant of 15 min implies

$$g \approx 1-\frac{0,5}{15\cdot 60} = 0,9995$$

6. Filtering of outliers

Every system is more or less exposed to noise and disturbances, specially radio transmission links. This generates outliers in the wave records.

Data causing prediction errors

$$|\varepsilon(t)|>3\sqrt{E(\varepsilon^2(t))} \tag{10}$$

are rejected and replaced by the predicted value. This means that 1 of 300 rejected values could be a true one.

7. Choice of model set

In many physical systems there is a natural choice of the model. Finding the order of the ARMA process for waves from wind speed and direction, fetch, waterdepth etc is however an impossible task. Some kind of statistical measure must be used.

Akaike [8] has suggested a minimum information crieteria AIC based on the maximum likelihood estimate of the ARMA-parameters

$$AIC(m,n) = N(E(\varepsilon^2(t)))+2(m+n) \tag{11}$$

where

E expectation value operator

N number of samples

If several models are possible the one that minimizes AIC is preferred.

8. Examples

The wave data logging system is outlined below

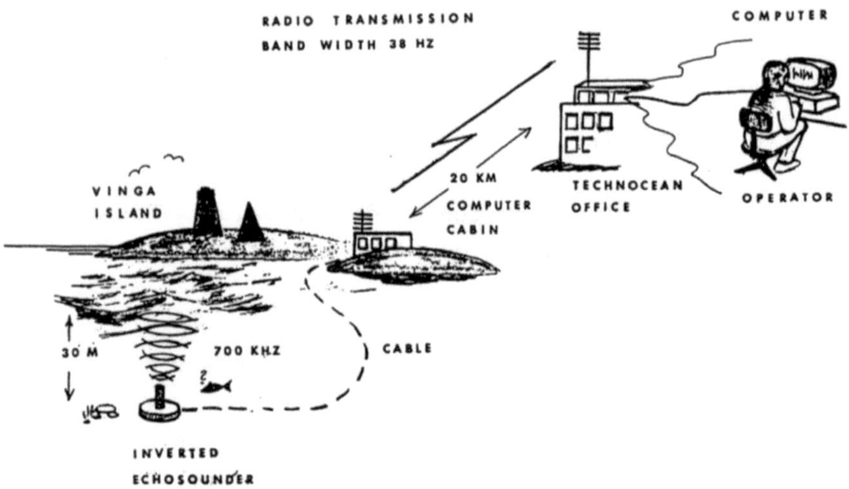

Fig.1. Wave recording system

The absence of too much noise is crucial to a succesful start of the parameter estimation. Otherwise the noise will be identified instead of the wave data. Figure 2 demonstrates the transient behavior of the recursive algorithm together with initial values.

```
Date and starttime    1 Jul 1985  14:49:12
Control code of the voltmeter is F1T1R2D.0001SN0001S
Number of AR-parameters=  4   Number of MA-parameters=  3
Forgetting factor=  .9995     Sampling time=  .5
Inital values for 4 AR-parameters and 3 MA-parameters in order
a( 1 )=  0          a( 2 )=  0          a( 3 )=  0          a( 4 )=  0
c( 1 )=  0          c( 2 )=  0          c( 3 )=  0
Initial values of the trace in the 7 x 7 covariance matrix,
suitable values 10<Pii<100
P( 1 , 1 )=  10     P( 2 , 2 )=  10     P( 3 , 3 )=  10     P( 4 , 4 )=10
P( 5 , 5 )=  10     P( 6 , 6 )=  10     P( 7 , 7 )=  10
Initial values of 4 old wave sampels and 3 old innovations
y( 1 )= -.6         y( 2 )= -.6         y( 3 )= -.6         y( 4 )= -.6
e( 1 )=  0          e( 2 )=  0          e( 3 )=  0
```

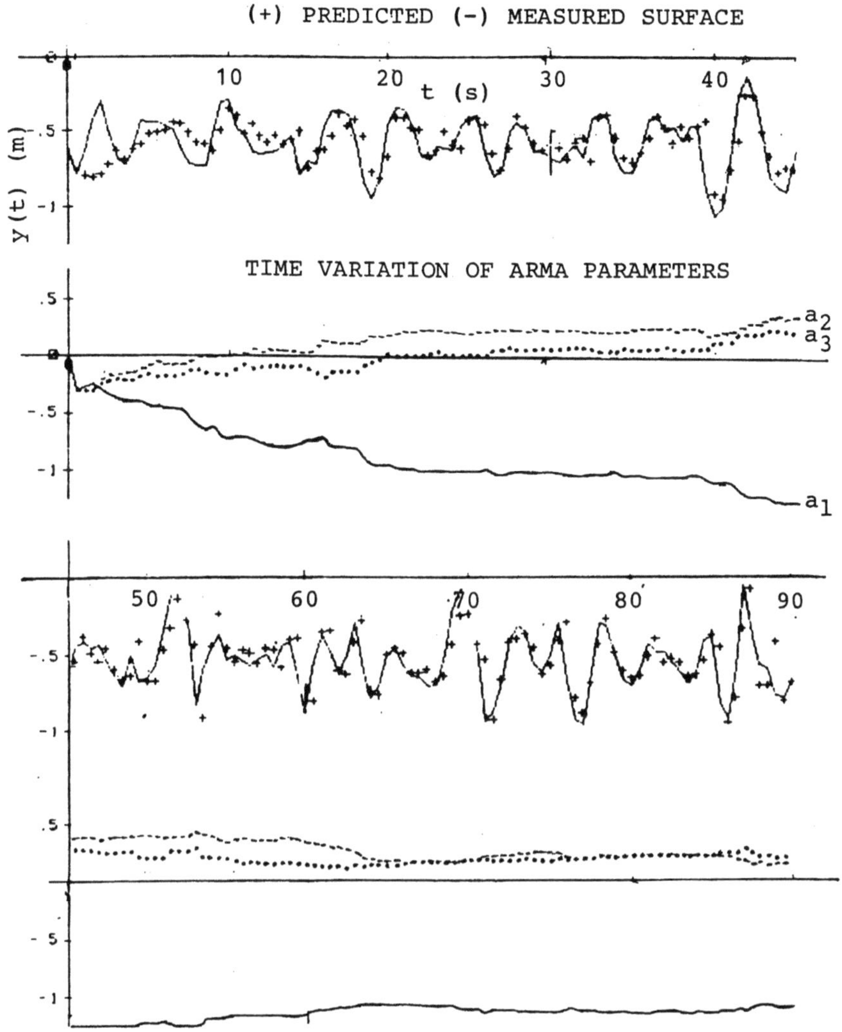

Fig.2. Initial transient behavior

After 10 minutes the parameter variations are small as shown in figure 3.

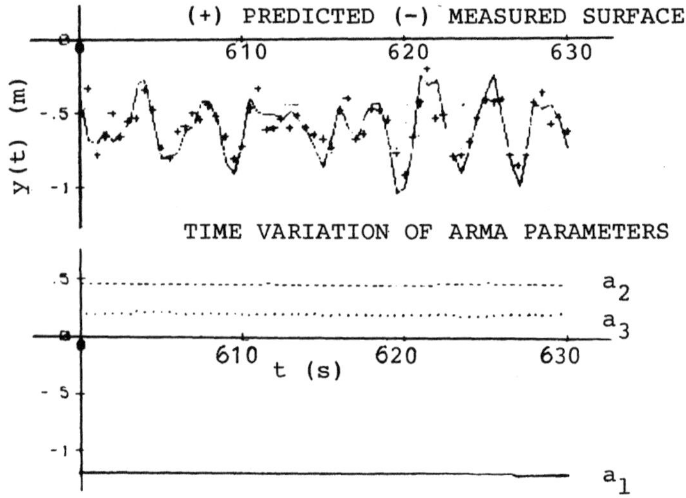

Fig.3. Stationary conditions

Figure 4 exemplifies the long term ARMA-parameter variations followed by the algorithm by means of the forgetting factor g. The records are from a storm in october 1983.

References

1. Åström, K.J.: Introduction to stochastic control theory. New York, London: Academic Press 1970.
2. Koopmans, L.H.: The spectral analysis of time series. New York, London: Academic Press 1974.
3. Goodwin, G.C.; Payne, R.L.: Dynamic system identification: Experiment design and data analysis. New York, London: Academic Press 1977.
4. Jones, R.H.: Identification and autoregressive spectrum estimation. IEEE Trans. Automat. Contr. No. 6 (1974) 894-897.
5. Overvik, T.; Houmb, O.G.: On the statistical modelling of ocean waves by parametric methods. NTH Trondheim, Norway.
6. Johnsen, M.H.: Signalprocessering i böyer, estimering av bölgeparametre. CMI Nr 821156-3 Bergen, Norway 1982.
7. Ljung, L.; Söderström, T.; Gustavsson, I.: Counterexamples to general convergence of a commonly used recursive identification method. IEEE Trans. Automat. Contr. No. 5 (1975) 643-652.
8. Akaike, H.: A new look at the statistical model identification. IEEE Trans. Automat. Contr. No. 6 (1974) 716-723.
9. Källström, C.G.: A dynamic positioning system for semi-submersibles. Swedish Maritime Research Center, SSPA, Gothenburgh, Sweden 1983.

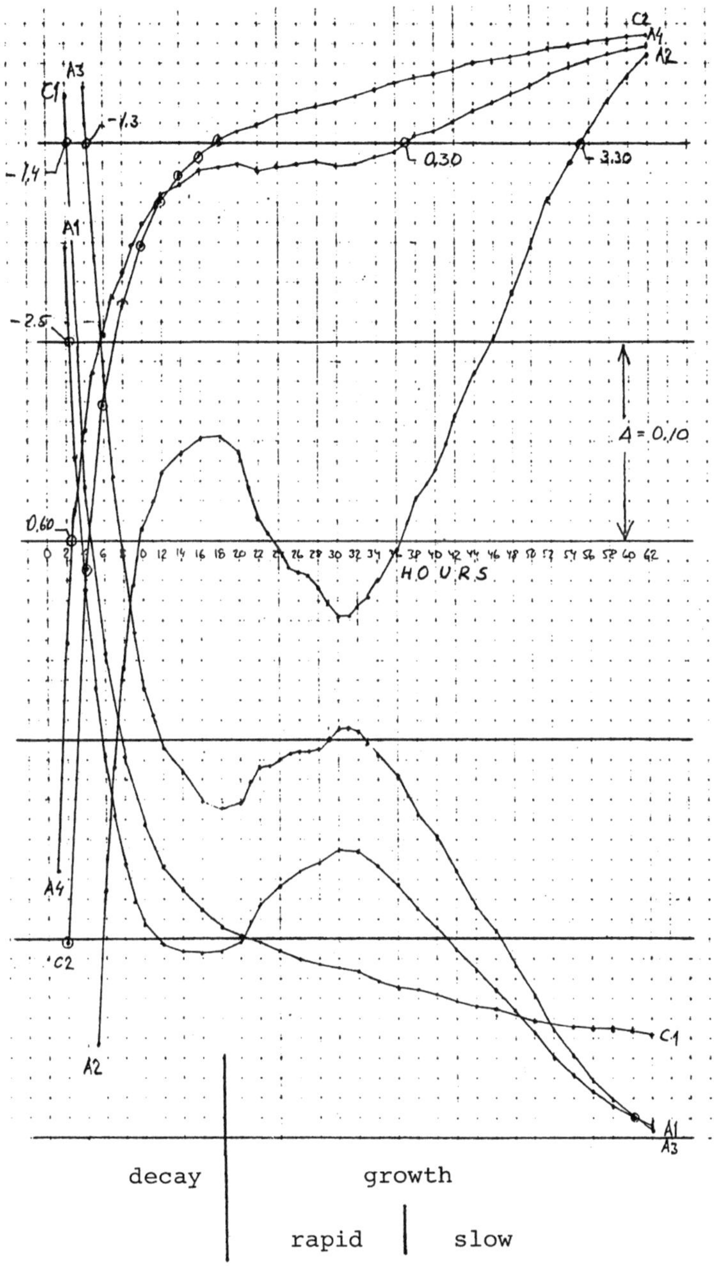

Fig.4. Long term variations of the ARMA-parameters

Extreme Wave Heights along the Norwegian Coast

MEIXIN ZHU

Nanjing College of Navigational Engineering
No. 6 Back Changjiang Road
Nanjing
People's Republic of China

Summary

The Gumbel distribution (first asymptote) is used with the annual largest significant wave heights Hs to calculate design wave heights. The formula $H=Hs\sqrt{\frac{1}{2}\ln N}$ has been derived from the extreme distribution. The reasonable values of τ_{Hs} and N are illustrated. In this paper directional wave statistics have been taken by using hindcast wave data. The worst directions and the worst directional design wave heights are obtained along the Norwegian coastal area.

1. The Calculation of Design Wave Heights

Design wave criteria for offshore and coastal structures are based on occurence probabilities of extreme wave conditions. Design wave height is a function of return period Rp (usually Rp=100 years)

$$Rp = \frac{Hs}{1-G(X)} \cdot \tag{1}$$

Here we use the Gumbel extreme distribution (first asymptote)

$$G(X) = e^{-e^{-y}}, \quad y = \alpha(X-\mu),$$

where X represents the annual largest significant wave height Hs, τ_{Hs} is the average time interval between two successive Hs, and μ is the mode of extreme Hs,

$$\left.\begin{aligned} &\mu = X - \frac{\overline{y(n)}}{\alpha}, \quad X = \overline{X} - \frac{\overline{y(n)}}{\alpha} + \frac{1}{\alpha} y. \\ &\frac{1}{\alpha} \text{ is a measure of dispersion } \quad \frac{1}{\alpha} = \frac{Sx}{\sigma(n)}, \quad Sx^2 = \frac{1}{n-1}\Sigma(X-\overline{X})^2. \end{aligned}\right\} \tag{2}$$

Being the functions of sample size n, the values of $\sigma(n)$ and $\overline{y(n)}$ are given in special table [5]. Here n should be larger than 20.

The derivation of $Hm = Hs\sqrt{\frac{1}{2}\ln N}$

If we have many records of wave height observation, many extreme wave height values can be obtained, each value coming from one record. These values $(H_N)_i$ form a long term series of H_N, where N is the number of waves in a given sea state Hs. The nonexceedance cumulative probability distribution is given by [5]

$$F(H_N) = e^{-Ne^{-2\left(\frac{H_N}{Hs}\right)^2}} . \qquad (3)$$

According to the general exponential extreme distribution

$$F(X) = e^{-e^{-y}} = \exp(-e^{-\alpha(X-\mu)}).$$

When X=μ, i.e. y=0, it is the most frequent extreme. If X represents extreme wave height, we call X=μ the most probable extreme wave height, then

$$F(X) = F(H_{N,m}) = \exp(-e^{-0}) = e^{-1}. \qquad (4)$$

Substituting (4) into (3):

$$e^{-1} = \exp(-Ne^{-2(H_{N,m}/Hs)^2})$$

$$H_{N,m} = Hs\sqrt{\frac{1}{2}\ln N} \quad \text{or simplified} \quad Hm = Hs\sqrt{\frac{1}{2}\ln N}$$

The adoption of the value of τ_{Hs}

In the literature, τ_{Hs} is often taken from 12 min. to 3 hrs. From eq. (1), the smaller the value of τ_{Hs}, the higher the security of the same X will be, but we can never take the τ_{Hs} value as small as possible only, ignoring the relationship with other factors. Using eq. (5), we assume that the 100-year return period most probable extreme wave height $H_{m(100)}$ can be obtained from the same return period significant wave height $H_{s(100)}$

$$H_{m(100)} = H_{s(100)}\sqrt{\frac{1}{2}\ln N} , \qquad (6)$$

where N is the number of waves for a given sea state Hs.

Since the shorter the τ_{Hs}, the smaller the N during the τ_{Hs}, it is sure that there exists a τ_{Hs} and its corresponding N ($N \leq Hs/7$, 7 sec. being the average time interval between two individual waves in the North Sea) which will form a highest value of the 100-year return period most probable wave height $H_{m(100)}$.

If

$$F_L(Hs) = 1 - \exp(-(\frac{Hs-0.5}{1.9})^{1.44}),$$

according to eq.. (1)

$$100 \times 365 \times 24 = 87600 = \frac{\tau_{Hs}}{\exp(-(\frac{H_{s(100)}-0.5}{1.9})^{1.44})}$$

$$H_{s(100)} = 0.5 + 1.9(\ln(\frac{87600}{\tau_{Hs}})^{\frac{1}{1.44}}).$$

Using τ_{Hs}=7N and eq. (6), we can get $H_{m(100)}$ which varies with τ_{Hs}

Table 1

τ_{Hs}	N	$H_{s(100)}$	$H_{m(100)}$
(hr)		(m)	(m)
0.2	100	13.1	19.9
1	500	12.2	21.5
2	1000	11.8	21.9
3	1500	11.5	22.0
6	3000	11.1	22.2

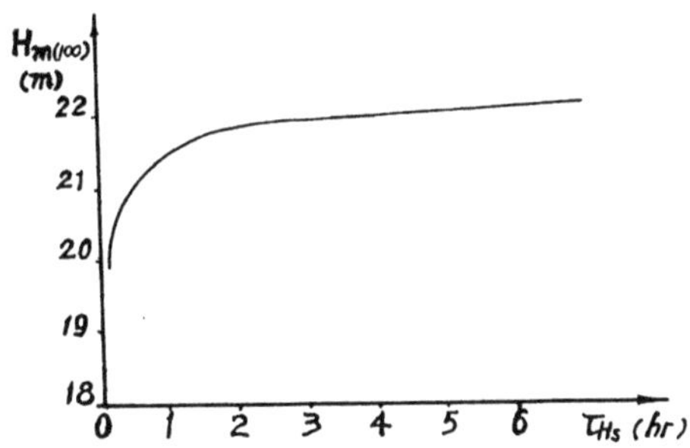

Fig. 1

From table 1 and fig. 1, it can be seen that the highest 100-year wave height does not occur at the lowest value τ_{Hs}=0.2 hr (12 min.) and we can adopt τ_{Hs}=2 hrs and N=1000 or τ_{Hs}=3 hrs and N=1500. The difference in $H_{m(100)}$ between these two sets is only 0.45%. Here the set of τ_{Hs}=2 hrs and N=1000 is adopted.

2. Prediction of Extreme Wave Heights (Omni-directional)

Description of data

The choice of a statistical model is not as important, especially in extreme statistics, as the need for an adequate data base. One can imagine, with or without the highest recorded wave height in the Norwegian coast — 27 metres — how large the difference will happen in curve-fitting and extrapolation of 100-year return period wave height. Since the reliable measured data in the Norwegian water area only started in 1974, we have to extend the length of wave data series by using available weather data. A large wave data base, timewise and areawise, can be created by hindcasting. In this paper, the hincast data of wave height are given by NMI (Norwegian Meteorology Institute) model. Having no practice of wave hindcasting, the author cannot estimate the uncertainties in that process, but scientists who did it [7] concluded that the wave heights hindcasted are, in average, a bit greater than the corresponding actual wave heights. So we took the hindcast wave data from the NMI model as a conservative estimate, and used 25 years of data (from 1955.07.01 to 1980.06.30) for 6 grid points (13:09), (14:12), (15:15), (16:16), (18:20) and (20:20) (see fig. 2).

In this method, data series are much longer than others and there exist strict formulas to fit the straight line with the empirical points. Avoidance of fitting lines by eye is an extra advantage. No tendency for $H_{m(100)}$ to change with latitude can be seen from tables 2 and 3. On the average, $H_{m(100)}$ along the Norwegian coast equals 31.60. This result comes from the omni-directional statistics. Furthermore, if there are enough data to carry out the directional wave statistics, taking the worst directional $H_{m(100)}$ as the design criterion will be more resonable than the omni-directional statistics in this paragraph (see paragraph 3).

3. Directional Wave Height Statistics

The normal procedure followed in extreme event design for off-shore installations is to assume that the design wave can aapproach in any direction. But this is not true. Especially in off-shore areas, the local topography influences the wind fetch, and hence influences the size of waves, so that there exist one or more directions from which waves come. Up until now, wave height statistics take into account all waves occuring

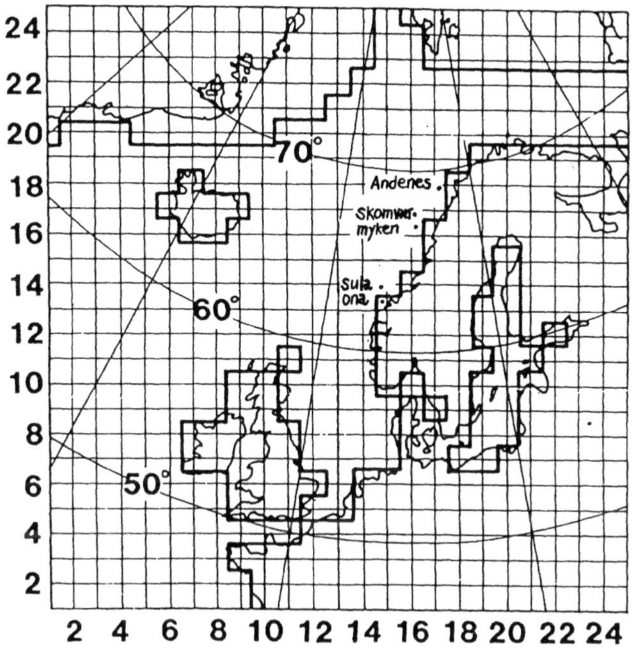

Fig. 2 Grid net of hindcast model

Ekofisk (13:09)	Halten (15:15)
Utsira (14:11)	Traenabank (16:16)
Troll (14:12)	Tromsøflakt (18:20)
Kråkenes (14:13)	Fruholmen (20:20)

$H_m(100)$ values predicted by using Gumbel distribution (first asymptote) and NMI hindcast model wave data for yearly largest Hs.

Table 2

Station	(13:09)	(14:12)	(15:15)	(16:16)	(18:20)	(20:20)
$H_s(100)$ m	18.28	17.95	16.38	18.08	17.60	17.33
$H_m(100)$ m	34.53	33.36	30.44	33.60	32.71	32.21
n* years	25	25	25	28	25	25

* n=25 (1955.07.01-1980.06.30)
n=28 (1955.07.01-1983.06.30)

Table 3 n=(1948–1977)

Station	(14:11)	(14:13)	Ona/Sula	Myken/Skomvar	Andenes	(20:20)
$H_s(100)$ m	16.25	16.00	16.68	16.90	16.04	16.00
$H_m(100)$ m	30.19	29.73	31.00	31.40	29.81	29.73

in different directions, without regarding the directions of waves. In fact, either for the having-occured or the coming waves, each one is associated with a certain direction. The design wave height should have a higher value and corresponds to a direction of propagation which is called the worst direction. So performing only omni-directional wave statistics is not enough; obviously directional wave statistics are needed. In directional wave statistics, wave height data base is split into eight compass sectors, No. 1, 2,...,8, representing eight structure response directions respectively, i.e. 1 represents the wave coming from south, and 5 from north. The directional statistics are only based on the data of waves coming from each respective compass sector, each direction having only 10-20% data of the total base.

If the data length is large enough, in every direction, to represent the population, and in the absence of curve-fitting and extrapolation errors, there still exists a difference between the worst directional and omni-directional design values, because they come from different populations. At first sight, the data length for performing omni-directional and individual direction extreme wave height statistics seems to be the same, but,according to the general statistics definition, the 100-year return period extreme value in omni-directional statistics should be numerically larger than that in individual directional statistics, if the standard deviations of these distributions are the same. A simplified

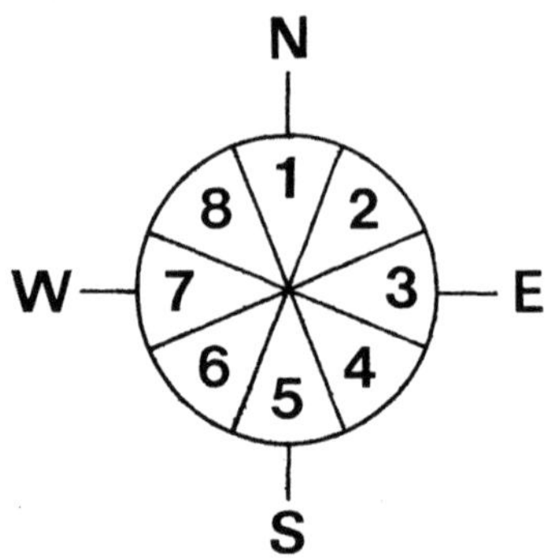

Fig. 3

theoretical example can be used to explain this view. Assume we have a representative 100-year wave data series from a particular site, where the wave directions are distributed evenly in north and south sectors. In each of these two sectors the annual largest wave height exceeding 30 metres occurred only once in the representative 100-year period. However, from the omni-directional view-point, the wave height which is equal to or larger than 30 metres occurred twice. In this simple example one can see that the 100-year return period wave height in directional statistics is 30 metres or so each, while that value in omni-directional statistics may be 31.8 metres. The difference will have the largest value when the waves are distributed evenly in all directions. In fact, curve-fitting and extrapolation errors cover up this difference and it is impossible to find the worst directional 100-year design wave height higher than the omni-directional one. What is the reason for this? The author deems that for extreme value statistics, corresponding to the directional individual wave height, the directional extreme wave height distribution belongs to an extreme distribution. Similarly, corresponding to the eight directional extreme wave heights, the omni-directional extreme wave height distribution belongs to another extreme distribution, and it is a super-extreme distribution. The standard deviation σ_{Hs} and mean value $\overline{Hs}$ in the worst directional extreme statistics are larger and smaller than those in the omni-directional extreme statistics, respectively (see table 5).

Table 5. Mean value $\overline{Hs}$ and standard deviation σ_{Hs} for different stations along the Norwegian coast

Station		(13:09)	(14:12)	(15:15)	(16:16)	(18:20)	(20:20)
$\overline{Hs}$	omni-direction	11.42	11.11	10.24	12.70	11.28	10.74
	worst direction	9.96	10.35	9.12	10.29	9.87	9.70
σ_{Hs}	omni-direction	1.91	1.54	1.63	1.45	1.66	1.76
	worst direction	2.30	2.52	2.35	2.23	2.32	2.47

Although, on the average, the mean value $\overline{Hs}$ in the worst directional extreme statistics is a bit smaller (12%) than that in the omni-directional statistics, the standard deviation σ_{Hs} is much larger (43%) than the corresponding values in omni-directional statistics. That is exactly the σ_{Hs} which determines the extrapolated design value $H_{s(100)}$. The standard deviation σ_{Hs} is the main factor which leads to the higher value of $H_{m(100)}$ in the worst directional extreme statistics. Values of σ_{Hs} in table 5 also prove that the omni-directional extreme wave height distribution is a "super extreme distribution". The results of 100-year $H_{m(100)}$ in eight directions for 6 stations are listed in table 6.

Table 6. Values of directional and omni-directional $H_{m(100)}$

Station \ Sector	1	2	3	4	5	6	7	8	omni
(13:09)	19.51	21.46	25.33	32.89	34.70	18.58	20.25	23.00	34.53
(14:12)	25.46	30.40	29.98	32.62	36.80	17.56	14.87	22.30	33.36
(15:15)	11.76	16.93	27.88	33.23	23.00	21.09	21.65	18.14	30.44
(16:16)	26.00	32.71	32.89	34.38	32.71	21.00	14.87	18.12	33.60
(18:20)	15.83	29.74	29.25	34.14	29.55	25.31	23.51	19.51	32.71
(20:20)	12.04	19.51	28.47	33.04	25.80	21.04	21.84	19.05	32.21

From table 6, one can see that there exists one worst direction for each grid point. For the Norwegian coastal area, the worst directions are sector 5 (in south part) and sector 4 (in north part). Approximately, the distinctive line is latitude 62°N. The change can be explained by topographic reasons. To the south of 62°N, the coastline makes an angle of 90° with the latitude line, and in this region the sea (North Sea) is bounded by Norway on the east, by Great Britain on the west, and by the European mainland on the south. Therefore the worst-wave-generating wind direction (the one one with the longest fetch) is north, and so the worst wave direction is in sector 5 (waves coming from north). The part of the coastline lying north of 62°N makes an angle of 45-60° with the latitude line. It is in the norwegian Sea, and from the Norwegian coast to the west there is a very wide water area, and so the worst wind generated wave direction is not strictly northern, but northwestern in sector 4.

The mean value of the worst direction 100-year return period wave height for the six stations is 34.38 m, whereas the omni-directional one is 32.81 m. This means that, on the average, the worst directional 100-year return period wave height is approximately 5% higher than the omni-directional one. It is more reasonable and conservative to take the worst directional extreme value as the design criterion for all structures affected by waves. The time is coming when enough date will be available to perform the directional statistics. For the average value of omni-directional $H_{m(100)}=31.60$ m (see paragraph 2), we can take the worst directional wave height $H_{m(100)}=33.18$ m as the design value in the Norwegian coastal water area.

4. Conclusion

(1) The formula $H = Hs\sqrt{\frac{1}{2}\ln N}$ can be derived from the extreme distribution. Since the shorter the τ_{Hs} the smaller the number of N during the τ_{Hs}, there exists a τ_{Hs} which will form a highest value of $H_{m(100)}$. Through analysis it is proposed that $\tau_{Hs}=2$ hrs and N=1000 be used to contribute to the larger value of $H_{m(100)}$.

(2) Up till now, according to the data length, it is suggested that using hindcast wave data and the Gumbel extreme distribution (first asymptote) may be the better method to predict 100-year return period wave height $H_{m(100)}$.

(3) Because any occuring wave only has one direction, adopting the worst directional extreme wave height instead of omni-directional one in design work is more reasonable and conservative than adopting omni-directional one when the data are enough to do the directional statistics.

(4) Due to the influence of local topography in the Norwegian coastal area, the worst wave propagation is northern in the area south of latitude 62°N, and nothwestern in the area north of latitude 62°N.

(5) The omni-directional extreme wave height distribution is a "super extreme distribution" which has a lower standard deviation than that of the worst direction, so the worst directional 100-year return period wave height is higher than the omni-directional one.

(6) According to the analysis of hindcast wave data, for the Norwegian coastal area the mean value of the worst directional $H_{m(100)}$, on the average, is 5% larger than the traditional omni-directional $H_{m(100)}$. It is suggested that the $H_{m(100)}$ of the worst directional 100-year return period wave height along the Norwegian coast equals 33.18 m.

References

1. Gumbel, E.J.: Statistics of extremes. Columbia University Press, New York 1958.
2. Battjes, J.A.: Long term wave height distribution at seven stations along the British Isles. National Institute of Oceanography, Godalming, England, Report No. A44, 1970.
3. Ramesh, B.A.; Houmb, O.G.: Comparative study of wave hindcast models. 1982.
4. Sarpkaya, T.; Isaacson, M.: Mechanics of wave forces on offshore structures. Van Nostrand, New York, 1981.
5. Brunn, P.: Port engineering. 3rd ed. 1981.
6. Graham, C.G.: Problems with the design statistical approach in extreme value. 1982.
7. Earle, M.D.; Baer, L.: Effects of uncertainties on extreme wave heights. 1982.

Wave Forces on a Horizontal, Submerged, Spinning Cylinder

K. BUDAL and P. M. LILLEBEKKEN

Department of Physics and Mathematics,
University of Trondheim,
N-7034 Trondheim-NTH. Norway.

Introduction and Summary

A submerged, fixed object is subjected to wave forces. If the object has an axis of rotational symmetry and is rotated around this axis with constant angular velocity, the wave force on the body may be changed substantially. The phenomenon may be explained as a Magnus effect in waves.

The paper develops, within linear theory, an approximate theoretical expression for the wave force on a horizontal, submerged, spinning cylinder with axis parallel to the wave crest. Defining the spin factor σ as the ratio between the wave force with and without rotation, it is shown that the cylinder has a theoretical spin factor $\sigma \cong 1 + \omega_r/\omega$, where ω_r and ω are the angular frequencies of the cylinder and the wave, respectively. The spin factor may be both larger and smaller than 1 , depending on the sign of ω_r .

Experimental results are presented and compared with the theory. Within restricted limits of wave parameters and angular frequencies of the cylinder, theory and experiment agree. Spin factors as lage as 6 and as small as 1/3 were measured. Deviations between theory and experiment are discussed. Finally, some applications of the phenomena are mentioned.

Theory

A fixed cylinder (Fig.1) with radius a and length $\ell \gg a$ is submerged to a depth h below the still water level. Waves incident in the x-direction, normal to the axis of the cylinder, exert time varying force components F_x' and F_z' on the cylinder. In a plane incident wave with velocity potential

$$\phi_i = A\, e^{i(\omega t - kx)}\, e^{kz} \tag{1}$$

these forces are according to linear theory

$$F_x' = -\int_S p \cos\Theta \, dS = i\omega\rho \int_S (\phi_i + \phi_d) \cos\Theta \, dS \tag{2}$$

and

$$F_z' = -\int_S p \sin\Theta \, dS = i\omega\rho \int_S (\phi_i + \phi_d) \sin\Theta \, dS \tag{3}$$

Here S is the surface of the cylinder, p is the dynamic pressure, ϕ_d is the velocity potential of the diffracted wave and ρ is the density of water.

Suppose now that the cylinder is being rotated around its own axis with constant peripheral speed v_a . Then the wave induced forces will be different from those given by the expressions (2) and (3) . An estimate of these force components is as follows:

Let the velocity potential with fixed cylinder be

$$\phi_o = \phi_i + \phi_d \tag{4}$$

Furthermore, let the time-independent velocity potential due to rotation of the cylinder without an incident wave be ϕ_{rot} .

The fluid velocity due to the rotation is

$$\vec{v} = \nabla\phi_{rot} \tag{5}$$

On the surface of the cylinder

$$(\nabla\phi_{rot})_S = v_a \hat{\Theta} , \tag{6}$$

where $\hat{\Theta}$ is a unit tangential vector. The fluid velocity has its maximum velocity on S . In an infinite fluid the velocity v_r at a distance r from the center of a long, spinning cylinder (see Fig. 1) is such that

$$v_r \cdot r = v_a \cdot a = \text{constant} = C \tag{7}$$

or

$$v_r = \frac{v_a \cdot a}{r} = \frac{C}{r} \qquad (8)$$

Eq. (8) does not apply in our case since the fluid is not infinite. However, under the assumption of deep water and deep submergence, $h \gg a$, a crude estimate of the maximum fluid velocity on the free water surface is

$$v_{max} \approx \frac{v_a \cdot a}{h} = \frac{\omega_r a^2}{h} , \qquad (9)$$

where ω_r is the angular frequency of the cylinder. If then the variables in eq. (9) are such that v_{max} is sufficiently small, the dynamic free-surface condition can be linearized, i.e. the term $1/2\ \rho v^2$ in Bernoulli's equation is of second order and can be neglected at the water surface.

Consider now the rotating cylinder placed in the plane incident wave. The above arguments lead to the conclusion that the velocity potential

$$\phi = \phi_i + \phi_d + \phi_{rot} \qquad (10)$$

satisfies Laplace's equation and all boundary conditions, i.e. ϕ is, to the first order approximation, the correct solution to our problem.

We are now in the position to estimate the force on the rotating cylinder. The pressure on the surface of the cylinder is given by Bernoulli's equation

$$p = \left(-\rho\frac{\partial \phi}{\partial t} - \frac{1}{2}\ \rho v^2\right)_S = \left(-\rho\frac{\partial \phi}{\partial t} - \frac{1}{2}\ \rho (\nabla\phi)^2\right)_S \qquad (11)$$

Notice that the term $\frac{1}{2}\ \rho v^2$ cannot be neglected as compared to $-\rho\frac{\partial \phi}{\partial t}$ on the surface of the cylinder, since $|\nabla\phi_{rot}|_S = v_a$ is not necessarily a small quantity. This is not inconsistent with our previous assumption that $\nabla\phi_{rot}$, with properly chosen variables, is small on the free surface of the fluid.

By means of eqs. (1) and (10) and noticing that ϕ_{rot} is independent of time, eq. (11) gives

$$p = - i\omega\rho(\phi_i + \phi_d) - \rho\nabla(\phi_i + \phi_d) \cdot \nabla\phi_{rot} - \frac{1}{2}\rho[\nabla(\phi_i + \phi_d)]^2 - \frac{1}{2}\rho\cdot[\nabla\phi_{rot}]^2 \tag{12}$$

Here the first two terms are of 1st order. The third term which is of 2nd order, is symmetrical in Θ . The last term is a constant in time.

Hence, to the first order approximation the x-component of the wave-induced force is

$$F_x = - \int_S p \cos\Theta \, dS \cong i\omega\rho \int_S (\phi_i + \phi_d) \cdot \cos\Theta \, dS + \rho \int_S \nabla(\phi_i + \phi_d) \cdot \nabla\phi_{rot} \cdot \cos\Theta \cdot dS \tag{13}$$

Here, the first term is according to eq. (2) equal to F_x' The second term is evaluated as follows:

$$F_x'' = \rho \int_S \nabla(\phi_i + \phi_d) \cdot \nabla\phi_{rot} \cos\Theta \, dS = \rho \int_S \frac{1}{a} \frac{\partial(\phi_i + \phi_d)}{\partial\Theta} v_a \cos\Theta \, \ell a \, d\Theta = \rho\ell v_a \int_S \frac{\partial(\phi_i + \phi_d)}{\partial\Theta} \cos\Theta \, d\Theta \quad , \tag{14}$$

where ℓ is the length of the cylinder. By partial integration we obtain

$$F_x'' = \rho\ell v_a \int_S (\phi_i + \phi_d) \sin\Theta \, d\Theta \tag{15}$$

Now, from eq. (3) it is seen that this may be written

$$F_x'' = - i \frac{\omega_r}{\omega} F_z' \tag{16}$$

The x-component of the force on the cylinder is consequently

$$F_x = F_x' + F_x'' = F_x' - i \frac{\omega_r}{\omega} F_z' \tag{17}$$

Similarly it may be shown that the z-component of the force is

$$F_z = F_z' + F_z'' = F_z' + i \frac{\omega_r}{\omega} F_x' \tag{18}$$

Notice that F_x' and F_z' are 90° out of phase. Then, the second force term in eqs. (17) and (18) is either in phase or 180° out of phase with the first term, depending on the sign of ω_r (or the direction of the incident wave). Within linear potential theory F_x' and F_z' are related by the equation [1]

$$F_z' \cong i\, F_x' \tag{19}$$

Eqs. (17) and (18) may then be written

$$F_x \cong F_x' \left(1 + \frac{\omega_r}{\omega}\right) \tag{20}$$

and

$$F_z \cong F_z' \left(1 + \frac{\omega_r}{\omega}\right) \tag{21}$$

According to these equations the force on the cylinder is doubled if $\omega_r = \omega$, i.e. when the cylinder is making one revolution each wave period, and that the force is zero if $\omega_r = -\omega$ Thus: even moderate rotational speeds of the cylinder changes the wave induced force drastically.

Notice that reversing the rotation of the cylinder corresponds to reversing the direction of the wave. Consequently, a spinning cylinder will experience a widely different force if the direction of the incident wave is turned 180^0.

Another implication is that a properly spinning cylinder used as a one-mode wave generator will radiate highly asymmetric waves.

The physical explanation of the additional force when the cylinder is rotating is simply a Magnus effect [2]. A rotating cylinder placed in an oscillatory flow will experience an oscillatory lift force normal to the direction of the flow. This explains the mixing og x- and z-components in the expressions (17) and (18).

Let us consider the x-force given in eq. (17). The term F_x' is the well known force without rotation. The second term is due to the combined effect of the fluid velocity in the z-direction set up by the incident wave and the circulation imparted to the fluid by the spinning cylinder, resulting in a lift force in the x-direction. The components F_x' and F_x'' are according to eq. (20) either in phase or 180° out of phase, depending on the sign of ω_r. This is so because F_x' is proportional to particle displacement and particle acceleration in the x-direction whereas F_x'' is proportional to the particle velocity in the z-direction. These quantities happens to have their maximum values at the same instant.

Experiments

The experiment was carried out in a wave flume of length 12 m, depth 0.75 m and width 0.32 m. Since the theory corresponds to an infinitely long cylinder, it was necessary to keep the ratio between the length and the diameter of the cylinder sufficiently large. Preliminary experiments showed that the spin-effect was substantially reduced if this condition was not satisfied.

The maximum length of the cylinder was selected 300 mm, only slightly less than the width of the flume, and its diameter was 16 mm. This gives a length to diameter ratio of 19.

The cylinder was rotated in its submerged position by a driving belt connected to an electric motor. The vertical wave force F_z on the cylinder was measured by a load cell. In order to minimize the influence of noise and non-linear effects, the measured signals were treated by a frequency analyzer.

Wave measurements were made by standard two-wire resistive water-wave gauges. To avoid standing waves, all measurements were finished before reflections from the beach could reach the cylinder.

The heave forces were investigated for three different frequencies: 1 Hz, 1.5 Hz and 2.1 Hz. During each sequence of measurements wave frequency and wave amplitude were kept constant, while the rotational speed of the cylinder was altered.

The result of a set of measurements is shown in Fig.2, where the ratio $\sigma = F_z/F_z'$ is plotted as a function of ω_r/ω. Notice that the cylinder in this case was rotated in both directions, i.e. $\omega_r/\omega \lessgtr 0$. It is seen that for a certain region around $\omega_r/\omega = 0$ there is good agreement between theory and experiment.

It should be mentioned that the force F_z' on the fixed cylinder was found to be slightly smaller than expected from linear theory. In spite of that the spin factor $\sigma = F_z/F_z'$ came out properly.

It might be asked if the added mass of the cylinder is influenced by its spin. This was investigated by measuring the natural period of a spring supported cylinder as a function of rotational speed [3]. No change in the natural period was found.

Fig. 3 shows another experimental result for a different frequency. The dotted, straight line is a least mean square fit to the experimental points within the linear region. The gradient α of the line should according to eq. (21) be equal to one.

In Fig. 4a - 4c experimental values for α is plotted as a function of the wave amplitude η_0 for the different wave frequencies.

Discussion

The experimental results shown in Figs. 2 and 3 were characteristic for the measurements. The curves are linear in a certain region around $\omega_r/\omega = 0$. At the upper end of this linear region there is a marked elbow on the curve, followed by a new fairly linear region.

Fig. 4 shows that the experimental coefficient α approaches the theoretical value 1 for a limited region of wave amplitudes. Notice that the curves have their maximum for $\eta_0 \approx a = 8$ mm . It appears that the reduction in α for $\eta_0 > a$ is related to the Keulegan - Carpenter number becoming so large that vortex shedding takes place. The deviation for $\eta_0 < a$ is more obscure.

The position of the characteristic elbow on the curves in Figs. 2 and 3 seems to be an effect of the stagnation points leaving the surface of the cylinder [2].

A particularly interesting observation from these experiments is that the Magnus effect in waves, for some wave parameters is found to be in excellent agreement with theory. This is in contradistinction to observations in stationary flow [2]. The reason may be less vortex shedding in oscillatory flow.

In spite of the fact that the spin-induced forces fall short of the theoretical values for some wave amplitudes, the spin-effect is still substantial for a large range of amplitudes. This is a particularly important fact with regard to wave energy conversion, since wave energy absorption for small devices is almost proportional to the second power in the wave force.

A major problem in wave energy conversion is to establish a frame of reference against which the conversion mechanism can work. Spinning cylinders render a particular possibility to that end. If two parallel cylinders rotate in opposite directions, the spin induced wave forces on the bodies will also be of opposite sign. Hence, energy can be extracted from the relative motions of the objects, eliminating the need for an external reference.

Controlling forces on submerged objects may also find an application in construction of off-shore platforms. If the net force on the structure is minimized, its heave motion is necessarily reduced. However, instead of using spinning submerged objects, forced circulation of water around submerged fixed objects (pontoons) may be used to the same effect.

Notice that spinning objects to some extent have their (non-linear) analogue in submerged foils.

Here we have studied a horizontal cylinder parallel to the wave crest. Needless to say, spin-induced forces will be present for any orientation of the object.

References

1. D.V. Evans, D.C. Jefferey, S.H. Salter and J.R.M. Taylor, Applied Ocean Research, 1979, 1, 3.
2. See for instance: H.Rouse, "Elementary Mechanics of Fluids", John Wiley, New York 1946.
3. A.P. Bartholsen and D. Mortensen, Thesis, 1984. University of Trondheim (Unpublished).

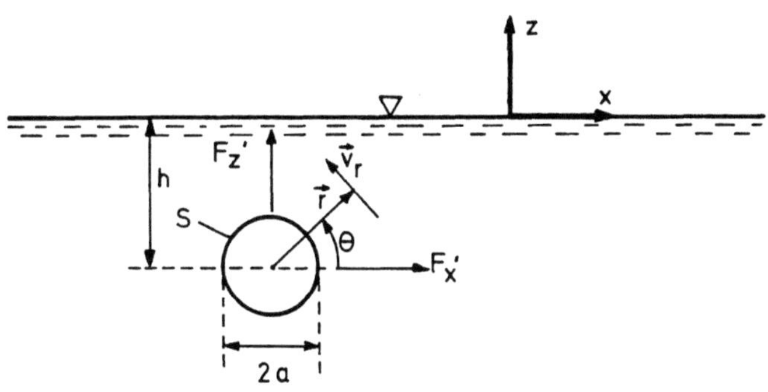

Fig.1. Submerged cylinder with axis parallel to the wave crest. The cylinder may spin around its axis with constant angular frequency ω_r. $\vec{v}_r$ is the spin induced particle velocity in the fluid. F_z' and F_x' are the components of the wave force when $\omega_r = 0$.

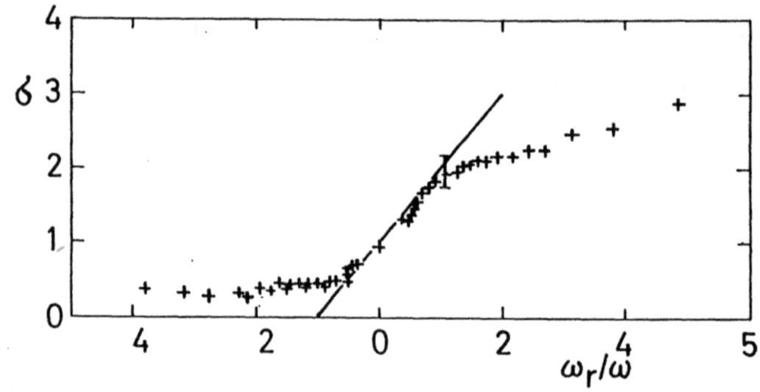

Fig.2. The ratio $\sigma = F_z/F'_z$ plotted as a function of ω_r/ω. (+): experimental points. (———): theoretical curve. Wave frequency $\nu = 2.1$ Hz . Wave amplitude $\eta_0 = 12.4$ mm . Depth $h = 70$ mm .

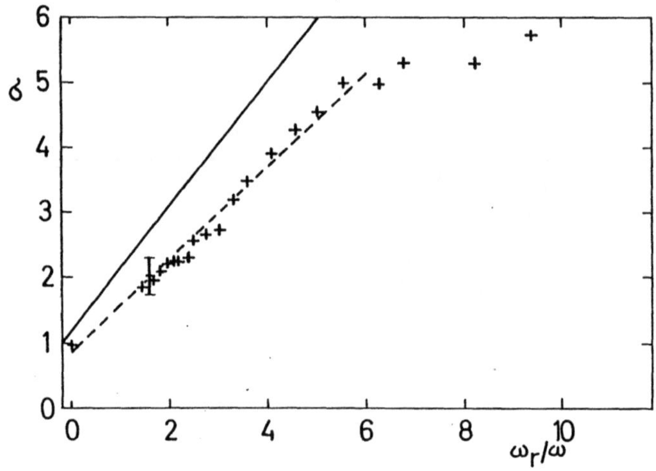

Fig.3. The ratio $\sigma = F_z/F'_z$ plotted as a function of ω_r/ω . (+): experimental points. (———): theoretical curve. (------): least mean square fit to experimental points from $\omega_r/\omega = 0$ to 6 . Wave frequency $\nu = 1.0$ Hz . Wave amplitude $\eta_0 = 18.9$ mm . Depth $h = 70$ mm .

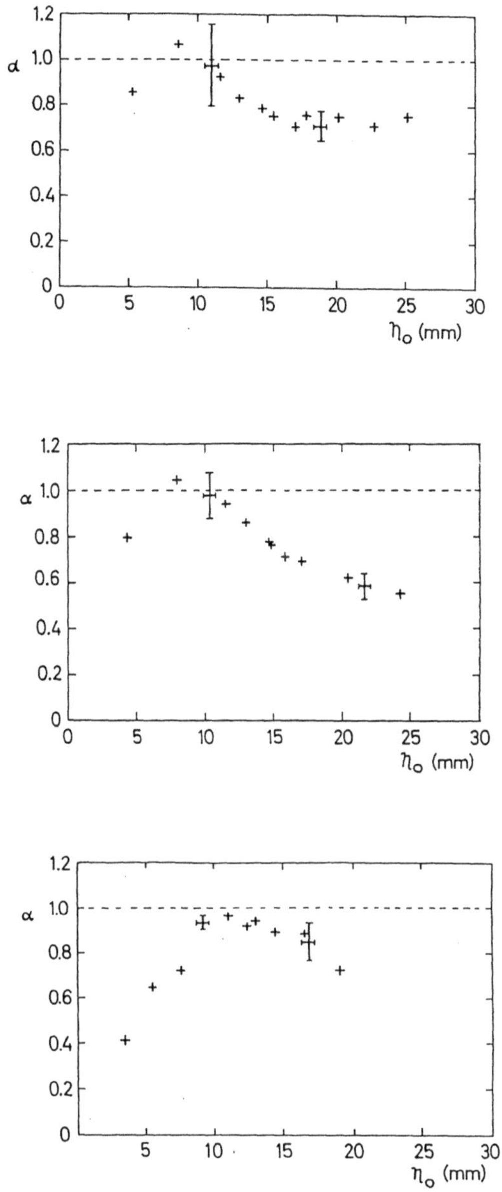

a)
ν = 1.0 Hz

b)
ν = 1.5 Hz

c)
ν = 2.1 Hz

Fig.4. Experimental values for the gradient α as a function of wave amplitude η_0 . Wave frequency: see a), b) and c). Depth h = 70 mm . (-----): theoretical values for α .

The Concept of a Bipartite Point Absorber

V. FERDINANDE* and M. VANTORRE**

(*) Dienst voor Scheepsbouwkunde, State University of Ghent
B-9710 Gent, Belgium

(**) National Fund for Scientific Research (Belgium)

Summary

A new wave power device is presented in the form of a bipartite point absorber. A simplified linear theory is used to demonstrate its satisfactory performance. An elaborate third-order theory however throws more light on the beneficial interactions between the composing parts, and on the strong nonlinear effects, which influence the power absorption in a negative way.

Basic considerations

The concept of the presented wave energy device takes into account severe environmental conditions and some particular requirements. The latter, and measures to be taken, are listed below. They suggest a primary outline of the design already.

Conditions and requirements	Solution
- rough sea area	springy point absorber
- considerable water depth	mooring lines
- large tidal amplitudes	oblique mooring line system
- efficiency independent of wave direction	vertical axisymmetry
- minimized slam damage risks	conical surface-piercing hull
- limited consequences of accidental loss of a device	array of small units
- no OWC-air turbines (option)	(nearly) fixed reference frame

Other characteristic features of the design are inspired by theoretical considerations, as given further.

Linear theory of wave power absorption

We limit the study to vertical and horizontal body motions in deep water, and assume that the diameter of the body waterplane is small compared to the wave length. The time-averaged extracted power, when the body motions are damped in a dashpot system on a fixed reference frame, is given in [1], and for a vertical axisymmetrical body

$$\frac{d\overline{E}}{dt} = (\frac{d\overline{E}}{dt})_1 + (\frac{d\overline{E}}{dt})_3 \qquad (1)$$

$$(\frac{d\overline{E}}{dt})_1 = \frac{\omega}{2}[\omega^2(\rho\forall + a_1)\cos\varepsilon_1 - \omega b_1 \sin\varepsilon_1] x_{A1}\zeta_A e^{-kz_1} - \frac{1}{2}\omega^2 b_1 x_{A1}^2 \qquad (2)$$

$$(\frac{d\overline{E}}{dt})_3 = \frac{\omega}{2}[(\omega^2 a_3 - \rho g A)\sin\varepsilon_3 + \omega b_3 \cos\varepsilon_3] x_{A3}\zeta_A e^{-kz_3} - \frac{1}{2}\omega^2 b_3 x_{A3}^2 \qquad (3)$$

($\forall$ = displacement volume and A = body waterplane surface ; a_i = added mass and b_i = hydrodynamic damping coefficient, i = 1 or 3, viz. horizontal or vertical mode ; x_{Ai} = motion amplitude in mode i and ε_i = phase angle with respect to wave motion, positive if leading ; ζ_A = wave amplitude ; k = wave number ; e^{-kz_i} is an averaged wave attenuation factor).

It is shown in [1], that for an axisymmetrical body, heave and surge motions may be uncoupled. x_{A1}, ε_1, and x_{A3}, ε_3 can be derived from equations based on the "relative motion hypothesis".

The extracted power is also

$$P_i = \frac{1}{2}\omega^2 d_i x_{Ai}^2 \quad (i=1,3) \qquad (4)$$

where d_i = damping coefficient in an energy absorbing dashpot, (damping force/velocity = constant).

It can be proved that the associated exciting wave forces are equal to those derived by means of a "long wave approximation", [2], taking account however of wave attenuation with depth. By this procedure, e^{-kz_1} and e^{-kz_3} in (2) and (3) can be evaluated.

Investigation of (2), (3), and the calculated body motions leads to following conclusions with respect to power extraction. z_i should be as small as possible. $\sin\varepsilon_3$ being always negative, and $\cos\varepsilon_3$ in any case rather small, waterplane area A ought to be large, and a_3 small. For a given wave spectrum, one should impose upon the device natural frequencies ω_{ni} corresponding with powerful wave components ; then A, m_i (mass of the device) and a_i can be regarded as variables, and m_3+a_3 is dependent on the choice of A. In a wave of $\omega \neq \omega_{n3}$ (non-synchronism), m_3 being fixed, x_{A3} and P_3 are smaller of course, but a_3 is of little importance in the range $\omega < \omega_{n3}$, whereas a_3 is severely detrimental to P_3 if $\omega > \omega_{n3}$ (in shorter waves). b_3 should be as small as possible, its contribution to P_3 being strongly negative at synchronism (if $b_3 d_3 < \omega^2 m_3^2$), although less detrimental when $\omega \neq \omega_{n3}$. Regarding the horizontal synchronous motion ($\varepsilon_1 = 0$), both $\rho\forall$ ($=m_1$) and a_1 should be as large as possible, ω_{n1} and the horizontal restoring force coefficient (springs) as well. In waves of $\omega \neq \omega_{n1}$, the decrease of P_1 is drastic in the range $\omega < \omega_{n1}$ ($\varepsilon_1 > 0$), whilst quite moderate in the range $\omega > \omega_{n1}$ ($\varepsilon_1 < 0$), where a_1 and m_1 are beneficial. Damping b_1 is very detrimental to P_1 (if $b_1 d_1 < \omega^2 (m_1 + a_1)^2$). The dissipative b_1 might be small, but viscous damping may be very harmful.

Especially because of the opposite influence of a_3 and a_1, the following can be concluded. Regarding the vertical motion, ω_{n3} of the device should be higher than the modal frequency ω_m of a representative irregular sea. A compromise must however be found, as the spectral densities diminish rapidly towards the higher ω. Regarding the horizontal motion, on the contrary, it will be beneficial for power extraction to choose a lower ω_{n1}, say near ω_m of the wave spectrum. These findings suggest the design of a bipartite device, with substantially different mass values in the vertical and in the horizontal mode.

In order to obtain an adequate natural frequency ω_{n3} in the vertical mode, i.e. corresponding to the ω of a desired component in the considered wave spectrum, the floating displacement body must be rigidly connected to an immersed flooded "water-mass vessel" of well determined volume. Without this vessel, the dimensions of the float would be excessive. The calculation of the compound body dimensions for a given ω_{n3} is straightforward.

The device is sketched in Fig.1. It consists of 1) a "vertical motion part" composed of a conical buoy B (displacement $\forall_B$) and a non-watertight "water-mass vessel" S (volume $\forall_S$), 2) a "reference part" composed of an immersed horizontal "inertia plate" P, held at a certain depth by (four) vertical buoyant columns C, which are braced at the top by traverses. The vertically moving part is guided inside the "reference part", through the central opening in the inertia plate and along slides on the columns. Due to the enormous inertia and drag of the "reference part" in

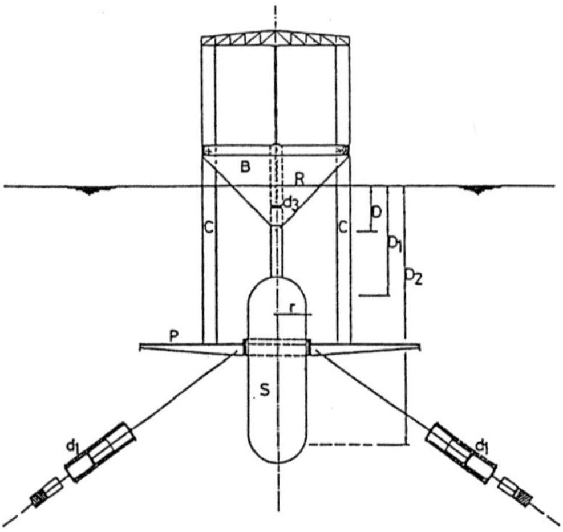

Fig.1. Schematic outline of bipartite wave energy point absorber

the vertical direction, the vertical wave induced motion of the inner float is near the relative motion, and can be damped in a dashpot d_3 between both parts, wherein wave energy is absorbed. This dashpot can be a piston pump, feeding e.g. a hydraulic swashplate motor with power generator. The simultaneous horizontal motions of the whole system are damped by the mooring lines attached to the "reference part", provided with spring restored piston pumps d_1. The latter produce additional power, and high pressure for the slide systems, which ought to function like hydrostatic bearings, and resist to transversal forces. No mooring lines are restraining the "vertical motion part".

According to linear theory, with separate wave attenuation factors for the buoy B and the supplementary body S, heave amplitude of the compound float in a regular wave (ζ_A, ω) is given by

$$\left(\frac{x_{A3}}{\zeta_A}\right)^2 = \frac{\left[(\rho g A-\omega^2 a_{B3})e^{-kz_{B3}}-\omega^2(\rho \forall_S e^{-kz_{S1}}+a_{S3}e^{-kz_{S3}})\right]^2+\left[\omega(b_{B3}e^{-kz_{B3}}+b_{S3}e^{-kz_{S3}})\right]^2}{\left[\rho g A-\omega^2(\rho\forall_B+\rho\forall_S+a_{B3}+a_{S3})\right]^2+\left[\omega(b_{B3}+b_{S3}+d_3)\right]^2} \tag{5}$$

It can be shown that approximately

$$e^{-kz_{B3}} = \frac{2}{kD}\left(1-\frac{1}{kD}+\frac{1}{kD}e^{-kD}\right) \tag{6}$$

for the cone with draught D, and

$$e^{-kz_{S1}} = \frac{e^{-kD_1}-e^{-kD_2}}{k(D_2-D_1)} \quad \text{and} \quad e^{-kz_{S3}} = \frac{1}{2}\left(e^{-kD_1}+e^{-kD_2}\right) \tag{7}$$

with D_1 and D_2 the respective equivalent depths of top and bottom of the water-mass vessel.

For a cone angle 45°, a_{B3} and b_{B3} can be derived from Fig.2, a_{S3} and b_{S3} from Fig.3. For λ_s' the experimental values should be used. They include viscous effects, and therefore are much

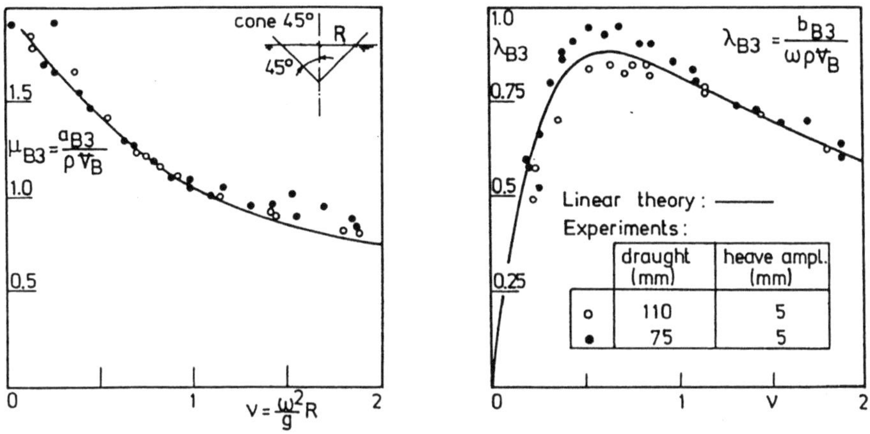

Fig.2. Nondimensional added mass and damping coefficients of cone 45°, vertical mode

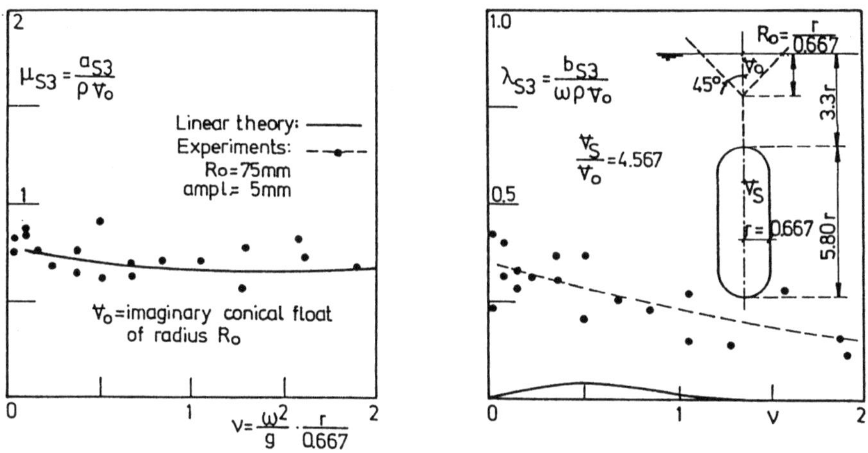

Fig.3. Nondimensional added mass and damping coefficients of immersed watermass-vessel, vertical mode

larger than the results from potential theory.

The formulas and data above have been used for the case of a device with waterplane radius R = 4.5 m ($\forall_B$ = 95.4 m^3), $\forall_S$ = 620m^3 (r = 3.37 m, height 19.6 m, D_1 = 15 m), i.e. according to ω_n = 0.8, and with different $\forall_S$ according to higher ω_n, in a seaway H_S = 3.25 m, T_m = 7s (Pierson-Moskowitz spectrum, modal frequency ω_m = 0.7). A linear superposition has been applied with respect to wave components $\Delta\omega = 0.05$. Computations have shown that the total extracted power is maximum for a value ω_n higher than ω_m, and for some d_3 much higher than b_3, viz. 130 kW for $\omega_n = 0.9$ and $d_3 = 250$ t/s. The explanation is that P_3 is maximum for $d_3 = [(\rho gA-(m_3+a_3)\omega^2)^2/\omega^2 + b_3^2]^{1/2}$, thus larger than b_3 for $\omega \neq \omega_n$. Introduction of $d_3 \cong 3b_3$ yields more power absorption over the ranges $\omega \gtrless \omega_n$ than $d = b$. Hence, a broader band of the wave spectrum contributes effectively to the power absorption. Moreover, resulting lower x_{A3}/ζ_A is desirable in view of non-linear effects, as will be seen further. High d corresponds to high pressure in the dashpot.

Unlike the vertical mode, the horizontal oscillatory motion involves the whole bipartite device. Linearity is not too a hazardous assumption, except for viscous drag. The motion equation contains, besides the inertial force related to the total mass, the horizontal components of the restoring force $-c_1x_1$ and the damping force $-d_1\dot{x}_1$ in the mooring lines, and the hydrodynamic forces. The hydrodynamic inertia forces are readily calculated : the columns and the water-mass vessel are regarded as long circular cylinders, and the conical float is ignored. The hydrodynamic dissipative damping forces are small compared to the viscous drag. The latter can be expressed in terms of plate friction and drag term in Morison's equation. It can be linearized by an "equal energy" method. The Froude-Krylov forces on the cylindrical parts are also readily calculated. The relevant wave attenuation factors are determined in a similar way as before. The calculation of x_{A1}/ζ_A however is less straightforward, since the several linearized damping coefficients b appearing in its formulation are function of x_{A1} and its phase angle themselves. An iterative procedure leads to the proper value of x_{A1}, and hence to the extracted wave power P_1, (4). The details of the calculation procedure can be found in [3]. With restoring force coefficient $c_1 = 750$ t/s^2, $\omega_{n1} = 0.75$, $d_1 = 200$ t/s, $P_1 \cong 70$ kW.

Presuming a mechanical efficiency 0.5, we may expect 100 kW in this sea state. However, considering the statistics of occurring sea states in an area like the west coast of Ireland, and possibilities of automation in adapting d, and even the water-mass in the immersed vessel to each sea state, the year averaged usable power could be much greater, not taking account yet of well known beneficial interaction in an array of point absorbers.

The fulfilment of strength requirements related to structure and mooring system for survival in extreme storms seems achievable at no prohibitive price.

Notwithstanding uncertainties in any evaluation, primary estimates let presume a cost per kWh not higher than other figures revealed so far, (viz. about 4 or 5 p).

Third-order theory

In view of the nonlinear character of the forces involved, the second author developed a nonlinear theory up to the 3rd order, [4]. The purpose is to see how far the credibility of a linear theory may be jeopardized. The theory aims at the computation of the forces acting on axisymmetrical floating and immersed bodies in a periodic heave motion

$$z(t) = \sum_{j=1}^{N} \mathrm{Re}[\bar{z}_j \, e^{i\omega_j t}] \tag{8}$$

Different from linear theory, the nonlinear Bernoulli equation is used, and the boundary conditions refer to the instantaneous body and free water surfaces. It is shown that the vertical force can be written as

$$\frac{F}{\rho g \forall_0} = \sum_{j=1}^{N} \mathrm{Re}\left[{}_1\bar{f}^{(\omega_j)} \frac{\bar{z}_j}{r_0} e^{i\omega_j t}\right] + 2 \sum_{\substack{k=-N \\ k\neq 0}}^{N} \sum_{j=|k|}^{N} A_{jk} \mathrm{Re}\left[{}_2\bar{f}^{(\omega_j+\omega_k)} \frac{\bar{z}_j}{r_0} \frac{\bar{z}_k}{r_0} e^{i(\omega_j+\omega_k)t}\right]$$

$$+ 6 \sum_{j=1}^{N} \sum_{k=j}^{N} \sum_{\substack{l=-N \\ l\neq 0}}^{j} B_{jkl} \mathrm{Re}\left[{}_3\bar{f}^{(\omega_j+\omega_k+\omega_l)} \frac{\bar{z}_j}{r_0} \frac{\bar{z}_k}{r_0} \frac{\bar{z}_l}{r_0} e^{i(\omega_j+\omega_k+\omega_l)t}\right] \tag{9}$$

where $\forall_0$ and r_0 are a characteristic volume and length respectively, and A_{jk}, B_{jkl} determined coefficients.

In the case of a harmonic motion with frequency ω (N=1), the 2nd-order terms in (9) consist of a second harmonic and a d.c.-term, while the 3rd-order forces contain harmonics with frequency 3ω and ω. The existence of the latter points to the importance of the 3rd-order terms with respect to power absorption. They influence added mass, damping and restoring force.

A general expression for the complex components ${}_n\bar{f}^{(\Sigma\omega)}$ in (9) has been found, and Söding's method of order reduction, [5], has been extended, so that the knowledge of only 1st and 2nd-order potential functions is required for the calculation of all components of (9). These potentials are computed by means of a boundary element method based on Kritis' theory, [6].

Computation results for several configurations (conical float, water-mass vessels and combinations of both) are in good agreement with experimental data for all frequencies of interest. They in general point to the importance of nonlinear terms. In particular, the 3rd-order forces associated with the water-mass vessel can be more important than the hydrostatic 3rd-order force due to the conical hull of the buoy ; the 2nd-order forces mainly concern the conical buoy alone. Moreover, the 1st-order (linear) solution reveals an interaction between parts B and S of the compound float. The hydrodynamic coefficients of a compound body can differ substantially from the sum of the coefficients of the individual parts. Due to phase differences of radiated waves, the immersed vessel counteracts the damping of the cone. Hence, (5) can be expressed more concisely, a_3 and b_3 being calculated directly for the complete body, by integration of the pressure components over the total wet body surface, and introducing attenuation factors for a_3 and b_3 separately, on a basis of the respective components of the depth-attenuated potential function.

In an irregular sea $\zeta = \sum \mathrm{Re}(\bar{\zeta}_j e^{i\omega_j t})$, $\bar{z}_j$ ought to be replaced by $\bar{z}_j - \bar{\zeta}_j e^{-k_j z'}$ in the force expression (9). The interactions will be multifarious. Let us restrict ourselves to a single sine wave. The heave response is not purely sinusoidal, but presented here as

$$z = \mathrm{Re}[\bar{z}_0 + \bar{z}_1 e^{i\omega t} + \bar{z}_2 e^{2i\omega t} + \bar{z}_3 e^{3i\omega t}] \qquad (10)$$

$\bar{z}_1$ is of first, $\bar{z}_0$ and $\bar{z}_2$ of second, and $\bar{z}_3$ of third order. They can be calculated by iteration, the corresponding force (9) having been split into its harmonics. The force 1st harmonic is influenced by nonlinear terms due to 1) interaction between motion 1st harmonic and body lifting d.c. term, $({}_2\bar{f}^{(\omega+0)})$, detrimental to power absorption ; 2) slight interaction between 1st and 2nd harmonics, $({}_2\bar{f}^{(2\omega-\omega)})$; 3) third powers of the 1st harmonic of relative motion, $({}_3\overline{f}^{(\omega+\omega-\omega)})$, favouring power absorption.

For instance, the nonlinear terms of the hydrostatic Froude-Krylov force on a conical buoy are represented in

$$F' = (-\rho g \pi R \tan\alpha)s^2 + (\tfrac{1}{3}\rho g \pi \tan^2\alpha)s^3 \qquad (11)$$

where $s = z-\zeta$. The terms are the respective forces due to ${}_2\overline{f}^{(\omega+0)}$ and ${}_3\overline{f}^{(\omega+\omega-\omega)}$. Their contribution to power absorption is determined by the relation

$$\delta P_{stat}^{(\omega+\omega-\omega)} = -\frac{1}{4}\,\delta P_{stat}^{(\omega+0)} = -\frac{\pi}{8}\rho g\omega \tan^2\alpha . z_{1A}\zeta_A \sin\varepsilon_3 (z_{1A}^2 - 2z_{1A}\zeta_A\cos\varepsilon_3 + \zeta_A^2) \qquad (12)$$

thus negative in its totality, since $\sin\varepsilon_3 < 0$. However, some nonlinear hydrodynamic forces have a counteracting effect on them. Only the cubic terms have a negative effect, which is reflected in corresponding apparent increments Δa_3 and Δb_3. As an example, Fig.4 is given, wherein for frequency parameter $\nu=0.4$ are shown the differences between linear theories with, and without body interaction, between results for various wave amplitude ratios ζ_A/r_0, and the influence of the different terms as mentioned above for a given ζ_A/r_0. It appears that the 1st-order theory (interaction between body parts) reveals higher power absorption than derived from (5), but that the nonlinear effects however lead to a considerable decrease.

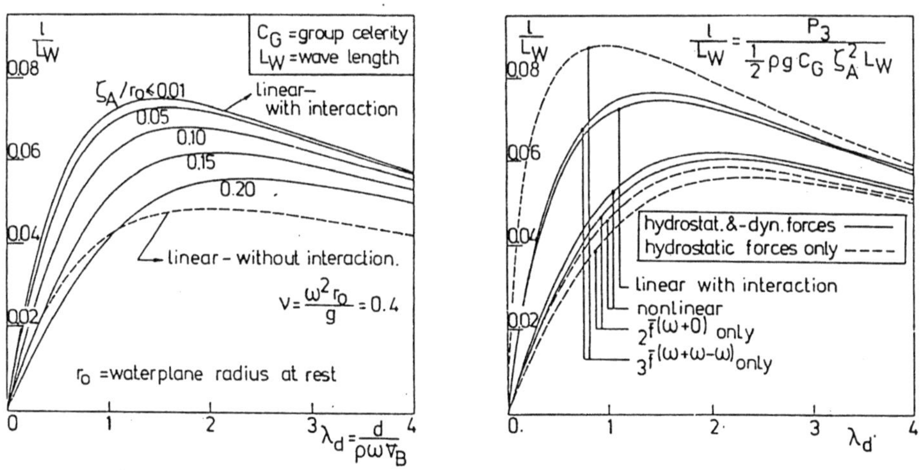

Fig.4. Power absorption length ratios according to linear and nonlinear theory - compound body, watermass-vessel with flat top ($T=2.2\ R_0$) and bottom ($T=6.067\ R_0$)

Conclusion

In the cases considered hitherto, the results of the linear theory as given by (5) are in fair agreement yet with those from the 3rd-order theory, at least for some reasonable finite wave heights. This agreement however is merely due to counteracting effects. Although elaborate, the 3rd-order theory will therefore be used for power absorption computations in the future. This theory moreover gives a clearer insight in the various interactions, and apparently provides a tool for designing more efficient configurations of this type of wave energy device.

Acknowledgments

The development of this third-order theory has been executed under the auspices of the Belgian NFWO (National Fund for Scientific Research) and the State University of Ghent.

References

1. Ferdinande, V. : On power absorption by floating devices in waves. International Shipbuilding Progress, Vol.32 - No.373, September 1985.
2. Newman, J.N. : Marine Hydrodynamics. Cambridge, London : MIT Press 1977.
3. Ferdinande, V. : A study of wave energy absorption by floating bodies (in Dutch). Gent : Dienst voor Scheepsbouwkunde RUG, Rep. 11-82, 1982.
4. Vantorre, M. : Third-order potential theory for the determination of hydrodynamic forces on axisymmetrical floating and immersed bodies in forced periodic heaving (in Dutch). Ghent: Doctor's thesis RUG 1985.
5. Söding, H. : Second-Order Forces on Oscillating Cylinders in Waves. Schiffstechnik, Band 23, Heft 114, November 1976.
6. Ferdinande, V. and Kritis, B. : An economical method of determining added mass and damping coefficients of axisymmetric floating bodies in oscillatory heaving motion. International Shipbuilding Progress, Vol.27 - No. 313, September 1980.

Thrust Generation by a Hydrofoil Driven by Waves — A Basic Aspect of Wave Devouring Propulsion

H. Isshiki[+], M. Murakami[+] & Y. Terao[++]

[+] Technical Research Institute
Hitachi Zosen Corporation, Osaka, Japan
[++] Department of Naval Architecture
Tokai University, Shimizu, Japan

Summary

A possibility of thrust generation through a utilization of wave power or Wave Devouring Propulsion (WDP) is discussed. Some inventions, observations and investigations related to the possibility have been reported [1-5]. The possibility may be applied not only to increase propulsive efficiency of a ship in waves but also to reduce mean and slowly varying components of wave drift force acting on a moored platform in waves. In the present paper, some experimental results on two basic models are reported. One is a two-dimensional model and the other a three-dimensional one. Some theoretical discussion is also given to the former.

1. Results on a Two-Dimensional Model

The experiments in this section are conducted in a tank of 25 m in length, 1 m in breadth and 0.71 m in water depth. The hydrofoil is a two-dimensional one of NACA 0015 section, and the chord 2ℓ and span b are 0.4 m and 0.96 m respectively.

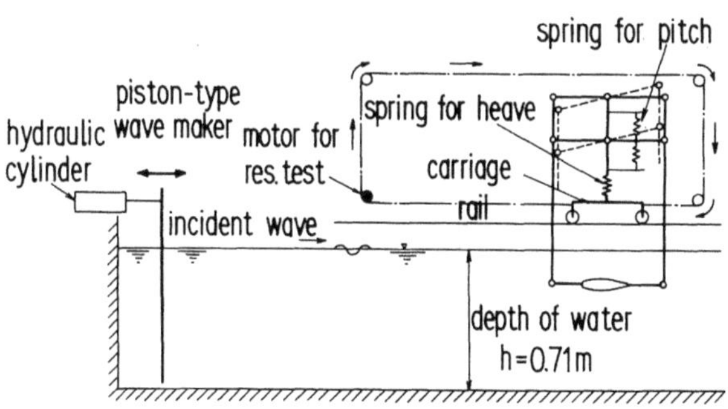

Fig.1. Scheme of a test system of a two-dimensional model. (Motor drive system is used when the resistance tests are carried out)

A foil motion mechanism is devised to make the foil resistance as small as possible as shown in Fig.1, since the resistance is quite big. Unfortunately, the foil motion mechanism has become very heavy. The results of resistance tests in calm water are shown in Fig.2. The draft at the midchord $d_{⊠}$ and attack angle α of the foil are 7.7 cm and -7.7° in the nose up direction respectively. The other particulars are shown in Fig.4.

Waves used in the present experiments are shown in Fig.3. Three kinds of wave height are used, that is, waves refered to as $(2a)_1$ $(2a)_{1/\sqrt{2}}$ and $(2a)_{1/2}$. The subscripts refer to the designed value of the wave height ratio between them.

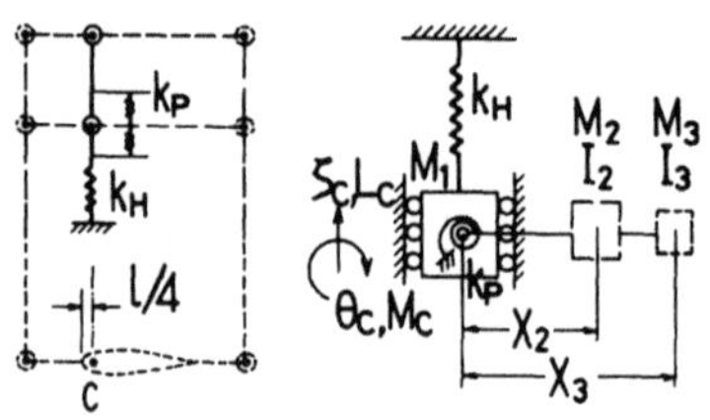

$M_1 g$ = 3.923 kgf	X_2 = 0.069 m
$M_2 g$ = 13.378 kgf	X_3 = 0.125 m
$M_3 g$ = 10.500 kgf	
$(M_2 X_2^2 + I_2) g = 1.452$ kgfm^2	$(M_3 X_3^2 + I_3) g = 0.888$ kgfm^2
k_H = 1.06×10^3 kgf/m	k_P = 13.188 kgf·m/rad

Fig.4. Dynamical model of foil motion mechanism

A dynamical model of the foil motion mechanism is given in Fig.4. The mechanism is composed of a foil, a heave guide, a pitch guide, a heave spring and a pitch spring. The rotational axis of the foil is denoted by C and located 1/8 chord length behind the leading edge of the foil. Let $\zeta_C(t)$ and $\Theta_C(t)$ refer to the heave and the pitch about C of the foil. The equations of foil heave and pitch motions may then be derived as

$$M\ddot{\zeta}_c-(M_2X_2+M_3X_3)\ddot{\theta}_c+k_H\zeta_c=L_c \tag{1a}$$

$$I\ddot{\theta}_c-(M_2X_2+M_3X_3)\ddot{\zeta}_c+k_P\theta_c=M_c \tag{1b}$$

respectively, where

$$M = M_1 + M_2 + M_3 \tag{2a}$$

$$I = (M_2X_2^{\ 2} + I_2) + (M_3X_3^{\ 2} + I_3) \tag{2b}$$

M_1 = mass of the heave guide
M_2 = mass of the pitch guide
M_3 = mass of the foil
X_2 = horizontal distance of the gravitational center of the pitch guide from C
X_3 = horizontal distance of the gravitational center of the foil from C
I_2 = rotational inertia of the pitch guide about its gravitational center
I_3 = rotational inertia of the foil about its gravitational center
$L_c(t)$ = hydrodynamic lift acting on the foil
$M_c(t)$ = hydrodynamic moment about C acting on the foil
k_H = spring constant of the heave spring
k_P = spring constant of the pitch spring

L_c and M_c are obtained on the basis of Wu's theory for an unsteady hydrofoil in a wave [2,5].

If the advance speed U is assumed, the heave and pitch motions ζ_c, θ_c are determined by solving Eqs. (1a) and (1b). The thrust may then be obtained[2,5]. This thrust is the apparent thrust $\overline{T}_{app.}$ and is explained below. The free running speed may now be estimated as the speed where the apparent thrust is equal to the resistance of a carriage with a foil in calm water.

When the advance speed is not zero, two kinds of definition are used with respect to the thrust generated by the hydrofoil. One is the above-mentioned apparent thrust $\overline{T}_{app.}$, and the other is the net thrust $\overline{T}_{net}$. The former is a thrust which does not take account of the foil resistance, and the latter is the difference between the apparent thrust and the foil resistance in calm water. The apparent and net WDP efficiencies $\eta_{W\ app.}$ and $\eta_{W\ net}$ may then be defined as

$$\left.\begin{array}{l}\eta_{w\ app.}\\ \eta_{w\ net}\end{array}\right\} = \left.\begin{array}{l}\overline{T}\ app.\\ \overline{T}\ net\end{array}\right\} U/(P_w^+ b) \tag{3}$$

where P_w^+ is the power of an incident wave per unit breadth of the wave and is given as

$$P_w^+ = \frac{\rho g a^2}{2}(c_g+U) \tag{4a}$$

ρ = density of fluid

g = gravitational acceleration

a = amplitude of incident wave

c_g = group velocity of the incident wave

$$= \frac{1}{2}\left[1+\frac{2kh}{\sinh(2kh)}\right]\sqrt{\frac{g}{k}\tanh(kh)} \tag{4b}$$

k = wave number

h = depth of water

According to a theoretical procedure discussed above, some numerical calculations have been conducted, and the results are compared with the experimental ones. In Fig.5, the apparent thrust coefficient is compared between the theory and experiment. The experimental results are those obtained by the free running test where the advance speed is changed by changing the wave height. Hence, they include the effects of the wave height.

The free running speed is shown in Fig.6. Three different wave heights for each wave length are used to study the effects of the wave height. In Fig.7, the heave and pitch motions are compared between the theory and experiment. There seems to be a relation between the free running speed and heave response, although the correlation in the experiment is not so clear as in the theory.

The above results seem to give a rather good correlation between the theory and experiment, if the very shallow draft of the hydrofoil is taken into accout. In the present theory, the hydrodynamic lift and moment acting on the foil are evaluated by Wu's theory[2] which neglects the free surface effect completely. This means that the present theory may not be correct in case of a very shallow draft of the foil. However, the above results seem to be better than expected.

The experimental results of the net WDP efficiency $\eta_{W\ net}$ are shown in Fig.8. The definition of $\eta_{W\ net}$ is given by Eq.(3). The net WDP effeciency $\eta_{W\ net}$ is less than 5%, when the ratio of wave length to chord $\lambda/(2\ell)$ is bigger than 10. The low WDP efficiency in a long wave range is rather disappointing to the authors. The increase of the WDP efficiency is considered to be one of the most important problems at present.

2. Results on a Three-Dimensional Model or Wave Drifting Force Free Model

In this section, some experimental results of a wave drifting force free model are shown. The circular model is prepared to investigate the motion and thrust generation in waves under the mooring and freely drifting conditions.

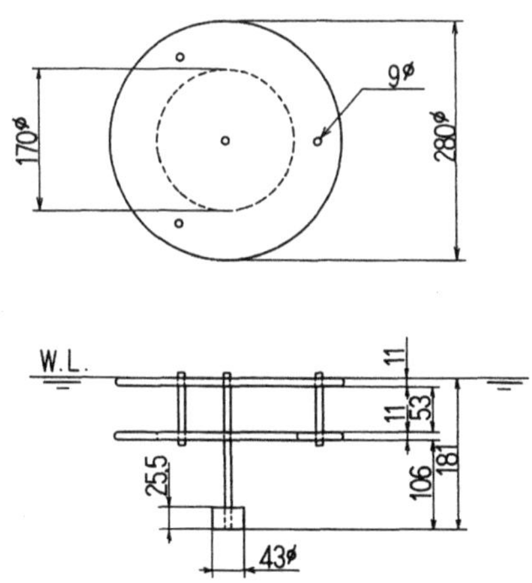

Fig.9. Wave drifting force free model

The model shape is shown in Fig.9. The model consists of two circular foils, a weight and a rod which connects the upper foil and weight. The circular foils have the same diameter D, but the lower one is a ring foil. This model has no directivity because of its circular form.

In the mooring condition, two soft springs are used as the mooring lines and set at the weather and lee sides of the model to give the initial mooring tension and to impose restriction on the surge motion. The mooring force discussed here is the difference of the tension from the initial condition.

Fig.10 shows typical records of the model motions and mooring force. The top figure shows the motions in case of freely drifting condition. The lower two figures refer to the results of mooring condition.

The model advance speed U is obtained by analyzing the record of surge in case of freely drifting condition and is shown in Fig.11. The positive speed means that the model moves against waves. The phenomenon is very important, since a usual floating body drifts in the direction of wave propagation. It is interesting that the phenomenon is observed in a wide range of wave length, that is, $2<\lambda/D<8$ and the advance speed has a sharp peak.

The results of the mean component of thrust acting on the model in case of mooring condition are shown in Fig.12. The vertical axis is the time average of thrust $\overline{T}$ normalized by $(1/2)\ \rho g a^2 D$. It is interesting that the thrust is "positive" and has "no sharp peak". The positive thrust means that the thrust component due to lift exceed the wave drifting force component.

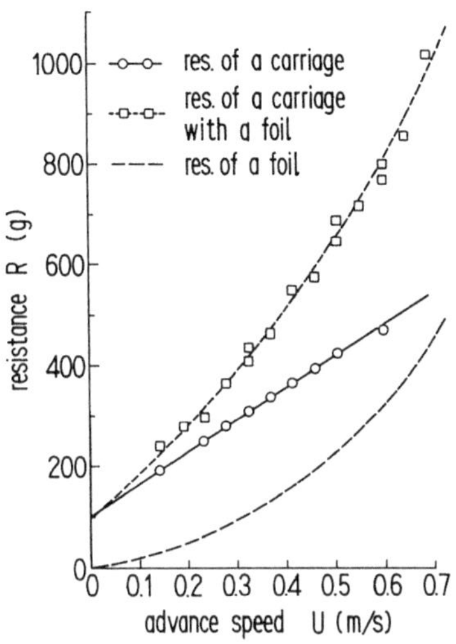

Fig.2. Results of resistance test. ($d_{\boxtimes}$=0.077 m, α=-7.7°)

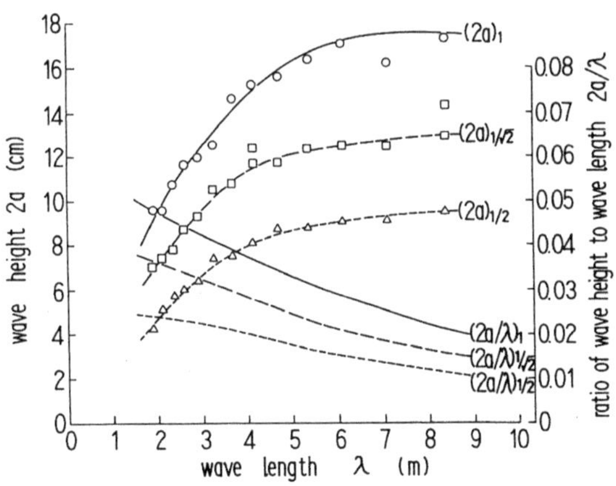

Fig.3. Waves used in experiments of free running test

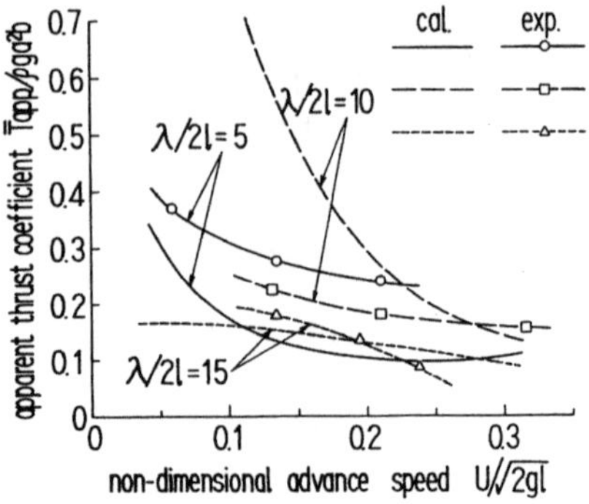

Fig.5. Comparison of apparent thrust coefficients between theory and experiment

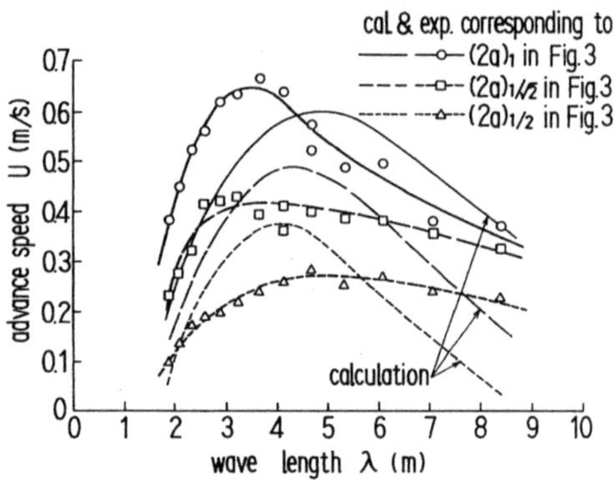

Fig.6. Effect of wave length and height on free running speed

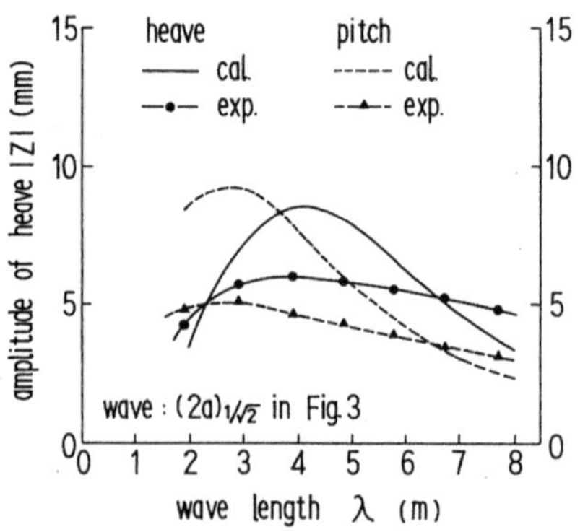

Fig.7. Effect of wave length on foil motion

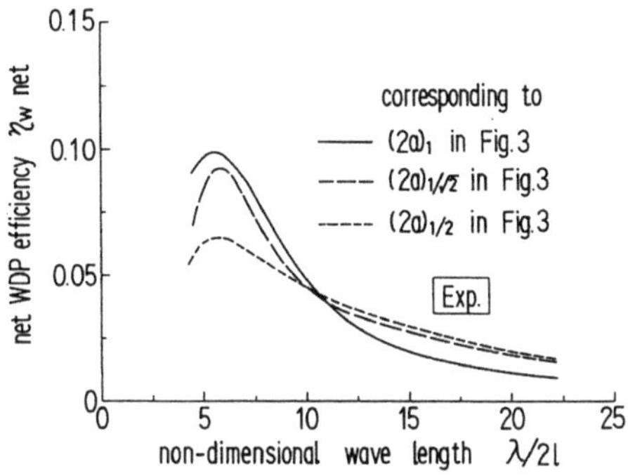

Fig.8. Effect of wave length and height on net WDP efficiency

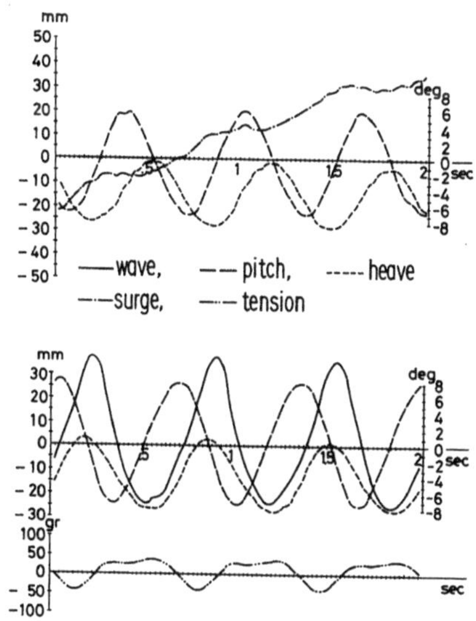

Fig.10. Typical records of model motion and mooring tension

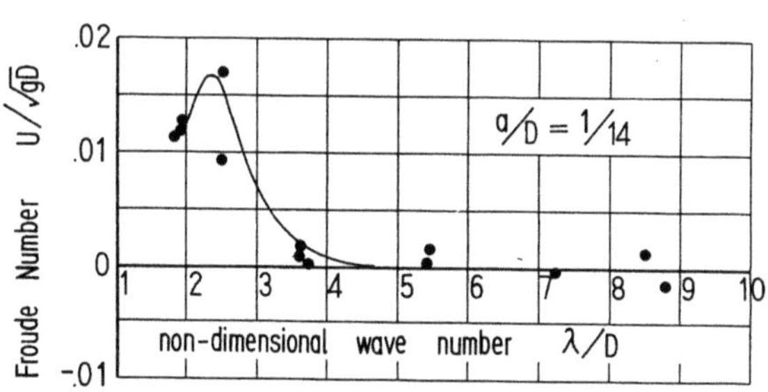

Fig.11. Model advance speed against waves in freely drifting condition

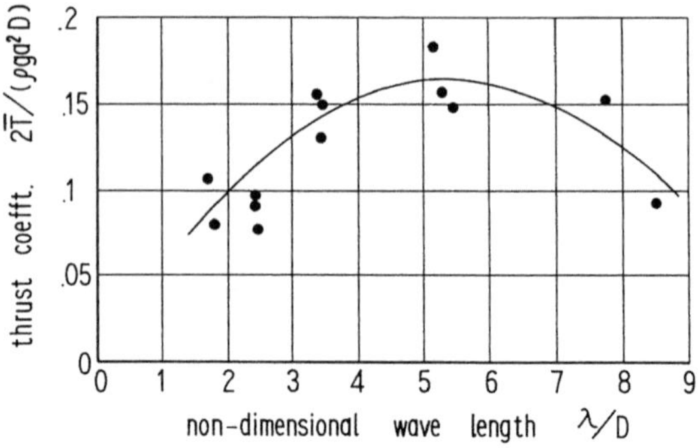

Fig.12. Thrust acting on the model in case of mooring condition

References

1. Wave energy for propelling craft - Nothing new. The Naval Architect (Nov. 1973) 239.

2. Wu, T.Y.: Extraction of flow energy by a wing oscillating in waves. Journal of Ship Research (Mar. 1972) 66-78.

3. Jakobsen, E.: The foil propeller, wave power for propulsion. 2nd International Symposium on Wave & Tidal Energy, BHRA Fluid Engineering (1981) 363-368.

4. Terao, Y.: A floating structure which moves towards waves (Possibility of wave devouring propulsion). Journal of the Kansai Society of Naval Architects, Japan No.184 (1982) 51-54.

5. Isshiki, H.: Murakami, M.: A theory of wave devouring propulsion (4th Report) -A comparison between theory and experiment in case of a passive-type hydrofoil propulsor-. Journal of the Society of Naval Architects of Japan No.156 (1984) 102-114.

Wave Energy Absorption Characteristics of Circular Air-Chamber for Use of Light Beacon Fixed on Sunken Rock

R. INOUE, M. IWAI

Ocean Engineering Division, Ship Research Institute
Ministry of Transport, Tokyo, Japan

M. YAHAGI and T. YAMAZAKI

Engineering Division, Ryokuseisha Co. Ltd., Tokyo, Japan

Summary

The light beacon which is installed on the sunken rock, etc.as navigation aid has great difficulty in securing power supply because of its installing location.
The research project concerns both with the experimental investigation on the performance of circular air chamber and that of air turbine, which means a technical extension of the light buoy technology which already has been realized as only case successfully utilizing wave energy.
This paper deals with the first stage of the investigation, i.e., the power absorbing characteristics of circular air chamber in order to squeeze wave energy efficiently even in the calm or moderate sea state throughout a year.
The circular air chamber is attached around circular base as shown in Fig. 1 and is divided into two or four chambers.
In the experiment a 1/5 scale model has been used in the Offshore Structure Experimental Basin of Ship Research Institute, and the wave power absorbing efficiencies for each air chamber have been measured for various wave directions, wave periods and heights, water-line levels and so on. As the results, it is clarified that this system can generate sufficient power for use of the light beacon. Furthermore, absorbed wave energy was calculated by use of the equivalent floating body concept and compared with experimental values.

Design Conditions of Air Chamber

Based on wave and tidal observation data in the sea area where the light beacon is to be constructed, the air chamber was designed under the following conditions:

			(Frequency of Occurrence)
Wave height	$H_{1/3}$	0.35 m	50%
Wave period	$T_{1/3}$	4 sec	
Sea level	ΔH	+1.0 ~ -0.6 m	80%

The necessary wave energy to be absorbed, which is able to produce electric power of 300W for light beacon, on the assumption that the air turbine efficiency is 50% and the generator efficiency 40%, was estimated roughly as follows ;

$$E_a = 300 / (\eta_H \cdot \eta_{\Delta H} \cdot \eta_T \cdot \eta_G) = 3.75 \text{ KW}$$

where,

η_H ; 50% (Frequency of occurrence of wave)

$\eta_{\Delta H}$; 80% (Frequency of occurrence of sea level)

η_T ; 50% (Air turbine efficiency)

η_G ; 40% (Generator efficiency)

Air Chamber Model

The circular air chamber model is illustrated in Fig.2. the model is constructed by the circular foundation section and the air chamber section located around it. The 1/5 scale model is made of 2.3 and 4.5 mm thick steel plates. Part of the top and side of the air chamber section is made of transparent acrylite plate so that behavior

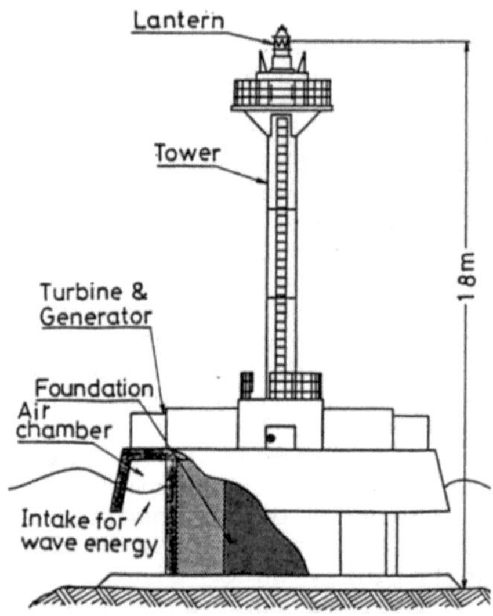

Fig.1 Wave activated generating system for light beacon

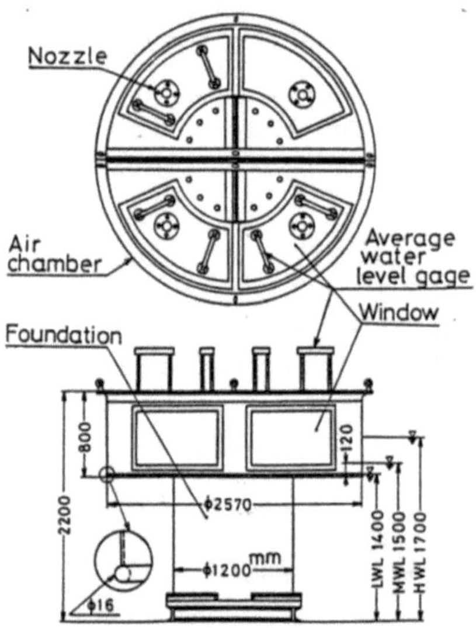

Fig.2 Detail of circular air chamber model

Fig.3 Circular air chamber model

of water surface in the air chamber can be visually observed. Also, in order to minimize the consumption of wave energy to be introduced into air chamber, a 16ϕ round steel bar was welded throughout the circumference at the lower end of the front part of the air chamber.

The air chamber section can be divided into two or four chambers in the direction of its diameter. Moreover, to change the incident angle of wave against the air chamber, the air chamber section can be rotated to any angle.

The load on the air chamber which takes the place of an air turbine or generator was changed by changing the area of the nozzle on the top of each air chamber.

The water plane area of each air chamber is 0.985 m^2 in 4 air chambers type model and 1.970 m^2 in 2 air chambers type model.

The photo of the model is shown in Fig.3.

Experimental Conditions

Using the model shown in Fig.2, experiments were conducted by changing the number of air chamber, the depth of front of air chamber d (change of sea level), the direction of incident wave χ , the ratio of nozzle to water plane area of air chamber R and so on.

Experimental conditions are shown in Table 1. The direction of incident wave to air chamber is shown in Fig.4 .

Table 1 Experimental conditions

Items			Conditions	
Number of air chamber			4	2
Direction of incident wave χ (deg)			0° 22.5° 45°	0° 45° 90°
Depth of front of air chamber		d (cm)	12 32 2	
Ratio of nozzle to water plane area in air chamber		R	1/100 1/150 1/270	
Wave	Regular	T (sec)	1.0 1.5 2.0 2.5 3.0	
		H_i (cm)	5 10	
	Irregular	T_m (sec)	1.2 2.0	
		$H_{1/3}$ (cm)	6.0 9.5	

Measuring Items

Measuring items included incident wave, water level and air pressure in air chamber, etc. The air pressure was measured by use of a miniature pressure gage on the top of air chamber. Also, variation of water level in air chamber was measured with capacitance type average water level gage.

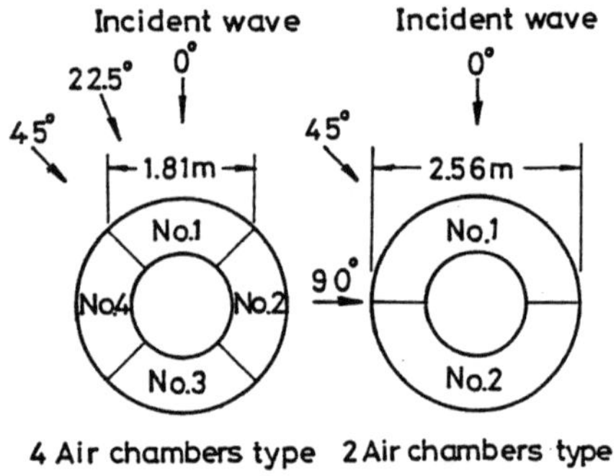

Fig.4 Direction of incident wave

Experimental Results and Considerations

Absorbed Wave Energy in Each Air Chamber

The absorbed wave energy calculated in terms of the full scale model by Froude's law is shown in Fig.5 and Fig.6 . In these figures, T and E_a indicate the incident wave period and absorbed wave energy in air chamber respectively. The values correspond to the case where direction of incident wave $\chi = 0°$, the depth of the front of air chamber $d = 0.6$ m , the ratio of nozzle to water plane area of air chamber $R = 1/100$ and the height of incident wave $H_i = 0.25$ and 0.5 m . Also, the air chamber output $E_a = 3.75$ KW that satisfies the generating output 300 W is indicated in these figures as a target value.

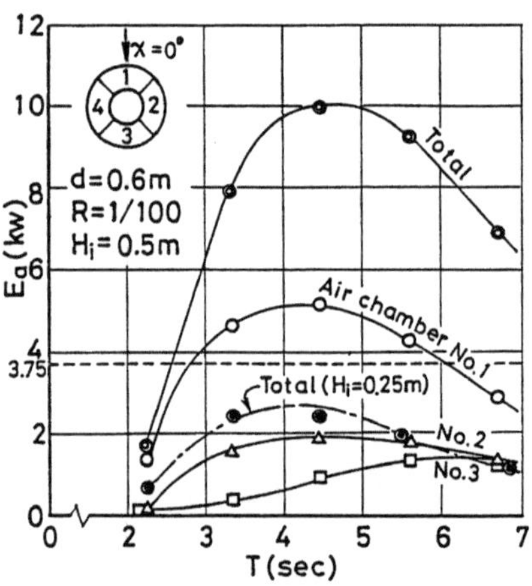

Fig.5 Absorbed wave energy in full scale model

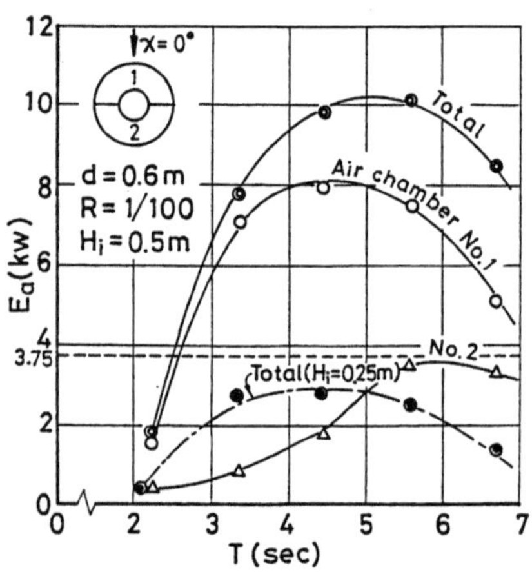

Fig.6 Absorbed wave energy in full scale model

In the 4 air chambers type model, the absorbed wave energy in the air chambers No.2 and No.4 is about 1/3 of that in the air chamber No.1 and that in the air chamber No.3 is about 1/5. With regard to the 2 air chambers type model, the absorbed wave energy in the air chamber No.2 is about 1/2 of that in the air chamber No.1 .

In either case, when $H_i = 0.25$ m, the total absorbed wave energy fails to reach the target value 3.75 KW. When $H_i = 0.5$ m, however, it fully satisfies the target value. Also, the total absorbed wave energy is almost similar in both cases even if the number of the air chamber is different.

Variation of Absorbed Wave Energy Owing to Ratio of Nozzle to Water Plane Area of Air Chamber

Fig.7 shows a variation in absorbed wave energy owing to the ratio of nozzle to water plane area of air chamber R . The wave energy absorption coefficient η_1 can be defined by the ratio E_a/E_i between the energy of incident wave E_i and the air energy absorbed in the air chamber E_a .

$$\eta_1 = E_a / E_i \qquad (1)$$

In such a three dimensional experiment, however, the incident width of wave energy B can not defined. Here, B is assumed to be the opening width of the air chamber No.1 when $\chi = 0°$ as shown in Fig.4 (in 4 air chambers type model, B = 1.81 m;

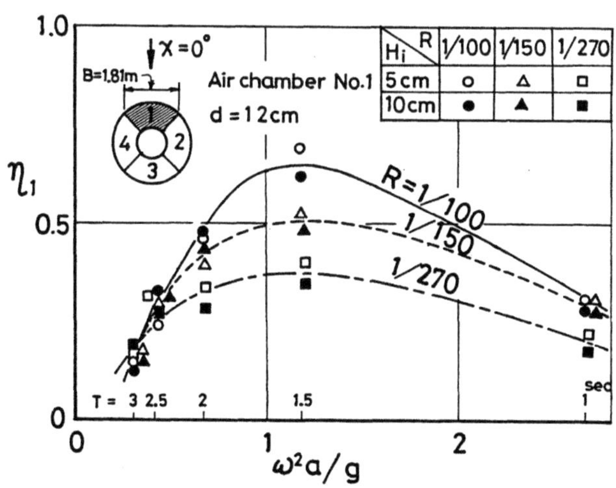

Fig.7 Variation of wave energy absorption coefficient owing to R

in 2 air chambers type model, B = 2.56 m).
Fig.7 shows variation of η_1 in the 4 air chambers type model. ω , a and g in the abscissa indicate circular frequency of incident wave, width of air chamber in its diameter direction (a = 0.66 m) and acceleration of gravity respectively.
The value η_1 is the largest when R = 1/100 . This is entirely similar to results of two dimensional experiment.

Variation of Absorbed Wave Energy Owing to Direction of Incident Wave

Fig.8 and Fig.9 show the variation of absorbed wave energy of the air chamber No.1 (A.C. No.1) and all air chambers in the 4 and 2 air chambers type full scale models in the case where the direction of incident wave χ is changed. In case of the 4 air chambers type model, there is no major variation of absorbed wave energy from whatever direction wave may come, as shown in Fig.8. In case of 2 air chambers type model as shown in Fig.9, the absorbed wave energy in the air chamber No.1 becomes about 2/3 and 1/3 of the value $\chi = 0°$ when $\chi = 45°$ and 90° respectively.

Variation of Absorbed Wave Energy Owing to Depth of Front of Air Chamber

Fig.10 shows the variation of absorbed wave energy in full scale model, corresponding to the variation of the depth of front of

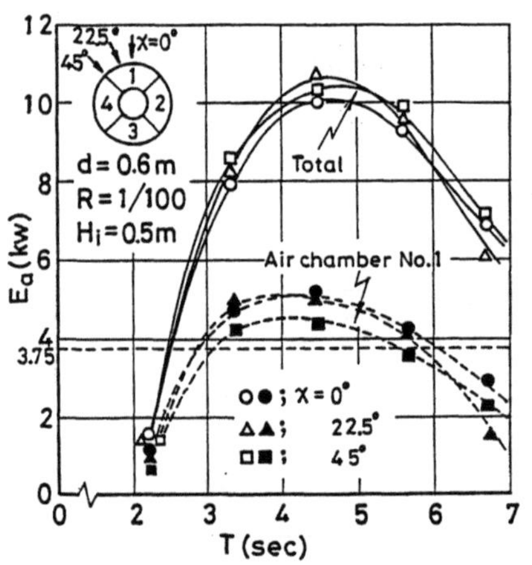

Fig.8 Variation of absorbed wave energy in full scale model owing to χ

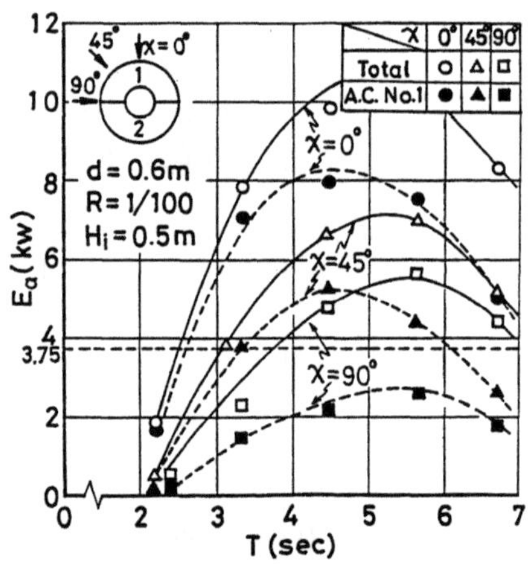

Fig.9 Variation of absorbed wave energy in full scale model owing to χ

air chamber in the 4 air chambers type model. The absorbed wave energy in the air chamber No.1 is the largest when d = 0.6 m (optimum depth of front of air chamber obtained from the results of two dimensional experiment), while the maximum value of total absorbed wave energy in all air chambers is showing a similar value when d = 0.6 and 1.6 m . When d = 0.1 m , the value is small because the lower end of the front part of the air chamber is exposed to the air. Also, the maximum value at d = 1.6 m is produced on the longer period side than that of others, and this is attributable to that the resonant period of a water column in the air chamber is larger.

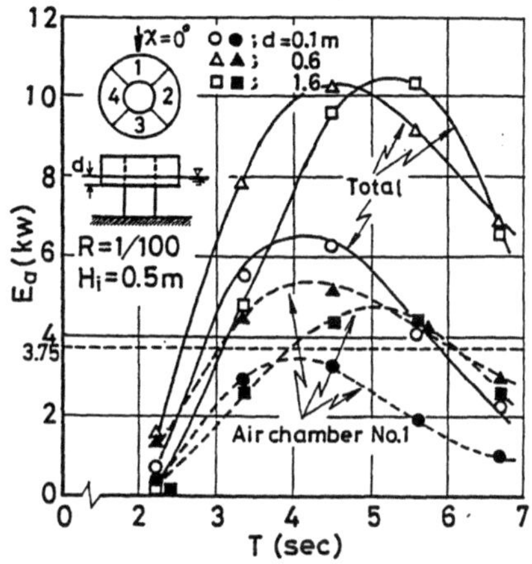

Fig.10 Variation of absorbed wave energy in full scale model owing to d

Experiment with Irregular Wave

The irregular wave used in the experiment was prepared with an ISSC spectrum. Its mean wave periods T_m = 1.2 and 2.0 sec . and significant wave heights $H_{1/3}$ = 6.0 and 9.5 cm.
Fig. 11 shows the value of wave energy absorption coefficient η_1 at $\chi = 0°$ in the air chamber No.1 of the 4 air chambers type

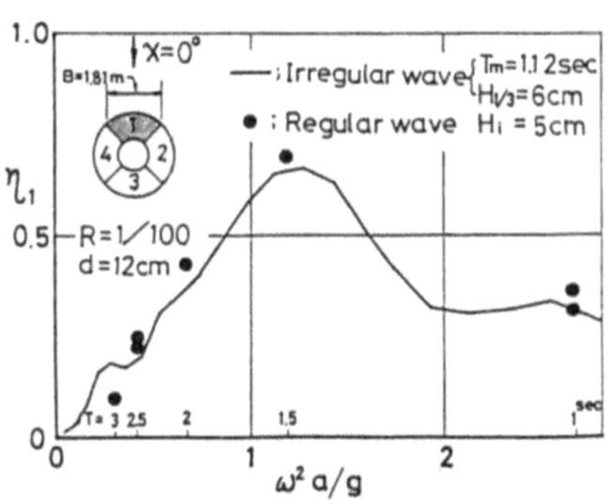

Fig.11 Wave energy absorption coefficient in irregular wave

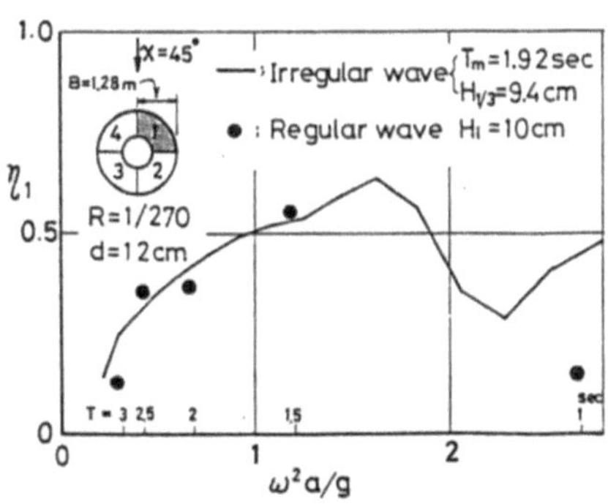

Fig.12 Wave energy absorption coefficient in irregular wave

model, as compared with the experimental results of regular wave. In Fig.12, the wave energy absorption coefficient η_1 at $\chi = 45°$ is shown. The incident width of wave energy for air chamber No. 1 at $\chi = 45°$ B is 1.28 m as shown in this figure.
From these results, the same results can be obtained in both irregular and regular wave experiments.

Calculation of Absorbed Wave Energy

The absorbed wave energy was calculated by the equivalent floating body concept, which is a method of making approximate calculation by replacing a water column in the air chamber with a floating body to be equivalent to it. The heaving motion of the water column in the air chamber is expressed by the following equation.

$$(m + m_z)\ddot{z} + (N + d_o)\dot{z} + \rho g A z = F_z e^{i\omega t} \qquad (2)$$

Where, m and m_z are mass and added mass of the equivalent floating body respectively, N is a wave damping coefficient, d_o is a linearized load damping coefficient, A is a water plane area of the equivalent floating body, and F_z is an amplitude of wave exciting force.
The values m_z and N were calculated by the Finite Element

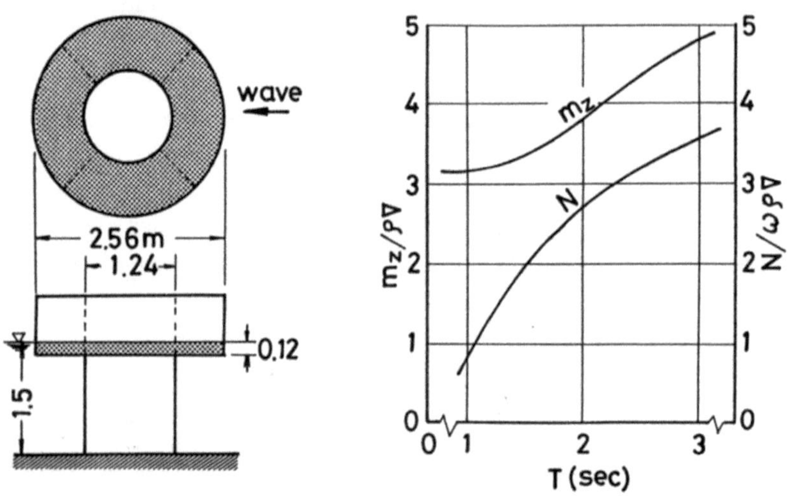

Fig.13 Water column in air chamber and calculated values of added mass and wave damping coefficient

Method with respect to the doughnut-shaped floating body having a core in the center as shown in Fig.13. The nondimensional values of their calculated results are shown in Fig.13. Also, F_z was obtained by integrating the pressure acting on the bottom surface of the floating body calculated by the Finite Element Method, over the water plane area of each air chamber. Since experimental values of load damping coefficient d_o shows almost constant against a wave period, this linearized value was used for calculation.

The experimental values of absorbed wave energy in the full scale 4 air chambers type model are compared with the calculated value in Fig.14. The tendencies of both values agree very well in each air chamber, though the experimental values are slightly larger than calculated ones in the vicinity of T = 3 and 4 sec. Fig.15 shows the values in the 2 air chambers type model, and indicates that the experimental values are larger than calculated ones on the whole, while both agree well qualitatively.

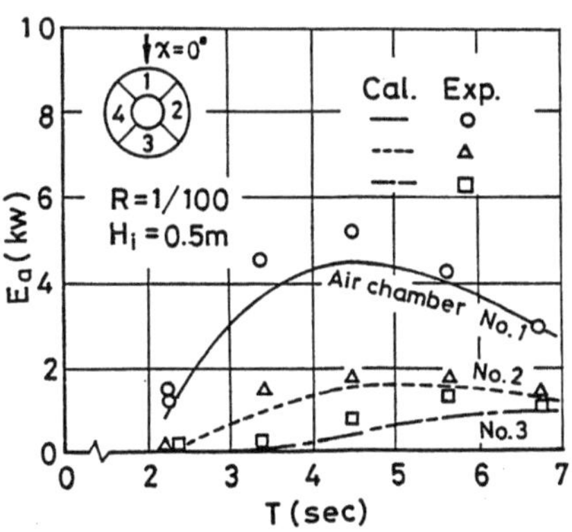

Fig.14 Comparison between experimental and calculated values

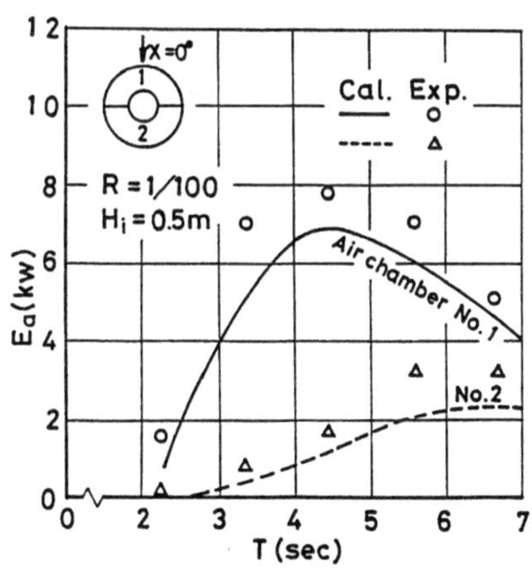

Fig.15 Comparison between experimental and calculated values

Conclusion

(1) The absorbed wave energy in the air chamber in front of wave is the largest. In the rear air chamber, absorbed wave energy is from about 1/2 to 1/5 of that in the front air chamber.

(2) The circular air chamber of 4 air chambers type can absorb

almost constant wave energy for any wave directions. But in 2 air chambers type, that varies with the wave directions.

(3) The absorbed wave energy becomes the largest in the depth of front of air chamber $d = 0.6$ m .

(4) The absorbed wave energy becomes the largest in the ratio of nozzle of water plane area of air chamber $R = 1/100$.

(5) Similar results can be obtained in both regular and irregular wave experiments.

(6) The absorbed wave energy in circular air chamber as well as two dimensional air chamber can be calculated by the equivalent floating body concept.

(7) Assuming the total absorbed energy of the air chamber in the design wave ($T_{1/3} = 4$ sec, $H_{1/3} = 0.35$ m) from the experimental value, it can be considered that the 4 air chambers type satisfies output target value 3.75 KW of the absorbed wave energy of the air chamber.

Acknowledgement

This work was carried out partly with the support of the Grant-in-Aid of Japan Foundation for Ship Building Advancement. The authors acknowledge the favor for this research received from Navigation Aids Department of Maritime Safety Agency and Japan Association for Aids to Navigation, and thank Mr. H.Kagemoto for the assistance of the calculation by Finite Element Method.

References

1) Inoue,R., Iwai,M., Yahagi,M. and Yamazaki,T.; Wave Energy Absorption Characteristics of Air Chamber for Use of Light Beacon, 1st Symposium on Wave Utilization in Japan, 1984.

2) Maeda,H., Kinoshita,T., Masuda,K. and Kato,W.; Fundamental Research on Oscillating Water Column Wave Power Absorbers, OMAE Symposium, 1984.

3) Akane,T. and Izumi,H.; New Type of Wave Activated Generator Light Buoy, The 3rd ODC Symposium, 1975.

Heaving Point Absorbers Reacting against an Internal Mass

M. J. FRENCH

R. BRACEWELL

Department of Engineering,
University of Lancaster,
Bailrigg,
Lancaster,
U.K.

Summary

A design study has been made of a wave energy converter in the form of a heaving buoy reacting against an internal moving mass. Phase control has been obtained by modulating the stiffness of the spring governing the moving mass. Promising results, agreeing closely with linear theory, have been obtained with two shapes of buoy. Mean annual captures of the order of 200W/tonne in a typical South Uist sea have been estimated.

Introduction

Except where natural formations are particularly favourable a major part of the cost of wave energy converters usually lies in the structure. This structural cost may be reduced either by using an inherently cheap and efficient structure [1], or by reducing the structural task by going downwards for a reaction from the sea, or by using an internal reaction against a moving mass (Evans, Tondl, Parks [2]). This paper studies a possible solution in this last group, in the form of a heaving buoy (Fig. 1)

Dynamics

The dynamics of such systems has been studied by Parks [2], but he used a fixed spring stiffness though he varied the damping. However, for the best results it is necessary to vary both the damping and the effective spring stiffness, to obtain quasi-resonance. With such an arrangement it is theoretically possible to obtain the maximum possible power for a point absorber (Evans [3]) over all but a part of the bandwidth (the 'hole') and even in that part the capture can approach the ideal. However, to obtain the ideal power may require impractically large values of the product of the reaction mass and its travel.

Our intention originally was to obtain quasi-resonance or phase control by latching, but another method was devised which seems superior, at least in this application (stiffness modulation).

The variation of the efficiency (actual capture/ideal capture) with period for a constant energy flux is typically of the form shown in Fig. 2, which also shows the power output. Even though the efficiency falls in long waves, because of the greater ideal capture width a reasonably level power output results.

Fig 1

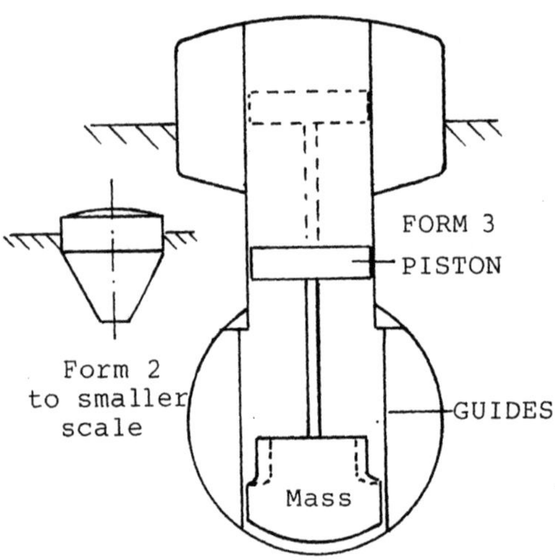

Fig 2

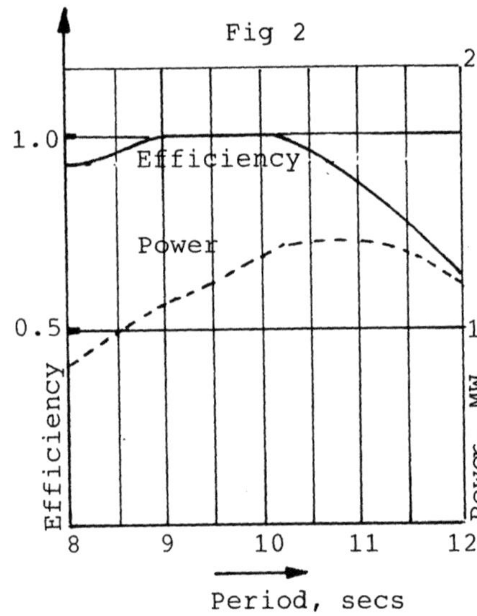

Fig 1 General form of buoy

Fig 2 Efficiency (capture/ideal capture) and power capture v. period

Stiffness modulation (SM)

In the stiffness modulation method of obtaining quasi-resonance, the spring alternates between two stiffnesses giving natural frequencies ω_1 and ω_2 at the extremes of the bandwidth to be covered. At an intermediate frequency ω, where $\omega_1<\omega<\omega_2$, the spring has its lower stiffness at smaller displacements and changes to its higher stiffness when the displacement exceeds a certain value: this arrangement is shown in Fig 3a. In contrast, in latching, the lower frequency is zero and obtains at large displacements.

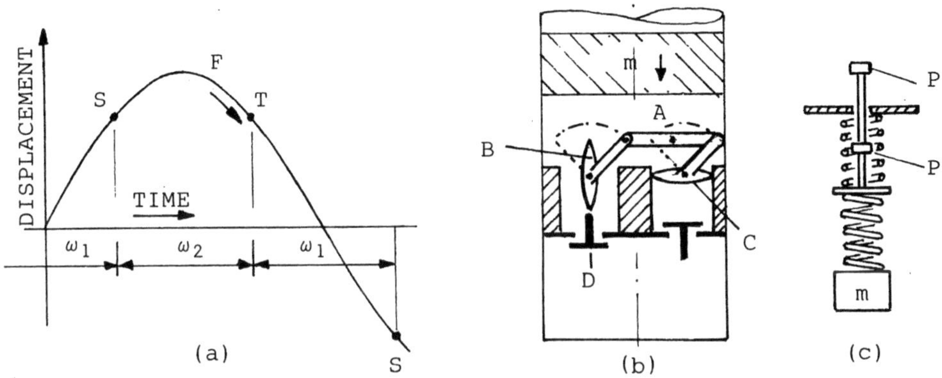

Fig 3 Stiffness Modulation

(a) Displacement/time plot

(b) Application in air spring, shown at point F, Fig 3a, approaching an automatic transition T. The butterfly B is open, and the mass m is descending, but the NRV D is closed because the pressure above it is less than that below. However, soon the pressure above D will rise above that below, it will open and all the spring will come into action, reducing the stiffness. At the next S transition, the link A will be moved to the left, closing the butterfly B and opening the butterfly C.

(c) Application to coil springs. At large displacements one or other of the stops P comes into operation so only the lower part of the spring is operational, giving the higher stiffness. This scheme also has automatic transition at T.

An important design feature in stiffness modulation is that the transition from the high to the low frequency regime, at points T in Fig 3a, can be made to take place automatically, without losses due to spring surge or 'ringing'. In practice, with an air spring, this is done by means of non-return valves, as shown diagrammatically in Fig 3b.

Generally, SM is easier to engineer than latching. It has also the advantage that it has no discontinuities in acceleration, only in jerk, with a corresponding large reduction in harmonic content.

Hydrodynamics

The behaviour of the device is represented by equations which have been derived elsewhere [4], but it is convenient to quote some of the results. They are best summarised in the force phasor diagram shown in Fig 4, where

	X_s	is the wave force on the buoy
	Y_1	is the buoy displacement
	r_I	is the ideal amplitude of the mass relative to the buoy
	m	is the moving mass
	ω	is the angular frequency
and	C	is a constant such that $m\omega^2 CY_1$ is the resultant force of the sea on the buoy due to its motion.

The solid lines show the ideal case where both the 'damping' of the mass-spring system, which is the power take-off, and the effective spring stiffness have been optimised.

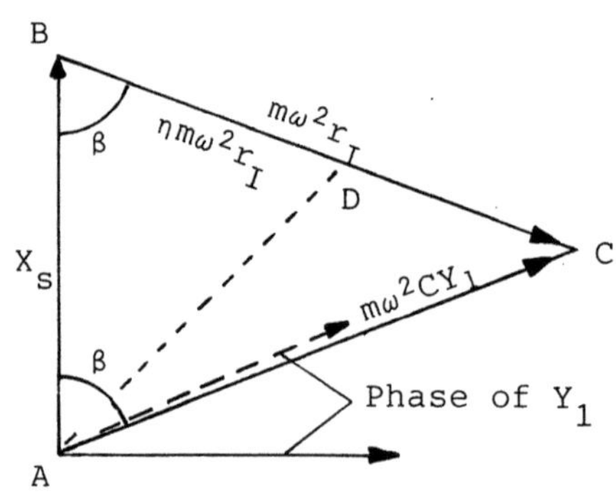

Fig 4 Phasor diagram: ABC; ideal case: ABD, case with less than ideal motion of the mass, r_I

In the ideal case, X_s and Y_1 are 90° out-of-phase, the phasor diagram is an equilateral triangle and $r_I = CY_1$. When ω is the natural frequency of bobbing or heaving oscillations, the angle β is zero and

$$mr_I = X_s/\omega^2 \qquad (1)$$

The longer the period, the greater the angle β and the larger the ideal value of

$$r_I = X_s \sec \beta/m\omega^2 \qquad (2)$$

For the ideal case, the power is $1/4\,\omega X_s Y_1$, which is equal to the radiated wavepower. However, it is not always practical to provide the required value of mr. If the available value of r is some fraction η of r_I and the power take-off is optimised, the power P as a fraction of the ideal power P_I is given by

$$P = (2\eta - \eta^2)\, P_I \qquad (3)$$

and the phasor diagram is the triangle ABD: the phase angle between X_s and r remains unchanged. A related result is given by Evans [5].

The result given in equation 3 is of great importance, for it shows that the power P falls off only slowly as η falls below unity. In fact, if internal losses were taken into account, the best power would probably be obtained with a value of η slightly less than unity.

Below the natural frequency of heave, the angle β is negative and there may be a range of periods for which optimisation is impossible without a negative spring rate, the 'hole'. However, even in this range only a few per cent of the ideal power is lost, and if the product of the non-dimensional radiation coefficient and the ratio of the moving to the total mass is less than one half, no such 'hole' exists.

Buoy Form

The choice of form for the buoy is constrained by a number of considerations, as follows:-

(a) In order that the power output should be large, the amplitude Y_1 must be large, which makes it desirable that the section piercing the sea surface should be steep-sided.

(b) In order that the power output should be large, the magnitude of X_s should be large, which requires that the area of the waterline section should be large. A deep cylinder, however, with the bottom too far below the surface, will not answer.

(c) The heave frequency should be suitable. At the time of writing of an earlier paper [4] it was thought that it should be in the middle of the bandwidth, but now a higher frequency is favoured.

(d) The form should be such as to accommodate a sufficient m and a sufficient r without instability or unfavourable interactions between heave and roll being set up because of the coupling caused by the moving mass. This favours a deep buoy.

(e) To keep costs down, the buoy should be small.

(f) Different forms have hydrodynamic coefficients (added mass and radiation coefficient) which vary in different ways with frequency, and these variations can have an important effect. There are several papers on the heaving of suitable forms, e.g. by Hulme [6], and it seems likely such studies could make useful reductions in size.

The form 3 shown in Fig 1 meets requirements (a), by the cylindrical section piercing the surface, (b), by the large cross-section and small draught of the upper portion, and (c), by virtue of the mass and added mass of the spherical lower portion. It also meets (d) well, the mass being located in the sphere, the piston of the air spring in the central cylinder and the sphere and central cylinder acting as pressure vessels. Form 2 is slightly inferior in yield/displacement.

Experiments

Experimental work has been done on three principal models. Hydrodynamic coefficients of the shapes were measured from free oscillations at various frequencies. Tests were conducted in regular waves, with sliding masses ranging between 0.18 and 0.5 of the displacement and periods in the region 0.8 to 1.6 secs, corresponding to linear scale factors in the range 60 - 100.

Three kinds of power-generating tests were carried out, with tuned springs, with latching and with stiffness modulation. The latching was achieved by means of an electromagnetic brake acting on a shaft turned by the moving mass. Stiffness modulation was achieved by the means shown in Fig 3c, which worked well but required very careful adjustment in long waves.

Two forms of loading were used. The first was a simple half-turn band-brake with a lost-motion arrangement enabling it to work with either rotation. The ends of the band were attached to strain-gauged spring-steel cantilevers which enabled accurate measurements of the load to be made. Unfortunately, the load could not be adjusted from the shore and its Coulomb nature inhibited the build-up of quasi-resonant vibrations, so a second type of load has also been used lately. This is an electric generator, in fact, a small D.C. permanent magnet motor, which has large frictional and electrical losses but makes testing easier.

No tests have been done with irregular waves but a computer simulation using a simple control strategy showed little difference in the results. Because phase and amplitude of all the phasors were measured, there were multiple internal checks in many of these experiments, and consistency was generally very good.

Results

The general conclusion from all the tests except those with

latching is that the behaviour of the system was very closely in line with that calculated by the simple theory given above using the experimentally-determined hydrodynamic coefficients. In particular, no evidence was found that non-linearity was important at the amplitudes used, which frequently reach the scale equivalent of 5m at the buoy. With latching the results were not as good, but this may only have been due to lack of perseverance in setting up.

Based on these results, assessments have been made of the average annual power capture at various scales, using the wave climate taken by the consultants to the U.K. Department of Energy in assessing a hypothetical device ('Mermaid') [7]. A few results are given in the table below.

Form	Displacement tonnes	mr tonne m	Diam m	mean annual power kW	Rating kW
2	1494	2390	15	313	700
3	1490	2390	12.8	342	700
2	3040	6160	19	510	1300
3	3020	6120	16.2	502	1300

Only a sketchy costing has been done, but we think this device is more promising than others we know of.

We feel that there may be other point-absorbers reacting against internal moving masses, more promising even than this one. Very small studies have been made of two of these.

References

1. Carter, R.G. and Hurdle, D.P.: Wave power absorption by a small body moving in surge. Applied Ocean Research 1983, Vol. 3, No. 1, pp 24-29.

2. Parks, P.C.: 'Wedges, plates and waves', pp 257-63 in Power from seawaves. ed. B. Count, London: Academic Press 1980

3. Evans, D.V.: A theory for wave-power absorption by oscillating bodies. Journ. Fluid Mech., 1976, Vol. 77, pp 1-25.

4. French, M.J. and Bracewell, R.: The reduction of structural costs in obtaining energy from sea-waves. Fourth (1985) Int. Offshore Mechanics & Arctic Engineering Symposium, A.S.M.E.

5. Evans, D.V.: Maximum wave power absorption under motion constraints. Applied Ocean Research, 1981, Vol. 3, No. 4, pp 200-203.

6. Hulme, A.: The wave forces acting on a floating hemisphere making periodic heaving oscillations. Journ. Fluid Mech., 1982, Vol. 121, pp 443-453.

7. Department of Energy: UK Wave Energy Programme. Consultants 1982 Report. (Not yet published)

Phase Control for the Oscillating Water Column

R.E. HOSKIN†, B.M. COUNT††, N.K. NICHOLS†, D.A.C. NICOL††

†Department of Mathematics,
P.O. Box 220,
University of Reading,
Reading, England.

††CEGB Marchwood Engineering Laboratory,
Marchwood,
Southampton, England.

Summary

In this paper we investigate the use of optimal control techniques for improving the efficiency of the power-conversion system in an oscillating water column device. A mathematical model of the system is developed and an optimal control strategy for power-generation is determined. An algorithm is described for solving the problem numerically given the incident wave form. The technique is applicable to irregular as well as regular waves. The results obtained show that the performance of the device can be significantly improved by means of a suitable control strategy.

1. INTRODUCTION

Theoretical studies of wave energy have generally concentrated on the computation of hydrodynamic forces acting on a structure, and on the analysis of energy absorption under idealised conditions [Evans (1976), Count (1982)]. Incorporation of a power-conversion system into the design has led to the development of wave-energy devices which are sub-optimal from a theoretical point of view - with an energy capture less than ideal. In this paper we investigate control mechanisms for improving the efficiency of power-conversion systems and increasing energy production. With such mechanisms the delivered cost of the energy is reduced, at the expense of a (hopefully) modest increase in capital investment, and wave-power can be made more attractive.

It has been recognised [Budal and Falnes (1978)] that the explanation for the reduced energy absorption achieved by early designs lies in the relative phase difference between the device motion and the wave-exciting force. For a small device, with a higher natural frequency than the incident wave force, the velocity of the device leads the wave in phase. For a single point absorber it has been demonstrated that if the motion of the device is halted for certain periods of time during the wave-cycle, then a more advantageous phase relationship can be obtained, with a subsequent increase in energy capture [Evans (1976)].

Here we examine similar phase control techniques for an oscillating water column. The device consists of a structure enclosing a column of water which is excited into motion by incoming waves, and thereby acts like a large piston, driving air through a turbine assembly [Count (1982)]. The power-conversion system is controlled by a valve which regulates the flow of air into the turbine. Consequently, when power-generation is halted, the motion of the water column - unlike that of the point absorber - persists, since the air in the system does not provide an infinite restoring "spring" force. Although the water-column device is a more complicated system than the point absorber, we show here that with an appropriate control strategy the natural phase of the oscillating column can be related advantageously to that of the exciting wave. Given certain information about the wave-force, the performance of the device can thus be significantly improved for waves in the complete spectrum.

The problem is formulated mathematically in §2. The dynamic behaviour of the device is described by a third-order system of ordinary differential equations, which determine the motion of the surface of the water column and the pressure of the air enclosed by the structure. A control variable is introduced to represent the volumetric proportion of flow permitted through the valve. The model does not take into detailed account the mechanical and electrical characteristics of the turbo-generators, but specifications of such properties can be incorporated. The problem is to determine the control which maximises the energy produced by the system, subject to the dynamic equations being satisfied, and certain constraints on the control. Necessary conditions for the solution are derived in §3 using optimal control theory, and properties of the optimal solution are developed. A numerical procedure for determining the optimal control strategy is described in §4 and numerical results are presented in §5. It is demonstrated that, for irregular waves, phase-control of the oscillating water column is feasible and that energy-capture can be significantly increased by applying a flow-control in the power-conversion system.

2. MATHEMATICAL FORMULATION

The mathematical model consists of three equations of state. The first equation describes the motion of the surface of the water column, [Cummins (1962)] and is given by

$$M\ddot{x} + \int_0^t \kappa(t-\tau)\dot{x}(\tau)d\tau + \rho_w g A_0 x + (p-p_0)A_0 = f_d, \qquad (2.1)$$

where $x(t)$ is the displacement of the surface from equilibrium; $p(t)$ and p_0 are the pressure of the air enclosed within the column and atmospheric air pressure respectively; A_0 is the cross-sectional area of the column; ρ_w is the density of water and g is gravitational acceleration. The constant M incorporates both the column-mass inertia and the frequency-independent 'added mass' component of the wave radiation force. The 'added damping' is described by

$$\kappa(t) = \frac{2}{\pi}\int_0^\infty N(\omega)\cos \omega t,\ d\omega \quad , \qquad (2.2)$$

where $N(\omega)$ is the frequency-dependent component of the radiation force, determined by the immersed shape of the column. The wave diffraction force acting on the column is given by the function $f_d(t)$.

It is assumed that the air flow from the column across the generating turbines is controlled through a valve. If $m(t)$, $\rho(t)$, $V(t) = V_0 - A_0 x(t)$ denote the mass, density and volume of the enclosed air, and ρ_0, V_0 denote the density and enclosed volume at atmospheric pressure, then the mass-flow equation is given by

$$\dot{m} \equiv d(\rho V)/dt \equiv \dot{\rho}(V_0 - A_0 x) - \rho A_0 \dot{x} = -uG(p-p_0)\rho \ , \qquad (2.3)$$

where the turbine discharge $G(p-p_0)$ is a function of the pressure difference across the turbine, and the control function $u(t)$, representing the proportionate valve opening, is constrained to satisfy

$$0 \leqq u \leqq 1 \ . \qquad (2.4)$$

The adiabatic pressure law is also assumed to hold, giving

$$p/p_0 = (\rho/\rho_0)^\gamma \ , \qquad \gamma = 1.44 \ . \qquad (2.5)$$

The equations (2.1) and (2.3) together with (2.5) fully describe the state of the system.

The instantaneous energy derived from the device is taken to be the product of the pressure difference and the flow across the turbine, and the total absorbed energy E is thus given by

$$E = \int_0^T u(p-p_0)G(p-p_0)dt \; . \tag{2.6}$$

The problem is to determine, for a given wave-diffraction force $f_d(t)$, the optimal control u which maximises E subject to state equations (2.1), (2.3) and (2.5), and the control constraint (2.4). To simplify the model, we assume that variations in air density are relatively small, and write

$$\rho(t) = \rho_0(1 + y(t)) . \tag{2.7}$$

From (2.5) and (2.7) we then obtain, to a first approximation,

$$p - p_0 = \rho_0 \gamma y \; . \tag{2.8}$$

We also assume that the turbine flow relations are linear, so that

$$G(z) = Kz \; , \quad K \text{ a positive constant} \; . \tag{2.9}$$

Neglecting second-order terms, the equations of state may then be written

$$\begin{aligned} \dot{x} &= v \\ \dot{v} &= \frac{1}{M}\left[f_d - \rho_w g A_0 x - \int_0^t \kappa(t-\tau)v(\tau)d\tau - p_0\gamma A_0 y\right] \\ \dot{y} &= \frac{1}{V_0}\left[A_0 v - uKp_0\gamma y\right] \; , \end{aligned} \tag{2.10}$$

and the total energy derived from the system becomes

$$E = Kp_0^2\gamma^2 \int_0^T uy^2 dt \; . \tag{2.11}$$

The system is assumed to be in equilibrium with $x = v = y = 0$ at atmospheric pressure.

The system equations (2.10) are now essentially bilinear in the state and control variables. A further simplification of the system equations can be achieved by replacing the convolution integral in (2.10) by a damping term $D\dot{x} \equiv Dv$. The model can then be interpreted as representing a simple, damped, mass/piston system with a finite spring restoring force.

3. THE OPTIMAL CONTROL PROBLEM

The optimal control problem is now to maximise E, given by (2.11) subject to the system equations (2.10) over a fixed period T where either

$$\text{(i)} \quad \underline{x}(0) = \underline{x}_0 \ , \quad \underline{x}_0 \ \text{given} \tag{3.1}$$

$$\text{or} \quad \text{(ii)} \quad \underline{x}(0) = \underline{x}(T) \ . \tag{3.2}$$

We observe that imposing boundary condition (3.2) results in a periodic solution. Admissible controls $u(t)$ are taken to belong to the set of measurable functions on $[0,T]$ satisfying the restraint condition (2.4). In order to establish the optimal solution to the problem, we examine variations in the Lagrange functional, $L(u)$, associated with the problem for admissible variations δu in the control u. In this case $L(u)$ is defined by

$$L(u) = E(u) + \int_0^T \lambda_1 (v - \dot{x}) + \frac{\lambda_2}{M}\left(f_d - \rho_w g A_0 x - \int_0^t \kappa(t-\tau)v(\tau)d\tau - p_0 \gamma A_0 y\right) + \frac{\lambda_3}{V_0}\left(A_0 v - uKp_0\gamma y - \dot{y}\right)dt \ , \tag{3.3}$$

where $\underline{\lambda}(t) = (\lambda_1, \lambda_2, \lambda_3)^T \in \mathbb{R}^3$ is a continuous vector of Lagrange multipliers. Using integration by parts, and constraining $\underline{\lambda}$ to satisfy the "adjoint equations"

$$\begin{aligned} \dot{\lambda}_1 &= \frac{\rho_w g A_0}{M} \lambda_2 \\ \dot{\lambda}_2 &= -\lambda_1 + \frac{1}{M}\int_t^T \kappa(\tau - t)\lambda_2(\tau)d\tau - \frac{A_0}{V_0}\lambda_3 \\ \dot{\lambda}_3 &= -2u \cdot Kp_0^2\gamma^2 y + \frac{p_0\gamma A_0}{M}\lambda_2 + u\frac{Kp_0\gamma}{V_0}\lambda_3 \end{aligned} \tag{3.4}$$

and the "transversality conditions", either

$$\text{(i)} \quad \underline{\lambda}(T) = \underline{0} \tag{3.5}$$

$$\text{or} \quad \text{(ii)} \quad \underline{\lambda}(T) = \underline{\lambda}(0) \ , \tag{3.6}$$

(corresponding to conditions (3.1) or (3.2) respectively), we find the first variation in the Lagrangian (3.3) to be

$$\delta L(u, \delta u) = \int_0^T (Kp_0^2\gamma^2 y^2 - Kp_0\gamma\lambda_3 y/V_0)\,\delta u \ dt \ . \tag{3.7}$$

Since, for an admissable control u to be optimal, it is necessary that $\delta L \leq 0 \ \forall \ \delta u$, [Gelfand and Fomin (1963)], it follows that u takes the piecewise constant form

$$u = \begin{cases} 1 \quad \text{if} \quad Kp_0\gamma y(p_0\gamma y - \lambda_3/V_0) & > 0 \\ 0 & \leq 0 \,. \end{cases} \tag{3.8}$$

Thus the optimal control takes values on the boundaries of the admissible interval $[0,T]$ - and it is therefore "bang-bang" in nature.

We remark that in the case of the simplified state equations, in which the convolution integral term in (2.10) is replaced by the damping term Dv, the adjoint equations are modified by replacing the convolution term in (3.4) by the corresponding term $D\lambda_2/M$. The form of the optimal control is unchanged.

4. NUMERICAL ALGORITHM

We have shown in §3 that the solution to the optimal control problem is a piecewise constant control u of the form (3.8), satisfying a two-point boundary value problem defined by the state and adjoint equations (2.10) and (3.4), with periodic boundary conditions (3.2) and (3.6), or non-periodic conditions (3.1) and (3.5). We can show [Lee and Markus (1967)] that such a solution exists, and we now present an iterative process for determining the optimal control.

4.1 Conditional Gradient Method

To approximate the optimal control we apply the following algorithm.

Algorithm 1

Step 1 : Set $E^o := 0$. Set $i := 1$, $\alpha := 1$.
Choose u^o, piecewise constant, such that
$u^o(t) = 0$ or 1, $\forall\, t \in [0,T]$. Set $\tilde{u}^o := u^o$.

Step 2 : Set $u^i := \alpha\, \tilde{u}^{i-1} + (1-\alpha)u^{i-1}$

Step 3 : Solve the state equations (2.10) with boundary conditions (3.1) or (3.2) for $\underline{x}^i = \underline{x}(u^i)$.

Step 4 : Solve the adjoint equations (3.4) with boundary conditions (3.5) or (3.6) for $\underline{\lambda}^i = \underline{\lambda}(u^i)$.

Step 5 : Evaluate $E^i := Kp_0^2\gamma^2 \int_0^T u^i\, {y^i}^2\, dt$.

Step 6 : If $E^i < E^{i-1}$; set $\alpha := \alpha/2$. Go to *Step* 2.

Step 7 : If $|E^i - E^{i-1}| < tol$, go to *Step* 10.

Step 8 : Set $\tilde{u}^i := \begin{cases} 1 & \text{if } Kp_0\gamma y^i(p_0\gamma y^i - \lambda_3^i/V_0) > 0 \\ 0 & \text{otherwise} \end{cases}$

Step 9 : Set $i := i + 1$, $\alpha := 1$. Go to *Step* 2.

Step 10 : Set $u := u^i$ and STOP.

In the case where a periodic result is required, the solution is achieved by repeated integration of the state and adjoint equations over time-intervals of length T, in *Steps* 3 and 4 of the algorithm, until the boundary values of the solutions are equal to within some specified tolerance. Algorithm 1 employs the conditional gradient method to generate a sequence of admissable controls $\{u^i\}$ and corresponding energy functionals $E^i = E(u^i)$. The sequence $\{E^i\}$ is bounded, and for α sufficiently small it is possible to ensure $E^i \geq E^{i-1}$ at each i. Thus, it can be shown [Gruver and Sachs (1980)] that the iteration process described by Algorithm 1 is convergent.

4.2 Discretization of Scheme

In order to discretise the procedure, the interval $[0,T]$ is partitioned into N steps of length $h = T/N$, and solutions are determined at the mesh points $t_j = jh$. The state and adjoint equations are solved using a finite difference technique, and the energy functionals E^i are evaluated using a quadrature rule. The trapezoidal scheme [Lambert(1973), Keller (1976)] is chosen to approximate the differential equations on the mesh, and the functionals E^i are approximated by the trapezium rule [Johnson and Reiss (1982)]. The convolution integrals in the state and adjoint equations and the values of the function $\kappa(t)$, given by (2.2), are also approximated using the trapezium rule. The state equations are integrated forward from the initial condition $\underline{x}(0)$, and the adjoint equations are then integrated backward from the final condition $\underline{\lambda}(T)$. Both of these numerical integrations can be shown to be absolutely stable. Since the system equations are essentially linear in $\underline{x}$ and $\underline{\lambda}$, the trapezoidal schemes reduce to 1-step explicit methods in the appropriate directions.

5. NUMERICAL RESULTS

Results have been computed for a water column with the following data:

$$A_0 = 1.2\ m^2, \quad V_0 = 0.6\ m^3, \quad M = 2619\ kg, \quad K = 0.5\times10^{-3}$$

$$\rho_w = 10^3 kg\ m^{-3}, \quad g = 9.81\ ms^{-2}, \quad p_0 = 10^5\ Nm^{-2};$$

the wave radiation and diffraction forces [Havelock (1929)] are chosen to correspond to the device shown in Figure 1. The turbine constant K has not been optimized for this device.

A typical water column response is shown in Figure 2 for an incident monochromatic wave. The results show that phase control gives a considerable improvement over the simple damping response. In this case the incident wave frequency is 1.0 $rads^{-1}$ as opposed to the natural frequency of the device of 2.12 $rads^{-1}$. In terms of energy absorption, the efficiency of the device when uncontrolled is 28%, whereas with phase control this is increased to 82%.

In Figure 3, device efficiencies in response to incident monochromatic waves are shown. The results are plotted against the parameter T_0/T where T_0 is the natural period of the device and T the period of the incident wave. Shown are the efficiencies for the uncontrolled device and the optimally controlled device. The improvement in the performance of the device with phase control is considerable. These results demonstrate that for incident monochromatic waves of non-resonant frequency, the power output of the water column can be significantly increased by phase control. For irregular waves in the natural sea spectrum, similar results are obtained. In Figure 4 the response of the column is shown for a particular example which has a spectrum with a mean frequency of 1.6 $rads^{-1}$. In this case the energy output of the device with phase control is increased by a factor of 1.6 over that of the uncontrolled device.

We remark that with the simpler mathematical model, using an appropriate linear damping term to approximate the 'added' hydrodynamic damping, the results are almost identical to those obtained from the more sophisticated model, and the same conclusion can be drawn.

6. CONCLUSIONS

A mathematical model of the oscillating water column is described here, and an optimal control strategy for power-generation from the device is determined. Numerical results are presented which demonstrate that, given certain information concerning the wave exciting force, it is possible to increase significantly the energy-capture of the device using phase-control in the case of both regular and irregular incident wave forms. The energy generated by the water column with the optimal control gives the maximum possible power output from the device and provides the target for practical control systems to achieve.

Implementation of a phase control mechanism requires forward prediction of the wave-exciting force in order to compute the optimal control strategy. It is expected that the acting wave force can be found from measurements of the system states and that a technique using Kalman filtering to estimate the future behaviour of the wave force over a short period from its recent history can be applied to obtain the necessary data. Further investigation of these techniques and of more realistic non-linear models accurately incorporating the turbo-generator characteristics is necessary.

ACKNOWLEDGEMENTS

This research has been supported by the U.K. Science and Engineering Research Council and the Central Electricity Generating Board.

REFERENCES

1. Budal, K.; Falnes, J.; Wave-power conversion by point absorbers. Norwegian Maritime Research, 6, (1978) (4), 2-11.
2. Budal, K.; et al. The Norwegian wave-power buoy project. Proc. of 2nd Int. Symp. on Wave Energy Utilisation, Trondheim, (1982).
3. Count, B.M.; On the physics of absorbing energy from ocean waves, Ph.D. Thesis, Univ. of Exeter, (1982).
4. Cummins, W.E.; The impulse response function and ship motions. Schiffstechnik 9, (1962), 101-109.
5. Evans, D.V.; A theory for wave-power absorption by oscillating bodies. J. Fluid Mech., 77, (1976), 1-25.

6. Gelfand, I.M.; Fomin, S.V.; Calculus of Variations. Prentice-Hall, (1963).

7. Gruver, W.A.; Sachs, E.; Algorithmic Methods in Optimal Control. Pitman, (1980).

8. Havelock, T.H.; Forced surface waves on water. Phil. Mag. 8, (1929), 51.

9. Johnson, L.W.; Reiss, R.D. Numerical Analysis. Addison-Wesley, (1982).

10. Keller, H.B.; Numerical Methods for Two Point Boundary Value Problems. Blaisdell, (1976).

11. Lambert, J.D.; Computational Methods in Ordinary Differential Equations. John Wiley & Sons, (1973).

12. Lee, E.B.; Markus, L.; Foundations of Optimal Control Theory. John Wiley & Sons, (1967).

13. Pontryagin, L.S. et al; The Mathematical Theory of Optimal Processes. John Wiley & Sons, (1962).

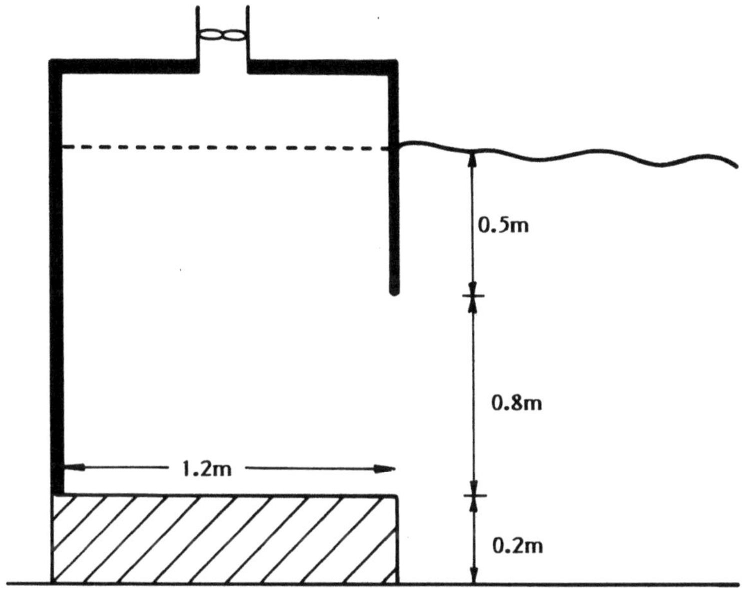

Figure 1. OWC used for tests

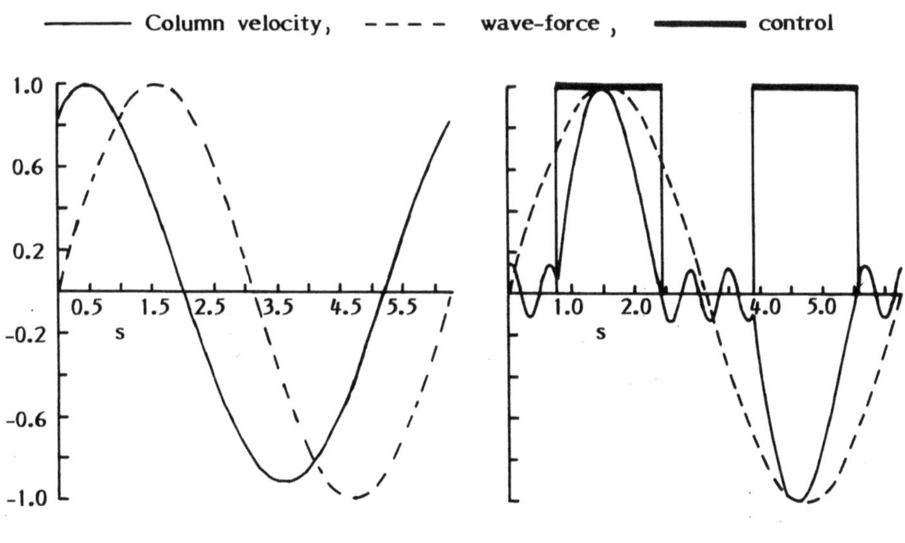

Figure 2a. Uncontrolled device

Figure 2b. Controlled device

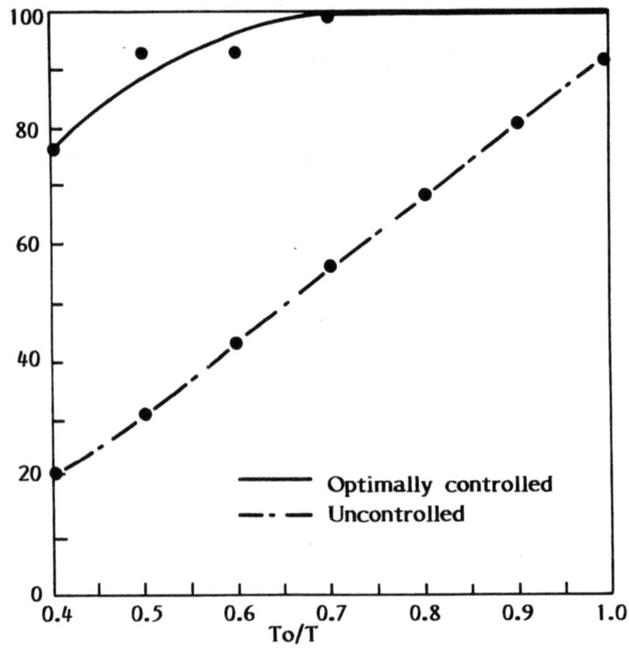

Figure .3. Efficiency

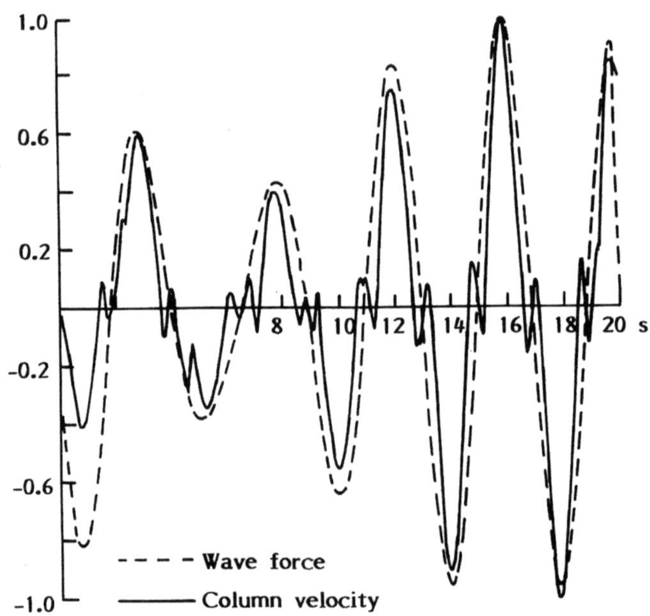

Figure .4. Response to irregular waves

Wave Energy Absorption in Irregular Waves by Feedforward Control System

SHIGERU NAITO , SHOICHI NAKAMURA

Osaka University, Faculty of Engineering, Department of Naval Architectue
2-1 Yamada-oka, Suita, Osaka, 565 JAPAN

Summary

A number of theoretical and experimental papers on ocean wave energy absorption have been presented over the past few years [1]. Many of them are concerned with the characteristics of absorbed wave energy and the condition of optimum control in regular waves. Practically this is not sufficient, and more interest must be focused on the problem in irregular waves. Absorbed wave energy in irregular waves obtained by a generating system which is composed of a floating body and external mechanism has two components, one is a steady component and the other is a fluctuating one. In this paper, some problems about the two components mentioned above are studied. One is the theory of optimum control to obtain the maximum mean energy of irregular waves related to the steady part, and the other is the formulation of the spectrum about absorbed wave energy related to the fluctuating part. The optimum conditions for complete absorption of wave energy in regular waves are well known. However, to establish the system of wave energy absorption in irregular waves, it is necessary to consider the problem on the basis of the time domain analysis, because the optimum conditions depend on wave frequency and irregular waves contain many component waves having different frequencies. A system, in which the incident irregular waves are measured at a certain distance in front of the floating body with the external mechanism, is controlled instantaneously to improve the efficiency by using the measured wave data, called "Feedforward Control System for Wave Energy Absorption" is proposed. Furthermore, the law of causation in the characteristics of wave propagation is discussed from the view point of practical approach, because the kernel function of wave propagation, which is used in this system, does not usually satisfy the law of causation.

From a practical point of view, the fluctuating component is not desirable for using the absorbed wave energy, so it is necessary to devise the method of smoothing it. The formula of the spectrum for gathering absorbed wave energy by the multi-floating bodies is obtained, in which the effect of the array on the fluctuation of absorbed wave energy is shown, and using the results, a suitable method to array the multifloating bodies to reduce the fluctuation of the absorbed wave energy is proposed. Some digital simulations with a two dimensional model of floating type wave absorber in irregular waves are carried out to confirm the validity of this formulations.

1 Theoretical Consideration on Wave Energy Absorption of Irregular Waves by Feedforward Control System

It is well known that the following two conditions must be satisfied for absorbing wave energy with high efficiency.
(1) The spring coefficient of external mechanism is selected so that a floating body synchronizes with incident regular wave.
(2) The damping coefficient of external mechanism is equal to the radiation wave damping coefficient of a floating body itself.
These two coefficients depend on the incident wave frequency. Therefore, to extract the energy of irregular waves composed of many wave components, it is necessary to make a time variable external mechanism. A general view of this system is shown in Fig.1.

We must consider three problems to realize this system, which are the external control force, the wave propagation property and the possibility on practical application.

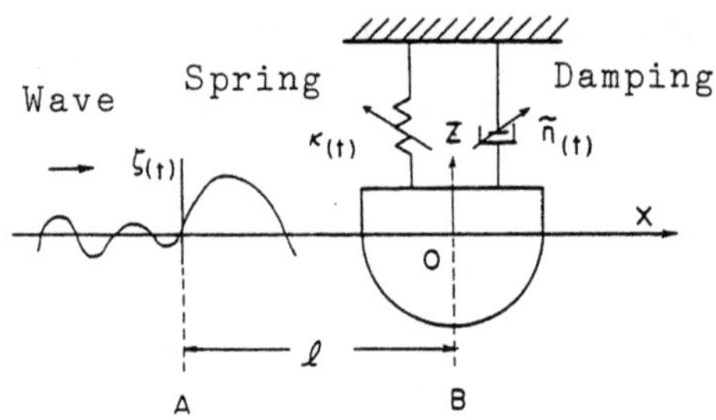

Fig.1 General view of waves energy absorption system

1.1 Optimum External Control Force in Regular Waves

In general, the equation of heaving motion Z is expressed by

$$[M+m(\omega)]\ddot{Z}+[N(\omega)+N_e(\omega)]\dot{Z}+[K_e(\omega)+C]Z = \rho g^2\frac{\bar{A}}{\omega^2}\zeta(t) \qquad (1)$$

where M: mass of floating body, $m(\omega)$: added mass, $N(\omega)$: wave damping coefficient, C: restoring force coefficient, $N_e(\omega)$, $K_e(\omega)$: damping and spring coefficients of external mechanism, $\bar{A} = i\bar{A}_H\exp(i\varepsilon_H)$, ($\bar{A}_H$, ε_H: amplitude ratio of progressive wave amp. to heaving motion amp. and its phase)

Let heave Z(t) and wave $\zeta(t)$ be expressed as follows,

$$Z(t) = Z_0\exp(i\omega t),\ \zeta(t) = \zeta_0\exp(i\omega t) \qquad (2)$$

Frequency response function of heaving motion $H_{\zeta z}(\omega)$ is given by

$$H_{\zeta z}(\omega) = \frac{Z_o}{\zeta_o} = \frac{\rho g^2\bar{A}/\omega^2}{K_e(\omega)+C-\omega^2[M+m(\omega)]+i\omega[N(\omega)+N_e(\omega)]} \qquad (3)$$

Then, the conditions to obtain the maximum energy of regular waves are given by [2]

$$K_e(\omega) = -C+\omega^2\{M+m(\omega)\} \quad , \quad N_e(\omega) = N(\omega) \tag{4}$$

1.2 Represetation of the Optimum Conditions in Time Domain

In irregular waves, it is necessary to satisfy the conditions shown by eq.(4) in the time domain instantaneously. We define the Fourier transformation as follows,

$$F(\omega) = \int_{-\infty}^{\infty} f(t)\exp(-i\omega t)dt, \qquad f(t) = \frac{1}{2\pi}\int_{-\infty}^{\infty} F(\omega)\exp(i\omega t)d\omega \tag{5}$$

Hereafter, let F^1 and F^{-1} denote the Fourier transformation and the inverse one respectively.

The impulse function L(t) of fluid is given by [3][4][5]

$$L(t) = \frac{1}{2\pi}\int_{-\infty}^{\infty} [N(\omega)+i\omega\{m(\omega)-m_\infty\}]\exp(i\omega t)d\omega, \; m_\infty = m(\infty) \tag{6}$$

L(t) is real function, so eq.(6) is transformed as follows,

$$L(t) = \frac{1}{\pi}\int_0^{\infty} N(\omega)\cos\omega t \; d\omega - \frac{1}{\pi}\int_0^{\infty} \omega\{m(\omega)-m_\infty\}\sin\omega t \; d\omega \tag{7}$$

Usually, an arbitrary function can be expressed by using of the even and odd functions, so let $L_e(t)$ and $L_o(t)$ denote the even and odd parts of L(t) respectively, then

$$L(t) = L_e(t) + L_o(t) \tag{8}$$

L(t) must satisfy the causality as follows,

$$\left.\begin{array}{ll} L(t) = L_e(t) + L_o(t) = 0, & t < 0 \\ L(t) = 2L_e(t) = 2L_o(t), & t > 0 \end{array}\right\} \tag{9}$$

At first, let $\kappa(t)$ be $F^{-1}[K_e(\omega)]$,then $\kappa(t)$ is given by

$$\begin{aligned}\kappa(t) &= F^{-1}[K_e(\omega)] = F^{-1}[-C + \omega^2(M + m_\infty) + \omega^2\{m(\omega) - m_\infty\}] \\ &= -C\delta(t) - (M + m_\infty)\delta''(t) + F^{-1}[\omega^2\{m(\omega) - m_\infty\}]\end{aligned}$$

$$L_o'(t) = -\frac{1}{\pi}\int_0^{\infty}\omega^2\{m(\omega)-m_\infty\}\sin\omega t \; d\omega \tag{10}$$

As the results,

$$\kappa(t) = -C\delta(t)-(M+m_\infty)\delta''(t)-L_o'(t) \tag{11}$$

Next, let $\tilde{n}(t)$ be $F^{-1}[N_e(\omega)]$, then $\tilde{n}(t)$ is written as

$$\tilde{n}(t) = F^{-1}[N_e(\omega)] = \frac{1}{\pi}\int_0^{\infty} N(\omega)\cos\omega t \; dt = L_e(t) = \frac{1}{2}L(t, \; t>0) \tag{12}$$

Eqs. (11) and (12) are the representation of the optimum conditions to extract the wave energy in the time domain. The phase control which is approached by Falnes and Budal[6] should be discussed relating to eqs.(11) and (12).

1.3 Equation of Motion under the Optimum Conditions

When the external force is acting on the floating body, the equation of motion becomes an integro-differential equation as follows,

$$(M+m_\infty)\ddot{Z} + \int_0^t L(\xi)\dot{Z}(t-\xi)d\xi + CZ + \int_{-\infty}^{\infty}\tilde{n}(t)\dot{Z}(t-\xi)d\xi + \int_{-\infty}^{\infty}\kappa(t)Z(t-\xi)d\xi = F_W(t) \tag{13}$$

Substituting eqs.(11) and (12) into eq.(13), it is rewritten as

$$\int_0^t L(\xi)\dot{Z}(t-\xi)d\xi + \int_{-\infty}^{\infty}\{L_e(\xi)-L_o(\xi)\}\dot{Z}(t-\xi)d\xi = F_W(t) \tag{13)'$$

The second term of left side is important, and it means a control force of external mechanism to extract the energy of irregular waves perfectly. This term is transformed by eq.(9) as follows,

$$\int_{-\infty}^{\infty}\{L_e(\xi)-L_o(\xi)\}\dot{Z}(t-\xi)d\xi = \int_{-\infty}^{0}\{L_e(\xi)-L_o(\xi)\}\dot{Z}(t-\xi)d\xi$$

$$= \int_0^{\infty} 2Le(\xi)\dot{Z}(t+\xi)d\xi = \int_0^{\infty} L(\xi)\dot{Z}(t+\xi)d\xi \tag{14}$$

This result means that the ideal system to absorb the energy of irregular waves perfectly can not be realized because it does not satisfy the causality. Eq.(13)' is rewritten as a final form by

$$\int_0^t L(\xi)\dot{Z}(t-\xi)d\xi + \int_0^{\infty} L(\xi)\dot{Z}(t+\xi)d\xi = F_W(t) \tag{13)''$$

Eq.(13)" can be transformed into the equation in the frequency domain easily as follows,

$$2\underset{\sim}{L}(\omega)i\omega\underset{\sim}{Z}(\omega) = \underset{\sim}{F}_w(\omega) \;, \quad H_{\zeta w}(\omega) = \frac{\underset{\sim}{\zeta}(\omega)}{\underset{\sim}{Z}(\omega)} = \frac{e^{i\varepsilon_H(\omega)}}{2i\bar{A}_H(\omega)} \tag{15}$$

This equation shows a frequency response function with optimum control of external mechanism. Let's consider the time mean of absorbing power $\bar{E}_T$ under the conditions mentioned above.

$$\bar{E}_T = \frac{1}{T}\int_{T/2}^{T/2} \dot{Z}(t)D(t)dt = \frac{1}{2\pi T}\int_{-\infty}^{\infty} F^1[D(t)]\dot{\underset{\sim}{Z}}^*(\omega)d\omega \tag{16}$$

where $D(t)$ is a force acting on the real energy absorbing part in the external mechanism, and * means a conjugate complex.

$$F^1[D(t)] = F^1[\int_{-\infty}^{\infty}\tilde{n}(\xi)\dot{Z}(t-\xi)d\xi] = \frac{1}{2}\underset{\sim}{L}(\omega)i\omega\;\underset{\sim}{Z}(\omega) \tag{17}$$

Substituting eq.(17) into eq.(16),

$$\bar{E}_T = \frac{1}{2\pi T}\int_0^{\infty}\omega^2\underset{\sim}{Z}(\omega)\underset{\sim}{Z}^*(\omega)\underset{\sim}{L}(\omega)d\omega \tag{18}$$

Using the eq.(15),

$$\bar{E} = \lim_{T\to\infty}\bar{E}_T = \frac{\rho g^2}{2}\int_0^{\infty}\frac{1}{\omega}\left[\lim_{T\to\infty}\frac{[\underset{\sim}{\zeta}(\omega)]^2}{2\pi T}\right]d\omega = \frac{\rho g^2}{2}\int_0^{\infty}\frac{S(\omega)}{\omega}d\omega$$

where $S(\omega)$ is a wave power spectrum. Usually, Pierson-Moskowitz type spectrum $S^{PM}(\omega)$ is used. Relation between both spectra is given by

$$S(\omega) = \frac{1}{2}S^{PM}(\omega) \;;\; 0 < \omega < \infty \tag{19}$$

As the results, $\bar{E}$ becomes

$$\bar{E} = \frac{1}{2}\cdot\frac{\rho g^2}{2}\int_0^{\infty}\frac{S^{PM}(\omega)}{\omega}d\omega = \frac{\text{(whole power of irregular waves)}}{2}$$

The theory mentioned above is based on the assumption that a floating body has a symmetric sectional form. In case of unsymmetric body, $\underset{\sim}{L}(\omega)$ and $\underset{\sim}{Z}(\omega)$ are changed as follows,

$$\underset{\sim}{L}(\omega) = \frac{\rho g^2}{\omega^3}(\overline{A^+}^2 + \overline{A^-}^2), \quad [\underset{\sim}{Z}(\omega)]^2 = \frac{\overline{A^+}^2}{\overline{A^+}^2 + \overline{A^-}^2}\cdot|\underset{\sim}{\zeta}(\omega)|^2 \tag{20}$$

In this case, $\bar{E}$ becomes

$$\bar{E} = \frac{\rho g^2}{2} \int_0^\infty \frac{S^{PM}(\omega)}{\omega} \cdot \frac{\overline{A^+}^2}{\overline{A^+}^2 + \overline{A^-}^2} d\omega \tag{21}$$

If we could make the unsymmetric body with a character of $\overline{A^-}(\omega) = 0$; throughout ω , eq.(21) is rewritten as

$$\bar{E} = \frac{\rho g^2}{2} \int_0^\infty \frac{S^{PM}(\omega)}{\omega} d\omega = \text{(whole power of irregular waves.)}$$

By the fact as described above, the theory on absorbing the energy of irregular waves perfectly can be clarified.

1.4 Propagation Property of Water Wave in Deep Water.

Transfer function of progressive waves in deep water without an attenuation from the point A to the point B in Fig.1, is presented by

$$H_{AB}(\omega) = \exp(-k\ell) = \exp(-i\frac{\ell}{g}\omega^2) \tag{22}$$

ℓ is distance between the point A and B.

The representation of eq.(22) in the time domain $h_{AB}(t)$ is known, but the equation does not satisfy the law of causation, and it is not useful because a high frequency vibration exists during a long period. Actual irregular waves consist of limited frequency waves($\omega_1 \leqq \omega \leqq \omega_2$), so the function $h_{AB}(t)$ can be modified as follows,

$$\tilde{h}_{AB}(t) = \frac{1}{\sqrt{2\alpha\pi}} \mathcal{R}_e[\exp(i\frac{t^2}{4\alpha}) \int_{\omega'-\eta}^{\omega''-\eta} \exp(-i\frac{\pi}{2}\xi^2) d\xi] \tag{23}$$

where $\eta = t/\sqrt{(2\alpha\pi)}$, $\alpha = 1/g$, $\omega' = \omega_1\sqrt{(2\alpha/\pi)}$, $\omega'' = \omega_2\sqrt{(2\alpha/\pi)}$

This integral is called Cornu integral. In case of $\omega_1 \to 0$ and $\omega_2 \to \infty$, eq.(23) agree with $h_{AB}(t)$. This function is more useful than $h_{AB}(t)$, but does not satisfy the causality yet. We must obtain a method to satisfy the causality.

Multiplying eq.(23) by rectangular pulse, and let it be $h_c(t)$.

$$h_c(t) = \tilde{h}_{AB}(t)\cdot q(t-t_0)\ ,\quad q(t-t_0) = \begin{cases} 1\ ; & 0 \leqq t \leqq 2t_0 \\ 0\ ; & t<0,\ t>2t_0 \end{cases}$$

$h_c(t)$ satisfies the causality, of course.

Now,

$$F^{1}[q(t-t_0)] = \frac{2\sin\omega t}{\omega}\cdot \exp(-i\omega t_0) \tag{24}$$

Using this result, the function of $WH(\omega) = F^{1}[\tilde{h}_{AB}(t)\cdot q(t-t_0)]$ is obtained as follows,

$$WH(\omega) = \frac{1}{2\pi} H_{AB}(\omega) * \frac{2\sin\omega t}{\omega} \exp(-i\omega t_0)\ ;\ (\text{* means a convolution integral.})$$

$$\equiv \exp(-i\omega t_0)\int_{\omega_1}^{\omega_2} \frac{\sin(\omega-\xi)t_0}{\pi(\omega-\xi)} \exp i(\xi t_0 - \alpha\xi^2)d\xi \tag{25}$$

This result is more useful than $\tilde{h}_{AB}$, because it satisfies the causality, so we can use it actually.

1.6 Feedforward Control System (refer to Fig. 1)

By using the wave elevation $\zeta_A(t)$, $\zeta_B(t)$ can be predicted as follows,

$$\zeta_B(t) = \int_0^t h_c(\tau_1)\zeta_A(t-\tau_1)d\tau_1 \tag{26}$$

A floating body is arranged at the place B. The heaving motion of the body is estimated by

$$Z(t) = \int_0^t h_{z\zeta}(\tau_2)\zeta_B(t-\tau_2)d\tau_2 \tag{27}$$

where $h_{z\zeta}(t)$ is F^{-1} of eq.(15), which is the frequency response function under the optimum conditions to absorb the wave energy. As the results, in order to extract the energy of irregular waves with high efficiency, the control force acting on the body is estimated as follows,

$$F_c(t) = \int_{-\infty}^{\infty} f_c(\tau_3)\dot{Z}(t-\tau_3)d\tau_3 \tag{28}$$

Substituting eqs.(26) and (27) into eq.(28), the theoretical external control force $F_c(t)$ is presented by

$$F_c(t) = \int_{-\infty}^{\infty}\int_0^{t-\tau_3}\int_0^{t-\tau_2-\tau_3} f_c(\tau_3)\dot{h}_{z\zeta}(\tau_2)h_c(\tau_1)$$

$$\cdot\ \zeta_A(t-\tau_1-\tau_2-\tau_3)d\tau_1 d\tau_2 d\tau_3 \qquad (29)$$

We can extend the theory mentioned above[7]. By using a floating body with external mechanism which is set at the end of a water tank, we can extract the energy of irregular waves perfectly by applying the method of images. Results of numerical simulation are shown in Fig. 2 and Fig.3 as the form of time histories. Fig. 2 shows a control force of external mechanism under the optimum conditions, and Fig.3 shows an extracted wave energy in which the apparent and the real energy components are mixed.

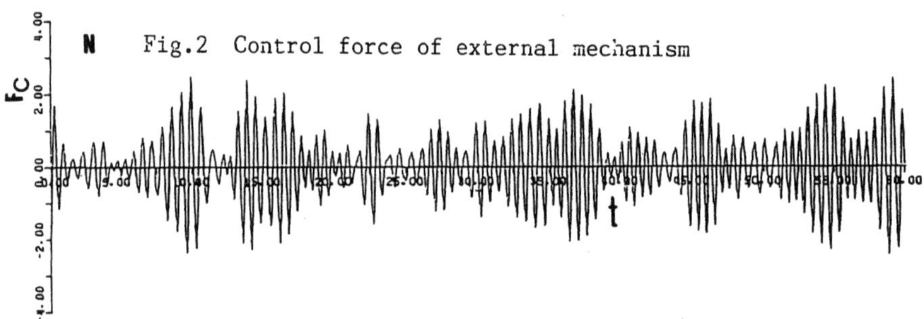

Fig.2 Control force of external mechanism

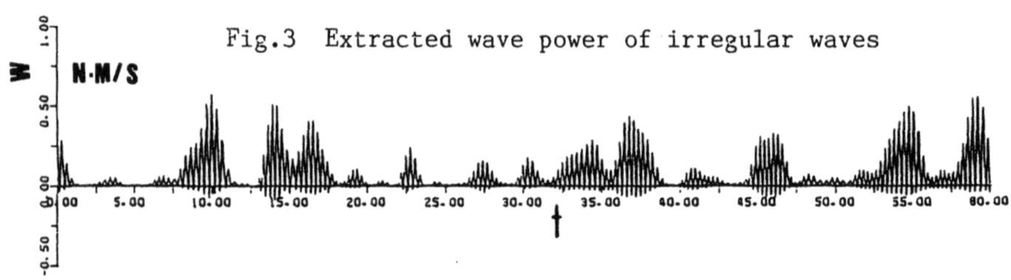

Fig.3 Extracted wave power of irregular waves

3 Spectrum of Absorbed Wave Energy Obtained by Multi-Floating Bodies

From the view point of practical purpose, fluctuating component of absorbed wave power is not desirable. So, we discuss the sum up effect of the energy to reduce the flucluation [8]. We assume that two floating bodies are placed at $x=x_j$ and $x=x_k$, and that the mutual interference between them does not exist. $R_{jk}(\tau)$ is a cross correlation function between them.

$$R_{jk}(\tau)=\lim_{T\to\infty}\frac{1}{2T}\int_{-T}^{T}Z_j^*(t)Z_k(t+\tau)dt=\int_{-\infty}^{\infty}S_z(\omega)d\omega\cdot e^{i\frac{\omega|\omega|}{g}(x_j-x_k)}e^{i\omega\tau} \tag{30}$$

$S_{jk}(\omega)$ is a cross spectrum defined by $F^1[R_{jk}(\tau)]$

$$S_{jk}(\omega)=\int_{-\infty}^{\infty}S_z(\tilde{\omega})e^{i\frac{\tilde{\omega}|\tilde{\omega}|}{g}(x_j-x_k)}\delta(\omega-\tilde{\omega})d\tilde{\omega} = S_z(\omega)e^{i\frac{\omega|\omega|}{g}(x_j-x_k)} \tag{31}$$

where the definition of F^1 is differ from eq.(5) by $1/2\pi$ and

$$L_{jk}(\omega) = e^{i\frac{\omega|\omega|}{g}(x_j-x_k)} = \begin{cases} e^{ik(x_j-x_k)} : & \omega>0 \\ e^{-ik(x_j-x_k)} : & \omega<0 \end{cases} \tag{32}$$

$L_{jk}(\omega)$ means a kind of transfer function converting $S_z(\omega)$ into $S_{jk}(\omega)$, and is very simple. So, eq.(31) becomes

$$S_{jk}(\omega) = S_z(\omega)L_{jk}(\omega) \tag{33}$$

In order to simplify this problem, let the coefficients of damping and spring in external mechanism be constant n_o and κ_o respectively.

We write the instantaneous value of energy as W(t) and replace the sum up energy $\Sigma W_p(t)$ by q(t).

$$q(t) = \sum_{p=1}^{N} W_p(t) = \sum_{p=1}^{N} [\kappa_o Z_p+n_o\dot{Z}_p]\,\dot{Z}_p \tag{34}$$

Auto-correlation function $R_{qq}(t)$ of q(t) is given by

$$R_{qq}(\tau)=\sum_{j}^{N}\sum_{k}^{N}[-\kappa_o^2\{R_{jk}(\tau)\ddot{R}_{jk}(\tau)+\dot{R}^2{}_{jk}(\tau)\}$$
$$+n_o^2\{\ddot{R}_{jj}(o)\ddot{R}_{kk}(o)+2\ddot{R}_{jk}^2(\tau)\}] \tag{35}$$

Let $F^1[R_{qq}(\tau)]$ be $S_{qq}(\omega)$ which is the spectrum of q(t).

$$S_{qq}(\omega)=\{Nn_o\ddot{R}_{zz}(o)\}^2\delta(\omega)+\int_{-\infty}^{\infty}\tilde{\omega}\{\kappa_o^2\omega+2n_o^2\tilde{\omega}(\omega-\tilde{\omega})^2\}S_z(\tilde{\omega})S_z(\omega-\tilde{\omega})$$
$$\cdot\sum_{j}^{N}\sum_{k}^{N} L_{jk}(\tilde{\omega})L_{jk}(\omega-\tilde{\omega})d\tilde{\omega} \tag{36}$$

The second term of right side in eq.(36) is very important, which indicates the characteristics on arrangement of floating bodies. We write this term as $T(\omega,\tilde{\omega})$, and consider one example

using this result.
As shown in Fig. 4, every floating body is arranged with the same interval d. In this case, $T(\omega,\tilde{\omega})$ becomes simple and is presented as follows,

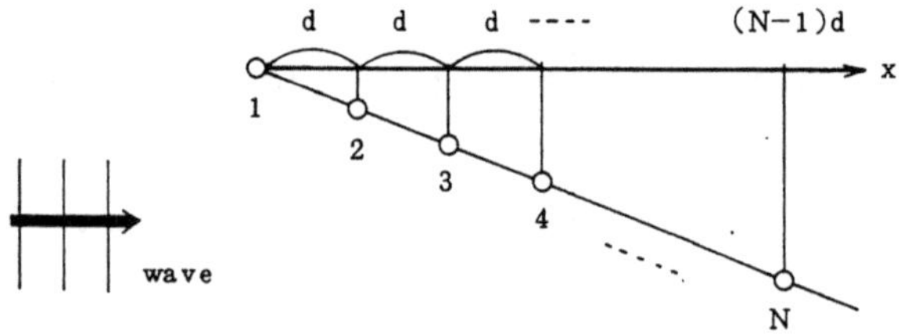

Fig.4 Arrangement of multi-floating bodies

$$T(\tilde{\omega},\omega) = \frac{1-\cos\gamma_k Nd}{1-\cos\gamma_k d}, \quad \begin{array}{ll} \gamma_1=\{\omega^2+(\omega-\tilde{\omega})^2\}/g & :\omega>\tilde{\omega}>0 \\ \gamma_2=\{\tilde{\omega}^2-(\omega-\tilde{\omega})^2\}/g & :\tilde{\omega}>0,\tilde{\omega}>\omega \\ \gamma_3=\{-\tilde{\omega}^2+(\omega-\tilde{\omega})^2\}/g & :\tilde{\omega}<\omega,\tilde{\omega}<0 \\ \gamma_4=\{-\tilde{\omega}^2-(\omega-\tilde{\omega})^2\}/g & :\omega<\tilde{\omega}<0 \end{array} \tag{37}$$

The Pierson-Moskowitz type spectrum of the energy obtained by multi-floating bodies arranged with the same intervals is expressed as

$$\begin{aligned} S_{qq}^{PM}(\omega)=&\{Nn_o\ddot{R}_{zz}(0)\}^2\delta(\omega)+\frac{1}{2}\int_0^{\infty}\{\kappa_o^2\omega^2+4n_o^2\tilde{\omega}^2(\omega+\tilde{\omega})^2\}S_z^{PM}(\tilde{\omega})S_z^{PM}(\omega+\tilde{\omega}) \\ &\cdot\frac{1-\cos\gamma_o Nd}{1-\cos\gamma_o d}\,d\tilde{\omega} \\ &+\frac{1}{2}\int_0^{\omega}\tilde{\omega}\{\kappa_o^2\omega+2n_o^2\tilde{\omega}(\omega-\tilde{\omega})^2\}S_z^{PM}(\tilde{\omega})S_z^{PM}(\omega-\tilde{\omega}) \\ &\cdot\frac{1-\cos\gamma_1 Nd}{1-\cos\gamma_1 d}\,d\tilde{\omega} \end{aligned} \tag{38}$$

where $\gamma_o=\{\tilde{\omega}^2-(\omega+\tilde{\omega})^2\}/g$
In case of N=1, this result agrees with Maeda's result [9].
The first term means the average energy, and the second and the third terms mean low and high frequency fluctuating components of the absorbed energy respectively. The term multiplied by κ_o gives the apparent energy and the term multiplied by n_o gives the real energy.
After all, by means of choosing the values of d(distance between two bodies) and N(number of bodies) reasonably, low and high frequency fluctuating components can be reduced. To confirm this fact some numerical simulations are carried out. Fig. 5 shows a spectrum of absorbed wave energy obtained by one

floating body. In this case, the spectrum of incident irregular waves has a peak at frequency of $\omega = 10(1/s)$. It is obvious at a glance that the fluctuations of low and high frequency cmponent are large. So, two bodies are arranged to reduce the fluctuation of $\omega = 20(1/s)$. The result is shown in Fig. 6. Furthermore, six bodies are arranged as shown in Fig. 6 to reduce the fluctuations of $\omega = 20(1/s)$ and $1(1/s)$. Distances d_1 and d_2 are decided to reduce the components of $\omega = 20(1/s)$ and $\omega = 1(1/s)$ respectively. We know from Fig.7 that the purpose is accomplished, and the agreement between the theory and the simulation results is fairly good.

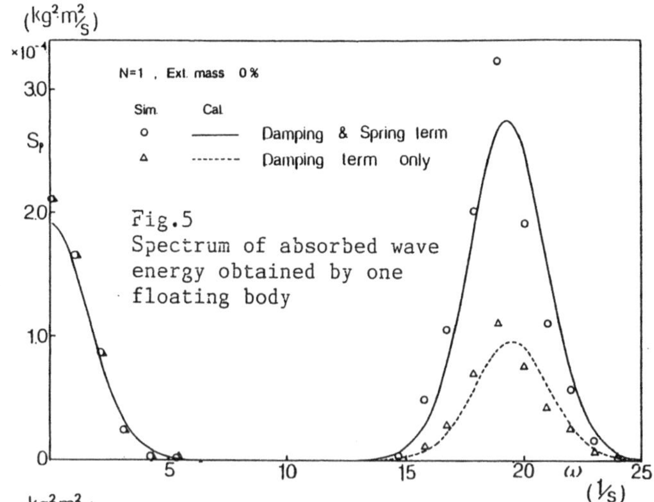

Fig.5 Spectrum of absorbed wave energy obtained by one floating body

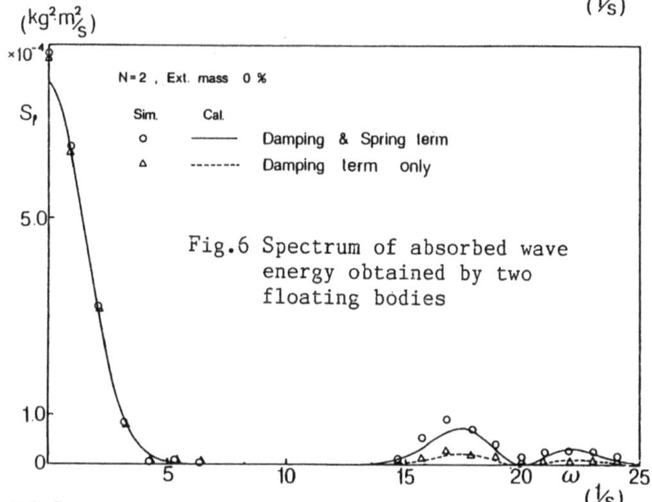

Fig.6 Spectrum of absorbed wave energy obtained by two floating bodies

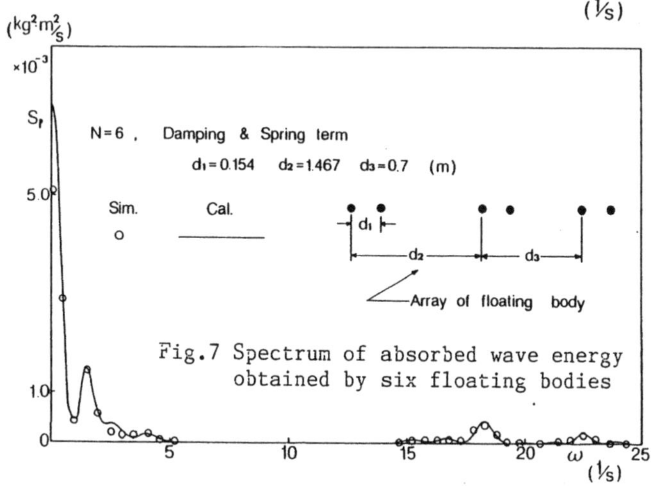

Fig.7 Spectrum of absorbed wave energy obtained by six floating bodies

4 Concluding Remarks

The results of this study are summarized as follows.

1) The optimum conditions in the time domain can be made clear to extract the energy of irregular waves perfectly. But the system consisted of a floating body and external mechanism can't be realized theoretically, because it does not satisfy the causality.

2) The equation on a spectrum of absorbed wave energy obtained by multi-floating bodies can be obtained. By considering the sum up effect term $T(\omega,\tilde{\omega})$, which is included in that equation, the fluctuation of the absorbed energy can be reduced.

3) The formulation obtained in this study has a possibility to apply into many other problems.

REFERENCE

1) Proc. 1st Symp. and 2nd Symp. on Wave Energy Utilization, Gothenburg,1979; Trondheim,1982

2) Bessho,M., : Feasibility Study of a Floating Type Wave Absorbr :, 34th JTTC, (Japan Patent No.S. 48-65359),1973

3) Cummins, W.E. : The Impulse Response Function and Ship Motions, Shiffstechnik, Bd.9, Heft 47, 1962

4) Ogilvie, T.F. : Recent Progress Toward the Understanding and Prediction of Ship Motions, Proc. 5th Symp. on Naval Hydrodynamics, Bergen, 1964

5) Count, B.M. and Jefferys, E.R. : Wave Power, The Primary Interface, Proc. 13th Symp. on Naval Hydrodynamics, Tokyo, 1980

6) Falnes,J. and Budal, K.: Wave-Power Conversion by Point Absorber", Norwegian Maritime Research, Vol.6, No.4, 1978

7) Milgram, J.H.: Active Water-Wave Absorbers, J.F.M., Vol.42, 1970

8) Thomas, G.P. and Evans,D.V.: Arrays of three-dimensional wave-energy absorbers, J.F.M., Vol.108, 1981

9) Maeda H., Kinoshita T., Kato S. and F.Suzuki : Fundamental Research on Absorbing Energy from Ocean Waves(3rd Report), Jour.of Soc. of Naval Arch. of Japan, Vol.150, 1981

Latching Control of an Oscillating Water Column Device with Air Compressibility

Richard Jefferys

Mechanical Engineering
University College
Torrington Place
London WC1E 7JE

Trevor Whittaker

Civil Engineering
The Queen's University of Belfast
Stanmillis Road
Belfast BT7 1NN

Abstract

A linearised model of air compressibility is developed and integrated with the conventional hydrodynamic model of an OWC. This is optimised in the frequency domain and the effects of compressibility are explored. Numerical predictions are validated by experiment. A latching control strategy is described and optimised by first harmonic methods. Simulations show that the approximate frequency domain analysis is accurate and that useful gains in efficiency can be realised by latching.

Introduction

Latching control can improve substantially the power absorbed by wave energy devices (ref. 1). In the oscillating water column (OWC) device, waves excite the water column which drives air through a turbine; while it is relatively simple to stop the air flow during part of the cycle, the compressibility of the air between the water surface and the latching mechanism allows the column to move even when air flow is halted. Since the column must be 'latched' to optimise the phase of the radiated wave, the control scheme should take account of the air compliance.

Some authors have ignored compressibility since 'speed of sound in air is high compared with chamber dimensions' but this only implies that pressure is the same everywhere in the chamber, not that the air is incompressible. In a typical OWC, the ratio of air to hydrostatic spring rates is about 2.5:1 so significant column and pressure transients will be generated when the air flow is latched.

In this study, we develop a linear model of the air column dynamics which represents the air spring and turbine damping. Optimal values of the air column characteristics are derived for the linear operating regime where the wave and resonant frequencies are close. The predictions of the mathematical model are compared with those obtained from wave tank tests on a model of a J-tube OWC. Variation of the air column length demonstrates the effect of air compressibility on performance.

A conventional latching algorithm controls the air flow; the switching times are optimised for sinusoidal waves by first harmonic analysis. Numerical computer simulations confirm the validity of the approximate frequency domain optimisation technique. The efficiency of the OWC in waves below the resonant frequency is improved significantly; transient column motions and air pressure histories are presented.

This paper overlaps to some extent but is mostly complementary to work reported by Sarmento and Falcao (ref. 2); they have investigated free surface column effects, compressibility, finite water depth effects and non linear power take off.

Compressibility

The exact modelling of the air compressibility needs a non linear thermodynamic model of some complexity (ref. 3). A linearised model is fairly accurate and yields physical insight to guide design.

Adiabatic flow is assumed; the air has little chance to exchange heat with the environment during a cycle and the modelling of the heat transfer processes would require detailed design information and a substantial program of research. These losses can be included in the turbine efficiency; lacking a detailed turbine and heat loss model, it is assumed that maximisation of energy absorption by the column corresponds to maximisation of overall efficiency.

The state equation for the air in the chamber is

$$PV^{\gamma} = M^{\gamma} P_0 V_0^{\gamma} / M_0^{\gamma} \tag{1}$$

where M is the instantaneous mass, P is the pressure and V is the volume; subscripted variables refer to undisturbed conditions. In the calculations which follow, the polytropic exponent, γ is set at the adiabatic value for air of 1.4.

The pressure change caused by the column velocity, $\dot{z}$ and the mass flow rate $\dot{M}$ is obtained by differentiating equ.1

$$\dot{P} V^{\gamma} + \gamma P \dot{V} V^{\gamma-1} = \gamma \dot{M} M^{\gamma-1} (P_0 V_0^{\gamma}) / M_0^{\gamma} \tag{2}$$

If the column walls are vertical, $\dot{V} = -A\dot{z}$, where A is the column area. The air flow through a Wells turbine is roughly proportional to turbine speed and pressure drop; for simplicity, turbine speed is assumed constant, leading to the following expression for mass flow rate.

$$\alpha \dot{M} = - (P - P_0)/P_0 \tag{3}$$

Hence

$$\dot{P} = \dot{z} \gamma P A/V - \gamma M^{\gamma-1} (P - P_0) (P_0 V_0^{\gamma}) / (\alpha P_0 M_0^{\gamma} V^{\gamma}) \tag{4}$$

Linearise the pressure and volume about the operating condition and denote the instantaneous values of the perturbed quantities by lower case letters:

$$p = P - P_0 \qquad v = V - V_0 \qquad m = M - M_0$$

Hence

$$\dot{p} = \dot{z} \gamma P_0 A / V_0 - \gamma p / M_0 \alpha \tag{5}$$

Rearranging, renaming and introducing the differential operator, D yields an expression for the pressure force acting on the column, f= p A.

$$f = kDz/(D + r)$$

where

$$k = \gamma A^2 P_0 / V_0 = \gamma A P_0 / L_\alpha \qquad r = \gamma A / M_0 \alpha \tag{6}$$

and $L_a = V/A$ is the air column effective length. Two parameters characterise the air column; k is the air spring, the pressure force per unit column motion when no air enters or leaves the chamber and r is the ratio of the air spring to the effective damping caused by the turbine.

The ratio of the air spring, k to the hydrostatic spring, K is

$$k/K = (\gamma A P_0/L)/\rho_w g A = \gamma P_0/\rho_w g L \simeq 14/L_a \qquad (7)$$

The final model of the air dynamics is shown in fig. 1. A spring, k and a damper, n in series represent the air compliance and the turbine energy extraction. The effective damping value n = k/r is determined by the turbine permittivity, α and other system parameters.

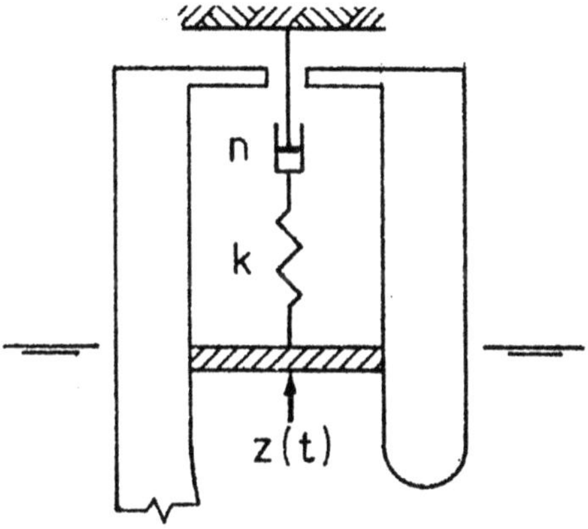

Fig.1 Linear model of air and turbine

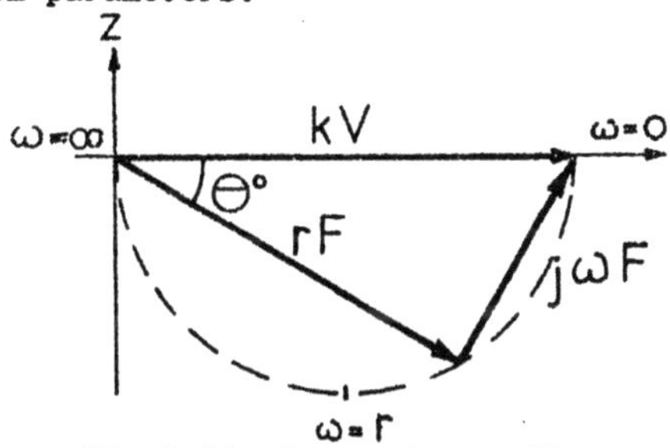

Fig.2 Air force phasor diagram

When the column oscillates in steady harmonic motion with complex amplitude, Z(jw), at frequency w the air force amplitude F(jw) is determined from equ.6.

$$F(jw) = jwk\,Z(jw)/(jw + r) = k\,V(jw)/(jw + r) \qquad (8)$$

This relation is shown as a phasor diagram in fig. 2. The pressure force lags the column velocity by a phase angle θ = tan (w/r)

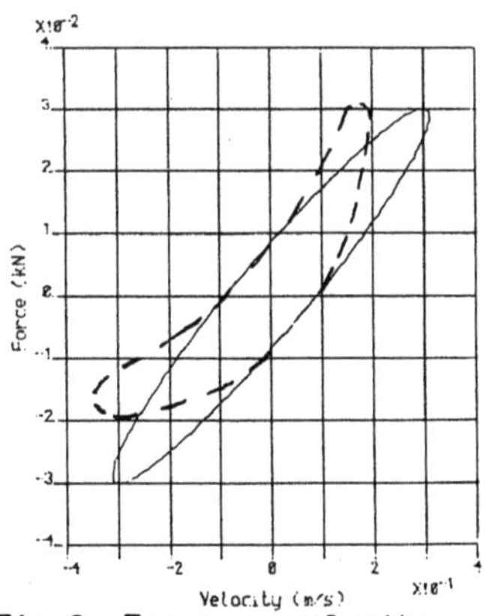

Fig.3 Force vs velocity

Figure 3 shows the plot of force against velocity when the column motion is harmonic, w/r=0.3; Penney et al. obtained very similar results from a full thermodynamic simulation for small to medium column motions. It would be easy to fit the parameters r and k to the curves generated by any thermodynamic model. If water column motions are a large fraction of the air column length, the air spring demonstrates a hardening (softening) characteristic for positive (negative) displacement, which cannot be represented by a linear spring; the dotted line in fig.3 shows the effect on the pressure - velocity characteristic.

In model scale experiments, air spring k must scale as X^2 with respect to a typical structure dimension, X. Hence A/L scales as X^2 since γ and atmospheric pressure are constant. The column area A must scale with X^2 to ensure hydrodynamic similarity so the effective column length above the water surface must be the same at full and model scale; of course, the desired result can be obtained by ensuring that the volume between the column and orifice is maintained at a value of AL. This effect was modelled in the trials of the Vickers submerged OWC (ref. 4) but seems to have been neglected in some subsequent work.

If the air is scaled incorrectly, the air stiffness and r will be large at optimal damping levels so the phase lag between column velocity and pressure force will be small at typical frequencies. The air/orifice system will look like a damper and the conventional analysis will apply. Heat transfer effects do not scale exactly so temperature fluctuations and the effective value of γ may not be modelled well at small scale. These phenomena must be investigated by further analysis or simulation.

Linearised system dynamics in the frequency domain

The column dynamics are modelled in the normal way, (ref. 5) assuming that the free surface remains plane; Evans (ref. 6) has shown that this is a good approximation so long as the incident wave is long with respect to the front to back dimension of the column. Intuitively, it is clear that the first 'slop' mode of the column must correspond to a wavelength near to twice the column dimension. Economically significant waves are usually longer than this for typical OWC designs; furthermore, slop degrades efficiency (ref.2) so the column dimensions must be chosen appropriately. High frequency pressure transients are bound to occur at slop mode frequencies but the uniform air pressure distribution cannot excite an asymmetric motion directly. Equally, asymmetric slop generates no net volume flux and cannot affect efficiency; at the first slop frequency, all the wave exciting force goes into slop generation so no net air flow results, leading to zero efficiency.

The conventional frequency domain equation of motion is

$$(- w^2 M(w) + jw\, N(w) + K).Z(jw) = R(w).H - A.P(jw) \qquad (9)$$

where M(w) and N(w) characterise the real and imaginary parts of the radiated wave force, the frequency dependent 'added mass and damping', R(w) is the exciting force per unit wave amplitude, K is the hydrostatic spring value, H is the incident wave amplitude and A.P(jw) is the air pressure force due to presssure P(jw) acting on the column area, A.

The pressure force F(jw) is substituted from equ.8 to express the amplitude of column oscillation Z(jw) in terms of the incident wave force.

$$[(rK - rw^2 M(w) - w\, N(w)) + jw(rN(w) + K + k - w^2 M(w))].Z(jw)$$

$$= (Re + jCo).Z(jw) = (r + jw).R(w).H \qquad (10)$$

The absorbed power is the average work done by the column.

$$\text{Power} = \frac{1}{2} \text{Re}\,[F(jw).\overline{jwZ(jw)}] = \frac{w^2 kr\, |Z(jw)|^2}{2\,(w^2 + r^2)} = \frac{w^2 kr\, |R(jw).H|^2}{2\,(Re^2 + Co^2)} \qquad (11)$$

Power is maximised at the frequency w_n, defined here as the resonant frequency, at which Re is zero.

$$K = w_n^2 (M(w_n) + N(w_n)/r) \quad (12)$$

$$w_n^2 = K/[M(w_n) + N(w_n)/r] \quad (13)$$

This tends to the conventional value as the air stiffness increases with respect to the damping and r becomes large. At this frequency, the absorbed power may be maximised with respect to r; power is proportional to

$$r/[\ rN(w_n) + K + k - w_n^2 M(w_n)\ ] \quad (14)$$

which is maximised when

$$r\ N(w_n) = K + k - w_n^2 M(w_n) \quad (15)$$

Substituting for M(w) from equ.12 and rewriting equ.15 in terms of turbine damping n = k/r yields a more familiar expression for the optimal damping.

$$n = N(w_n)/\ [\ 1 + w_n^2 n\ N(w_n)/k^2] \quad (16)$$

which again tends to the conventional value if k is large with repect to n. This implicit expression can be rearranged as a quadratic in n.

$$n^2 w_n^2 N(w_n)/k^2 + n - N(w_n) = 0 \quad (17)$$

The ultimate efficiency is unchanged by compressibility. These equations must be solved iteratively; compliance has a fairly small effect on the solution for short (less than 4m) air columns.

Comparison of Experimental and Numerical Results

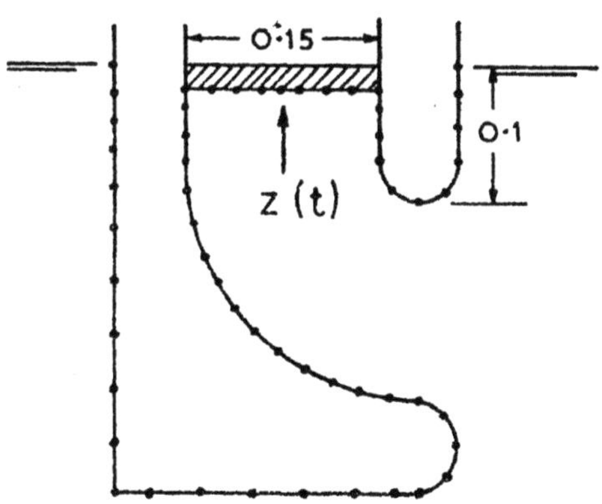

Fig.4 J column OWC

The device used to illustrate the theory in this paper and the subject of the experiments is a J column OWC, shown in fig.4.

Values of the added mass, damping and exciting force were calculated by a 2D diffraction program for the discretisation shown. A thin piston of water, which moves as a solid body is used to avoid numerical problems introduced when facets lie in the free surface; the added mass M(w) includes that of this piston.

Some authors put the boundary across the column mouth and add another term to the added mass. This technique is computationally economical but the fluid in the column does not all move with the same velocity in many OWC configurations; the column mass is not the required value.

The model OWC was tested in the narrow tank at QUB using monochromatic waves of constant steepness of 0.01 to avoid vortex losses at the duct entrance. Compressibility effects were investigated with two plenum chambers between the water surface and the adjustable orifice which models the damping. A 1:50 scale chamber gave a stiff spring while a chamber with an effective length of 5m represented the full scale air spring.

Measured and calculated wave exciting force are plotted in fig.5 and agree well. A resonant frequency was ascertained from the dynamic magnification of the undamped column. Applied damping was then adjusted at this frequency to determine the optimal 'stiff spring' damping. Pneumatic work done by the water surface was measured directly and used to define an effective damping. The OWC performance was then measured for both chambers at a range of applied dampings.

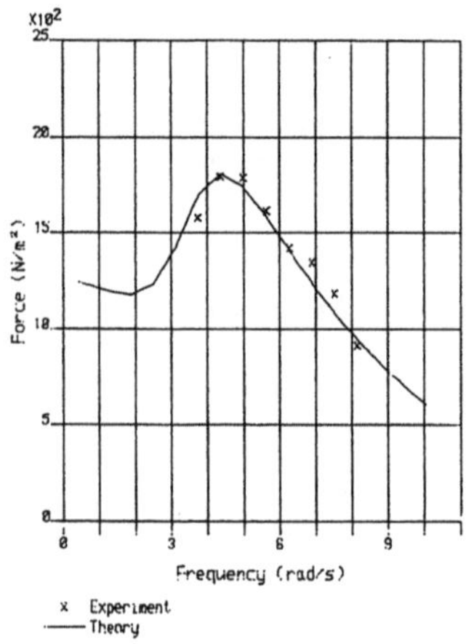

Fig.5 Measured and theoretical RAOs

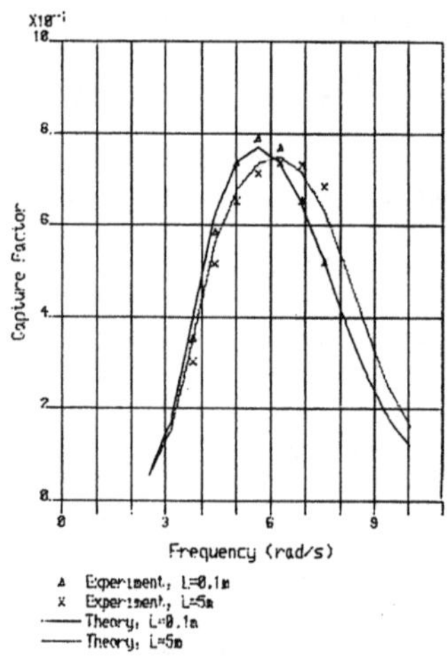

Fig.6 Theoretical and measured effect of compressibility

The performance with three times optimal damping at resonance applied to the air column was derived from the measured data and is plotted in fig.6. Continuous curves denote theoretical predictions and symbols indicate experimental points. Fair agreement was obtained and the predicted improvement in performance at high frequencies with the soft air spring was observed. The large column is still relatively stiff and the observed effects are not big enough to yield a value of effective isentropic exponent.

Latching theory and simulations including compressibility effects

Latching the air flow during the wave cycle might improve efficiency in the same way as the latching of the water column. An exact optimisation of the switching times is difficult but a simple approach based on the first harmonic gives accurate approximations and yields valuable physical insight.

Harmonic waves at frequency w excite the column. At some time during each cycle the air flow is halted by a valve whose action is assumed instantaneous. Optimal performance seems likely when the switching happens on both inflow and outflow alike since there is no asymmetry elsewhere in the system. The latching of the air will lead to a non harmonic column motion containing only odd multiples of the base frequency as the switching is symmetrical. The power is the sum of the products of the harmonics of velocity and force since the harmonics do not interact on the average.

$$\bar{P} = \sum_m \frac{2\pi}{mw} \int_0^{2\pi/w} F_m \sin mwt \times V_m \sin(mwt+\phi_m) dt = \sum_m F_m V_m \cos \phi_m / 2 \qquad (18)$$

The higher harmonics are smaller than the first, even for a substantially non sinusoidal motion; subsequent results show that the pressure force has typically 50% 3rd harmonic and that of the velocity is smaller, around 20%. Hence the 3rd harmonic contributes some 0.2x0.5 = 10% of the total power. A first harmonic analysis should yield switching times which are an excellent approximation to the optimal times.

The column motion is assumed cosinusoidal with an unknown amplitude Z_0; the air flow is halted at time $0<T_c<\pi/w$; it is subsequently released at T_0, where $T_c<T_0<\pi/w$ and the pattern is repeated on the second half of the cycle. Note that this algorithm is not completely general in that symmetry is enforced and only two latchings per cycle are permitted.

The pressure force obeys the equations

$$\text{(a)} \quad (D + r)\, f = k\, D\, z(t) \quad \text{or} \quad \text{(b)} \quad Df = k\, Dz \qquad (19)$$

These equations have the general solutions

$$\begin{aligned} &\text{(a) Open} && f(t) = A \exp(-rt) + B \cos wt + C \sin wt \\ &\text{(b) Latched} && f(t) = f(0) + k\,[z(t) - z(0)] \qquad (20) \end{aligned}$$

where A, z(0) and f(0) depend on the initial conditions and B and C depend on the column motion $z(t) = \mathrm{Re}[(Z + j0)\exp(-jwt)]$

$$B = k\, w^2 Z_0/(r^2 + w^2) \qquad C = -\, krw\, Z_0/(r^2 + w^2) \qquad (21)$$

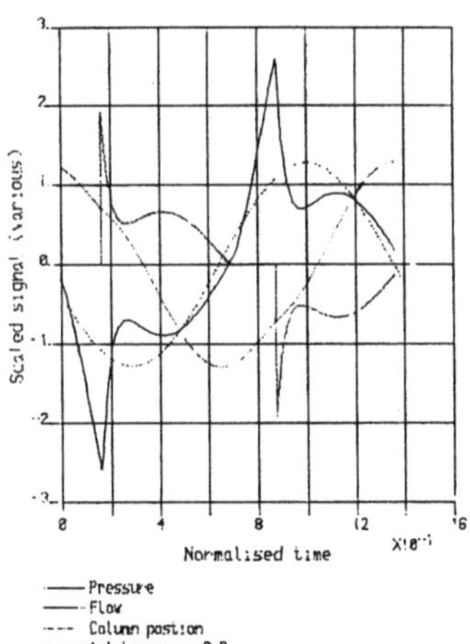

Fig.7 Column displacement, air flow and pressure signals

Values of f(t) during a half wave cycle are tabulated below in terms of these constants. The initial pressure force is unknown but the force at $t=\pi/w$ must be of the same size and opposite sign, which specifies f(0). The simulation of the OWC, discussed later, generated the values of pressure, its first harmonic, flow rate and column motion plotted in fig. 7. A harmonic wave of frequency 4.4 rad/s forced the column which was latched at the optimal times. The column moves nearly harmonically despite the latching; the signals generated assuming a harmonic column motion are almost indistinguishable from these. A relatively high damping of 140Ns/m, twice optimal, is needed to restrain the flow transients in a latched system.

Time t	Pressure force f(t)
$0 < t < T_c$	$(f(0) - B)\exp(-rt) + B\cos wt + C\sin wt \;=\; f(T_c)$ when $t = T_c$
$T_c < t < T_o$	$f(T_c) + kZ\,(\cos wt - \cos wT_c)$
$T_o < t < \pi/w$	$A\exp(-r(t - T_o)) + B\cos wt + C\sin wt \;=\; -f(0)$ when $t = \pi/w$

Some complicated algebra yields expressions for the unknowns f(0) and A.

$$f(0) = B - \frac{\exp(r(T_o - \pi/w))}{1 + \exp(r(T_o - T_c - \pi/w))}\,[(B - kZ_o).(\cos wT_o - \cos wT_c) + C(\sin wT_o - \sin wT_c)]$$

$$A = (B - f(0))\exp(r(\pi/w - T_o)) \qquad (22)$$

Fourier analysis yields analytic expressions for the 1st harmonic of f(t).

$$f(t) = P\cos wt + Q\sin wt \quad \text{where}$$

$$P = 2w/\pi \int_0^{\pi/\omega} f(t)\cos wt\,dt \qquad Q = 2w/\pi \int_0^{\pi/\omega} f(t)\sin wt\,dt \qquad (23)$$

The analytic expressions for P and Q, omitted here for brevity, were checked numerically by a simulation of a particular case and FFT of the resulting time history. The variables P and Q are linear in Z_o, the amplitude of motion; they may be calculated for a unit motion and substituted in equ.9 to yield an expression for the resulting amplitude.

$$[-w^2M(w) + jw\,N(w) + K).Z_o(jw) = R(w).H - (P - jQ)\,Z_o(jw)$$

$$|Z_o(jw)| = |R(w).H| / |(P + K - w^2M(w)) + j\,(wN(w) - Q)| \qquad (24)$$

The work done by the air column is equal to the component of the air force in phase with the column velocity.

$$\text{Power} = -0.5\,w\,Q\,|Z(jw)|^2 \qquad (25)$$

Given the parameters for a particular device, the power may be optimised numerically with respect to T_c, T_o. The NAG routine E04JAF was used in this application with the switching phases constrained between 0 and 2π.

Mechanical devices must usually be stationary when the latch is applied. The time T_c is not then a free parameter and must be found from the first equation of table 1. It is not clear that the column motion or excess pressure should be zero when the latch is applied; the results show that at frequencies below resonance, the optimal closure occurs when excess pressure is zero (see fig. 7); the valve remains closed so that the pressure restrains the column motion, effectively lowering the natural frequency. At higher frequencies, the valves close before maximum displacement is reached and open again when the column is moving down; the air spring stiffens the system and raises the effective natural frequency.

Optimal latching phases, referenced to the column motion are plotted in fig.8. They were evaluated in the frequency domain for an air column length of 5m and optimal damping; the applied damping has little effect on their value. At resonance, this latching technique cannot improve the response and the opening and closing times are indeterminate; so long as the closing and opening times are equal, they may take any value.

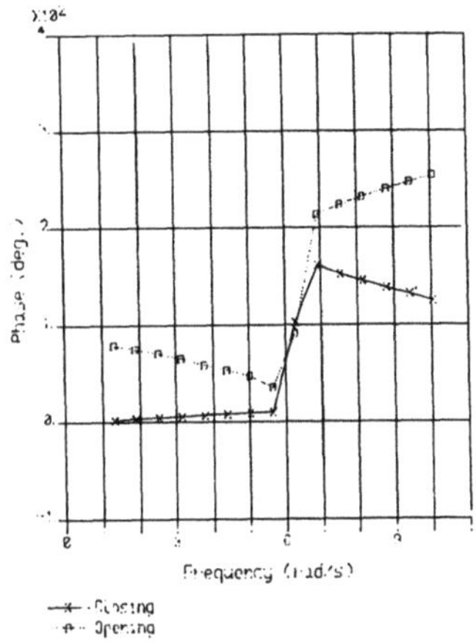

Fig. 8 Optimal latching phases

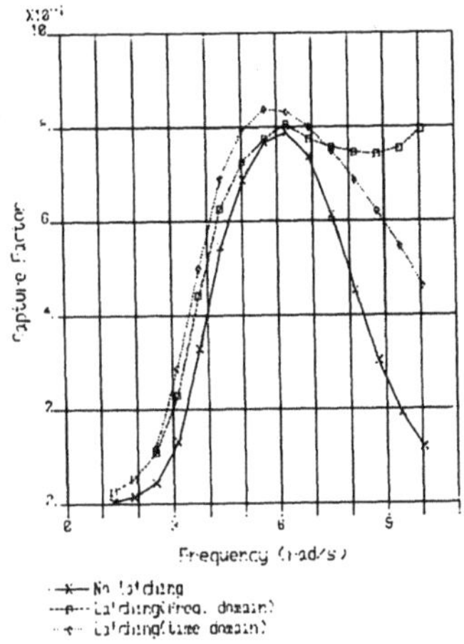

Fig.9 Performance gain from latching

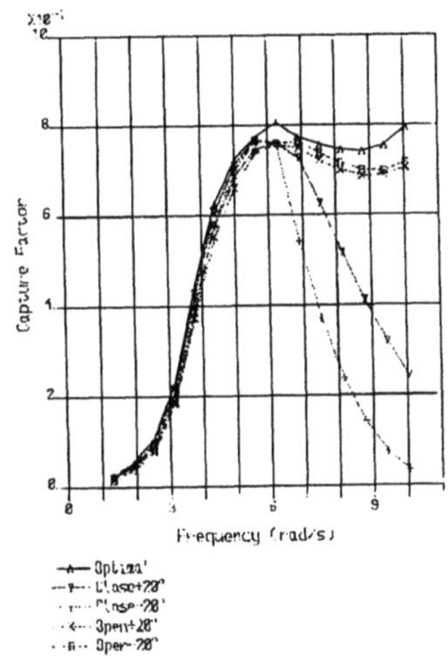

Fig.10 Sensitivity to latch phase

Capture factor with and without latching is plotted in fig. 9. Below resonance, the approximate frequency domain predictions agree well with the results of the simulation model, discussed below. At high frequencies, the higher harmonics ignored in the latching optimisation are more important and the agreement is less good

Changes of 20 deg. in the latching phases do not seriously affect performance in the important region below the device resonant frequency (fig. 10). It therefore seems likely that sub optimal latching algorithms based on heuristic conclusions about the optimal strategy will be able to realise most of the potential gains when the column is excited by random seas.

Simulation model

A digital simulation of the OWC was developed to validate the frequency domain model and derive time histories. Details of the technique are presented in ref.7; only the model is summarised here.

The motion of the column is due to the hydrostatic spring force, Kz, the exciting force, f_d, the radiation force (caused by the added mass and damping) f_r and the pressure force f which satisfies equ.20

$$d(z)/dt = \dot{z} \qquad d(\dot{z})/dt = (-K\, z - f_d - f_r - f)/ M(\infty) \qquad (26)$$

It is computationally convenient to model the radiation force by a subsidiary set of first order differential equations whose parameters are chosen to fit the added mass and damping curves (ref. 7).

$$\frac{d}{dt}\begin{vmatrix} x_1 \\ x_2 \end{vmatrix} = \begin{vmatrix} 0 & 1 \\ -19.5 & -4.86 \end{vmatrix}\begin{vmatrix} x_1 \\ x_2 \end{vmatrix} + \begin{vmatrix} 0 \\ 19.5 \end{vmatrix}\ddot{z} \qquad f = 20.22\, x_1 + 1.065\, x_2 \tag{27}$$

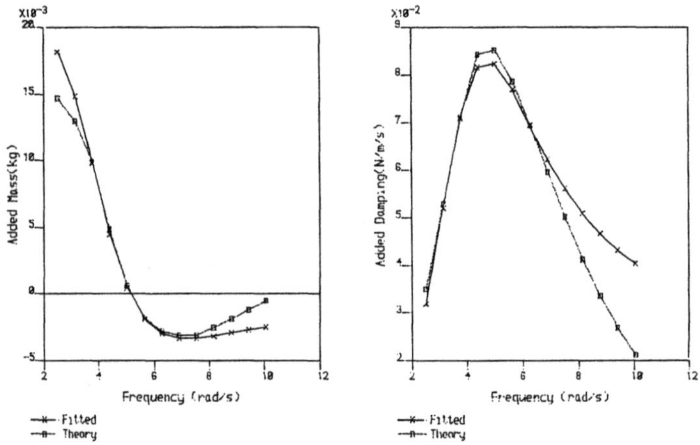

Fig.11a,b Correspondance between exact and fitted radiation force dynamics (a) Added mass, (b) Added damping

Figures 11 a and b show reasonable correspondance between the OWC added mass and damping (radiation force dynamics) and the real and imaginary part of the differential equation transfer function. More first order differential equations would yield a better fit. The model was validated by running simulations of the linear, unswitched system at check frequencies. Agreement between the responses and frequency domain predictions was better than 2%. It is hard to measure the phase of the first harmonic of the column motion since peaks and zero crossings are corrupted by higher harmonics; the latching times were therefore specified with respect to the known phase of the exciting force. Simulations and frequency domain predictions of efficiency agree well (fig.9)

Conclusions and further work

A simple model of air compressibility can guide more detailed simulations of OWC dynamics which include thermodynamic effects and a turbine model. Latching control could improve efficiency substantially but the algorithm will need careful optimisation to include realistic turbine and valve effects. The turbine must handle large transient flows efficiently.

Only a few results from the simple model have been presented here. Further exploration of the effects of compressibility and latching will be interesting.

The results suggest that simple sub optiomal control algorithms could yield substantial benefits in random seas; the best closing time occurs when pressure is zero in sinusoidal waves and this conclusion may extend to random seas. Given an estimate of the next wave force zero crossing time, an approximate opening time should be easy to calculate from the results shown in fig.8.

A more satisfactory approach may be to solve the full optimal control problem numerically; turbine characteristics and thermodynamic effects can be included and are likely to be important given the high transient flow rates and pressures present in a latched system. Such an optimal control algorithm for an OWC in a random sea has been developed by Hoskin, Nicholls, Nicol and Count (see this volume); the control law is not in feedback form so the optimal policy can only be calculated if the future of the exciting force is known, as has been assumed in this paper. In simulations or laboratory experiments, this is easily achieved but is rather more difficult at sea. Adaptive predictors (ref. 8) offer some hope of generating an estimate of the exciting force given the previous column motion, and perhaps, some wave pressure measurements from upstream of the device. Solution of the full optimal control problem in real time may be impractical but heuristic latching rules are likely to yield most of the potential gains, given the observed robustness of performance to switching phase.

Once a realisable control scheme has been formulated, there remains the formidable problem of engineering a reliable latching mechanism.

Acknowledgements

The authors would like to thank John Leitch of QUB for his help in running the experiments and diffraction programs.

References

1) Budal, K. and Falnes, J., 'Interacting Point Absorbers With Controlled Motion' in 'Power From Sea Waves' ed. B.M. Count, Academic Press, 1980.

2) Sarmento, A.J.N and Falcao, A.F. de O., 'Wave Generation by an Oscillating Surface-Pressure and its Application in Wave Energy Extraction', J. Fluid Mech. (1985), vol. 150, pp. 467-485.

3) Penney, T., McCormick, M. and Miles, J., 'Development Program Leading to an Optimised OWC Wave Energy Converter', Proc. AIAA 2nd Terrestrial Energy Systems Conference, 1981, Colorado Springs, USA.

4) 'Wave Energy Programme Feasibility Study on a Submerged Resonant Duct, Third Report', contract No. E/5A/con/1518, Vickers Limited, Engineering Group, Design and Projects Division, 1979.

5) Moody G.W., 'The NEL Breakwater - A Realistic Approach to Wave Energy Conversion', Proc. Marintec Offshore China, Shanghai 1983.

6) Evans, D.V., 'Wave Power Absorption by Systems of Oscillating Surface Pressure Distributions', J. Fluid Mech. (1982), vol.114.

7) Jefferys, E.R., 'Simulation of Wave Power Devices, Applied Ocean Research, 1984, vol.6, no. 1.

8) Jefferys, E.R. and Samra, B.S., ' Adaptive Prediction of the Motion of Marine Vehicles', Proc. 4th ASME OMAE conf., Dallas 1985.

Increase in Absorbed Wave Energy by the Phase Control for Air Flow on OWC Wave Power Device

H. HOTTA, T.MIYAZAKI, Y.WASHIO

Department of Ocean Development Research
Japan Marine Science and Technology Center
2-15, Natsushima-cho, Yokosuka-shi, 237, Japan

1. Introduction

As one of the methods to increase in efficiency of wave energy absorption by the point absorber wave power generator, the phase control was proposed by K.Budal et al.[1]. It is the method to make the float of point absorber oscillate as same as under the condition of resonance in any sea condition. Authors applied this method to the oscillating water column (OWC) type device and investigated the optimum control method, the increase in efficiency and the effect on an air turbine generator [2][3]. The principal results of these studies are reported in this paper.

2. Increase in air power

2.1 Mechanism of increase in air power

When the OWC wave power device is operated in waves, if the water column oscillates in a single degree of freedom (z;heaving), the equation of motion of it is expressed as follows.

$$(M+m)\ddot{z}+N\dot{z}+Cz+D(z,\dot{z},\dot{z}^2,\ldots) = Fw \quad (1)$$

where, M ; mass of the water column

m ; added mass of the water column

N ; wave making damping force coefficient

C ; restoring force coefficient

D ; damping force by the load device

Fw; wave exciting force

Since the force acting on the water column equals the hydrodynamic force ,it is calculated by next equation.

$$Fe = M\ddot{z}+Cz+D(z,\dot{z},\dot{z}^2,\ldots) \quad (2)$$

Mean value of absorbing power during one period ($\overline{E}$) is

$$\bar{E} = 1/Tw\int_0^{Tw} Re[Fe]\cdot Re[z]dt \qquad (3)$$

In this equation, Tw is wave period and Re[ ] means real part. If the load D is proportional to only z and $\dot{z}$, it can be expressed by the following equation

$$D(z,z) = d_1 z + d_2 \dot{z} \qquad (4)$$

We point out that the load by an orifice is proportional to $\dot{z}^2$ and the load by a Wells turbine is nearly proportional to $\dot{z}$.

When eq.(3) is calculated using eq.(4), term of mass and restoring force becomes zero and only damping term remains, therefore next equation is obtained.

$$E = 1/2\cdot\omega^2 d_N\cdot|Fw|^2 / [\{(C+d_1)-\omega^2(M+m)\}^2 + \omega^2(N+d_2)^2] \qquad (5)$$

where, $\omega = 2\pi/Tw$

In order to increase in $\bar{E}$, the following method is used. Firstly, it is a matter of course that Fw is increased. And secondly, there is a method to equalize N to d_2, which depends on the optimization of the air chamber and the load device. Further, there is a method to equalize $(C+d_1)$ to $\omega^2(M+m)^2$, which is the condition of resonance, and the phase control is applied by this method. In other words, the phase control is the method to make the water column oscillate just like resonance during $T_N/2$ (T_N; heaving natural period) while $0\sim\pi$ and $\pi\sim 2\pi$ of incident wave respectively, and restrained during another time.

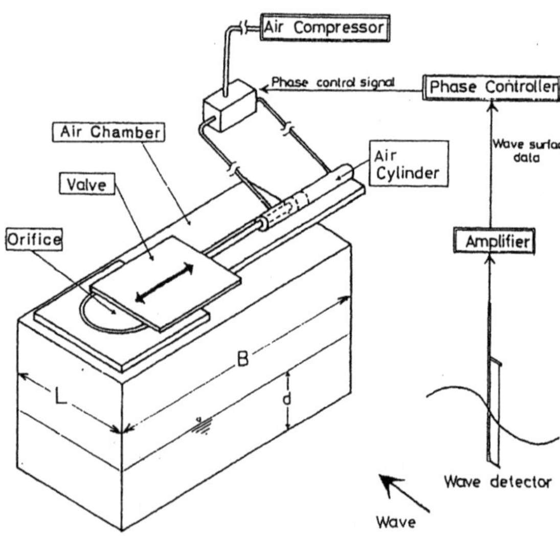

Fig.1 Scheme of the experiment model

This control is operated by following way: The wave detector set up at the side of the air chamber detects the incident wave surface and sends its data to the phase controller. This controller calculates and makes the signal to control (open and shut) the valve. This valve system is fixed on the air chamber as shown in Fig.1. When the valve receives this signal, it opens or shuts. This process is repeated so the motion of the water column becomes like the condition as shown in Fig.2 in regular waves.

2.2 Experiments and calculations

We carried out some experiments and calculations on the phase control with 2 and 3-dimentional air chamber where column oscillates in symmetrical motion in one degree. The width of the 3 -dimentional air chamber was about 1/5 of the width of the wave tank, and had an orifice as load. The width of the 2-dimentional air chamber is the same as the width of the wave tank, and the loaddevice was an orifice, too.

Since the motion of the water column is unsteady like this, the calculation was perfomed by the time domain equation of motion [4]. When the oscillation of the water column is only heaving (z-direction), the equation is

$$(M+m_\infty)\ddot{z}+\int_0^{t_c} Kz(t-\tau)\cdot\dot{z}(\tau)d\tau +(g\cdot\rho_w\cdot Aw)z+D = Fw . \qquad (6)$$

where, m_∞; heaving added mass at an infinite frequency

t_c ; convolution integral time

ρ_w; density of water

g ; acceleration of gravity

Aw; water line area of the water column

And Kz is the memory effect function of heaving water column, it is calculated by following equation.

$$Kz(t) = 2\rho_w\cdot g^2/\pi\cdot\int_0^{\infty}(\overline{A_H}^2/\omega^3)\cos(\omega t)\, d\omega \qquad (7)$$

where, $\overline{A_H}$; radiated wave height ratio by heaving

Some results of simulation on the motion of the water column controlled by the phase control was performed by working out of this equation using the newmark-β method which is one of the numerical integration. In this equation, the load D was calculated by following ways. When the valve is open and the water column is moving, the load equals the pressure drop produced by air flow at the orifice. And when the valve is closed and the water column is constrained, the load equals the restoring force produced by the compression stress of air in the air chamber.

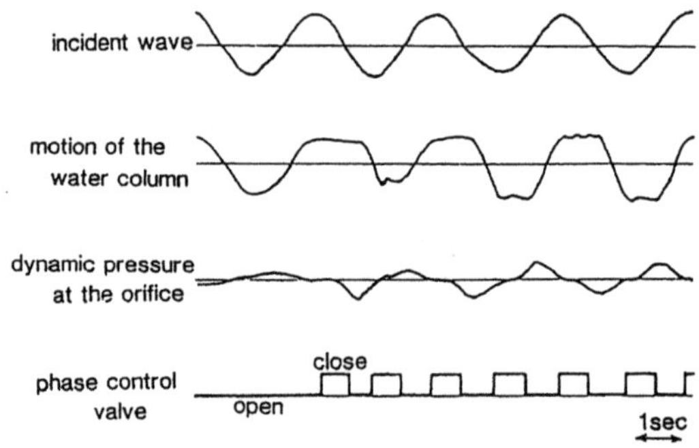

Fig.2 An example of Experiment data

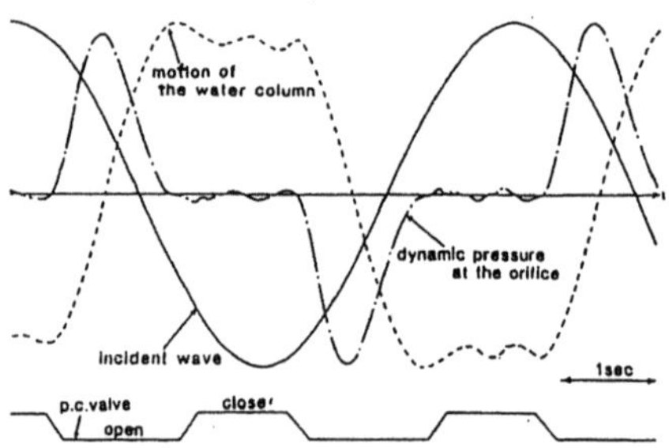

Fig.3 An example of calculated data

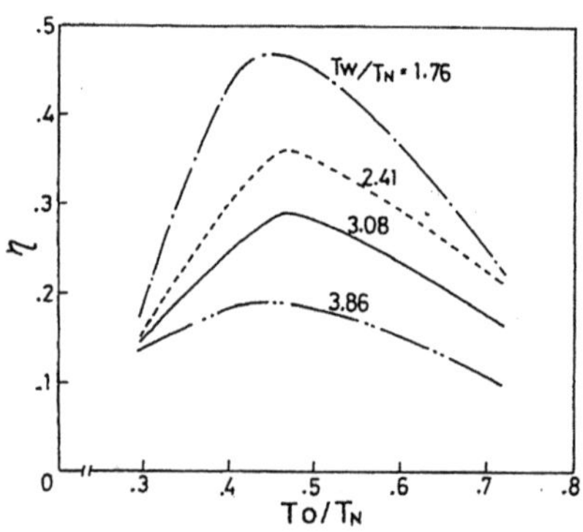

Fig.4 Optimum time during oscillation

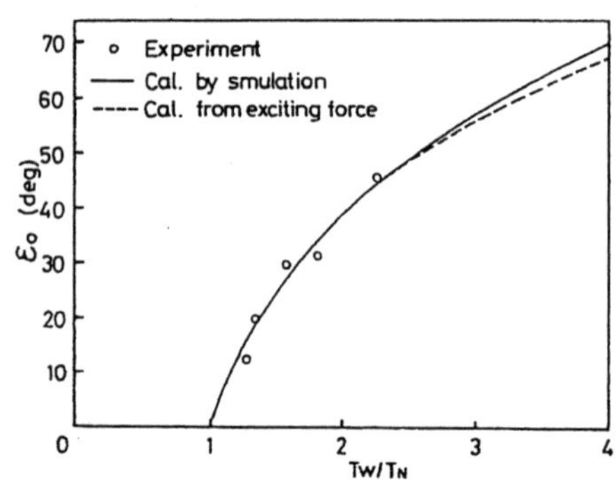

Fig.5 Optimum timing for releasing

In the calculation for the phase control, the load D is equals alternately those two forces. An example of simulation of motion by computation is shown in Fig.3.

2.3 Results

2.3.1 Optimum control timing

(1) time required for moving of the water column

As stated in 2.1, it is more effective to increase the motion of the water column in a state of resonance. So air output power increases due to a coincidence of the time requied for moving of the water column (To) and half period of the natural heaving period of it ($T_N/2$). Results of calculation in regular waves are shown in Fig.4. In this figure, η is wave energy absorption width ratio.

(2) phase of the released time of motion

In order to receive a wave exciting force as large as possible during To which is of constant value, the water column must be released at time Tp-To/2, where Tp is the time when the wave exciting force is maximum, and must be restrained at time Tp+To/2.

Results of experiments and calculations are shown in Fig.5. In this figure, ε_0 is phase angle between a zero cross time of incident wave level at the side of the air chamber and the released time of motion of the water column, and positive value means the phase of lag.

2.3.2 Increase in the air power

(1) 2-dimensional air chamber

The effect by the phase control for air flow on the 2-dimensional air chamber is shown in Fig.6. From this figure, it is clear that the air output power increases several times covering in a wide frequency band by the phase control in long wave period even on the OWC wave power device. This results are obtained in regular waves. But even in irregular waves, when the control goes well in response to each wave at optimum condition, there will be a hope for the improvement for the energy absorption ratio in such waves.

Though the calculated value was higher than the experimental value, it is a guess that there is an effect by eddys around the top of the plates of the air chamber.

(2) 3-dimensional air chamber

The results on the 3-dimensional air chamber are shown in Fig.7. Since the wave energy pours into the air chamber from the side, it was clear that the air output power increases remarkably in the waves of long period. But generally,the wave power device will be designed adjusting the natural oscillating period to the period of the wave which is most frequently observed at where the device will be set. Therefore, the scale of the OWC device is not so small, the phase control must not be operated in the waves whose period is under 2~3 times the natural oscillating period. So in order to clear the timing for control more exactly, the length of the air chamber must be smaller.

3. Effect on the turbine power

We considered some effects on an air turbine in the unsteady air flow caused by the motion of the water column as shown in Fig.3. When the turbine rotates in the air flow shut down by the valve,the turbine power will be lost by drag force acting on the turbine blade and frictional force acting on the bearing. Therefore, when we estimate the effect by the phase control, we must consider this loss. Now, we investigated the phase control air flow applied to the Wells Turbine in this paper.

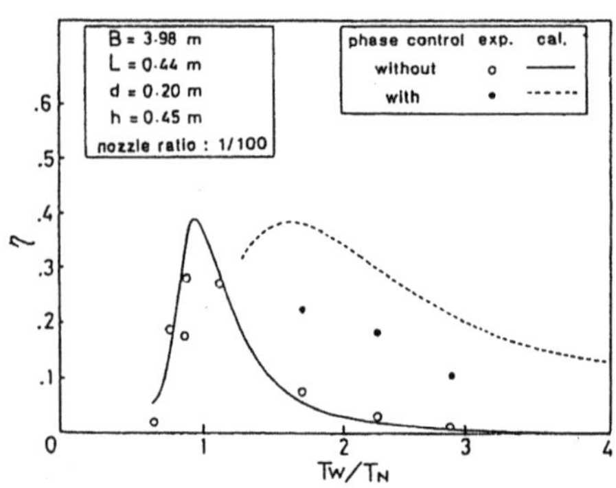

Fig.6 Efficiency by 2-dimensional model

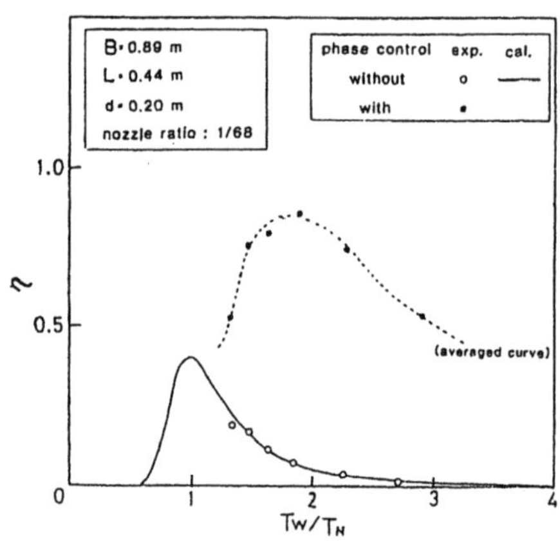

Fig.7 Efficiency by 3-dimensional model

3.1 Turbine-torque

The turbine-torque Tt gained in the air flow is

$$Tt = \rho a/2 \cdot Ct(\alpha) \cdot W^2 \cdot At \cdot Rt \quad . \tag{8}$$

where Ct; torque coefficient (tangential force coeff.)

α ; angle of attack [$\arctan(Va/(Rt \cdot \omega t))$]

Va; velocity of the airflow through the turbine

$W^2 = Va^2 + (Rt \cdot \omega t)^2$

ωt; angular velocity of rotation of the turbine

Rt; radius of the turbine

At; circular flow area of the turbine

Ct is a function of α , and is obtained by experiments. For instance, the value of Ct is shown in Fig.8. These are the data of the turbine set on "the Shore Fixed Wave Power Generator" [5] and also those presented by S.Raghunathan and et al.[6]. Va in eq.(8) depends on the incident wave height and period, and it is also altered by the phase control. When it is assumed that the oscillation of the water column and the variation of Va are sinusoidal, the peak value of α is calculated and an example of those results is shown in Fig.9.

The mean value of a turbine-torque in one period $\overline{Tt}$ is

$$\overline{Tt} = 1/Tw \int_0^{Tw} Ct(\alpha) \cdot \rho a/2 \cdot W^2 \cdot At \cdot Rt \quad . \tag{9}$$

When the turbine rotates at a constant high speed, W^2 nearly equals $(Rt \cdot \omega t)^2$. Therefore, $\overline{Tt}$ becomes

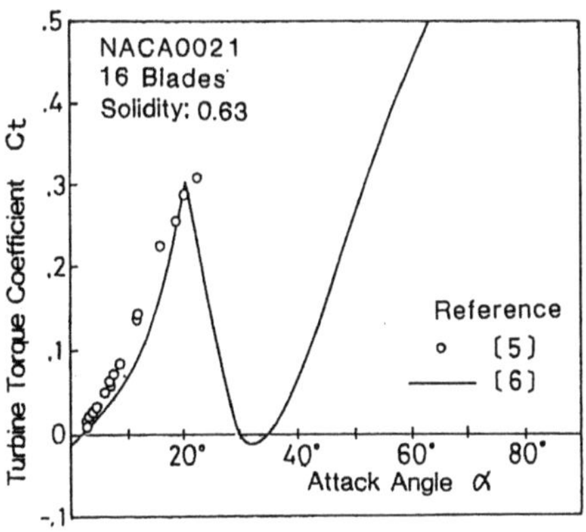

Fig.8 Turbine torque coefficient

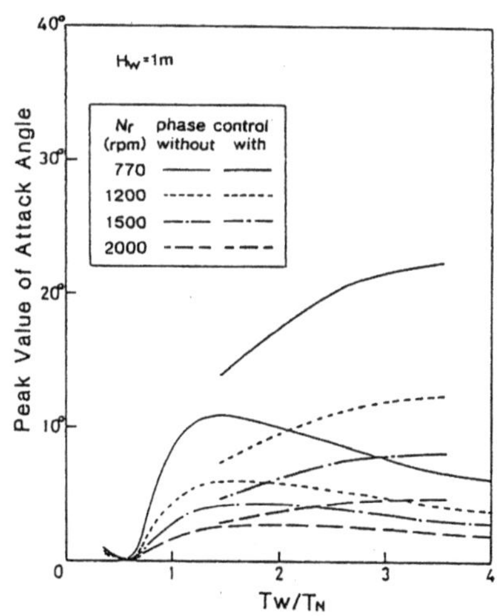

Fig.9 Peak value of attack angle

$$\overline{Tt} = \underbrace{\rho a/2 \cdot At \cdot Rt \cdot (Rt \cdot \omega t)^2}_{\text{dotted}} \cdot \underline{1/Tw \int_0^{Tw} Ct(\alpha)dt} \quad . \tag{10}$$

In this equation, since the part of a dotted line is constant value, we can estimate the effect on the turbine by only the mean torque coefficient $\overline{Ct}$ which is the result of the part of a solid line. When the phase control is operated and the valve is open during To and shut during Tc, $\overline{Ct}$ becomes

$$\overline{Ct} = 2/Tw \left\{ \int_{To} Ct(\alpha)dt + \int_{Tc} Ct(0^\circ)dt \right\}. \tag{11}$$

The second term on the right hand side of this equation is negative and is due to the phase control. Since the optimum value of To is equal to one half of the natural oscillating period of the water column, the time during Tc(α =0°) becomes long and $\overline{Ct}$ decreases in proportion to the wave period. As previously stated, $\overline{Ct}$ is a function of the rotational number of the turbine Nr(=$60 \cdot \omega t/2\pi$ (rpm)) and Va. Therefore, when we appraise effects on the turbine by the phase control, we must establish wave height (Hw), wave period (Tw) and Nr as parameter. Fig.10 is one of the results of calculation in regular waves. In this figure, in case of a upward line with an arrow, the control is efficacious. But in case of a downward line, it is inefficacious. Taking consideration of some results of these calculation, the turbine torque increases by the phase control under the condition of low wave height and long wave period.

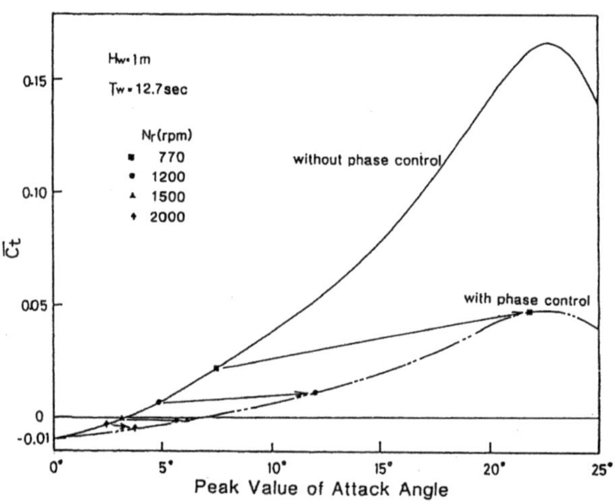

Fig.10 Effect on the air turbine by the phase control

4. Control in irregular waves

We proposed a new technical method to predict the release timing at the proper times, named "Wave propagation detection method" and the same experimental results with a heaving-buoy device model instead of an OWC device are examined [7]. Two wave gauges which measure the wave peak or trough were set in front of the device at distances of ℓ and $\ell+L$. The signals from these gauges were fed into the computer which calculated the propagation velocity and then gave out a trigger signal that released the buoy at the proper time. This method has the technical advantage of using the measurement of the real wave peaks which is not much affected by the device movement, and it deals with regular and irregular waves in the same manner

The phase control was quite successful in regular waves, the absorption efficiency being kept constant at 40% in the phase control region, and the incident wave width being twice the buoy width (diameter). On what concerns the phase control timing, the absorption efficiency was almost unchanged by taking a control timing ealier than the optimal timing, but the efficiency decreased as the control time was taken later. This means that the phase control timing is not strict in the early timing zone. For irregular waves, a result of maximum efficiency of 30% was achieved. The model test results were about 10% lower than the theoretical results in regular and irregular waves.

5. Conclusion

1) It was made sure that, even for the OWC type of wave power device, the wave energy absorption ratio increases with the phase control, this effect being especially remarkable in the 3-dimensional air chamber such as an isolated device.
2) The optimum phase control is achieved by making the timing to coincide with the time of appearance of the largest wave exciting force and the time of occurence of the maximum oscillating speed of the water column, and to make it oscillate during only half period of the natural heaving period in half period of a wave.
3) It was clear that the changes of airflow due to the phase control

control caused some losses on the air turbine, but it was effective in waves of long period and low wave height even on the air turbine.

4) In order to optimize the timing of the phase control, "Wave propagation detection method" was quite successful in regular and irregular waves in wave tank.

6. Acknowledgement

This work is one of the results of the research and development of the floating attenuator wave power device "Kaimei". We wish to thank Dr.T.Kinoshita and Dr. T.Arakawa, assistant professor of Tokyo Univ. and Dr.B.Count, head of special studies group for many valuable advices.

7. References

[1] K.Budal, et al.:The Norwegian Wave Power Buoy Project, Proceedings of the 2nd International Symposium on Wave Energy Utilization, 1982

[2] H.Hotta, et al.:Fundamental Study on Phase Control for Air-Flow to Improve on Wave Power Absorption (2nd report), Journal of the Kansai society of Naval Architects, No.194, September 1984 ,(Japanese)

[3] H.Hotta, et al.:Increase of Wave Energy Absorption Ratio by Phase Control, Proceedings of the 1st Symposium on Wave Energy Utilization in Japan, November 1984, (Japanese)

[4] B.M.Count :On the Physics of Absorbing Energy from Ocean Waves, Ph.Dr.Thesis, Dept. of Physics, University of Exerter, 1982

[5] Y.Washio, et al.:Full-Scale Perfomance Tests on Tandem Wells Turbine, Proceedings of the 1st Symposium on Wave Energy Utilization, November 1984, (Japanese)

[6] S.Raghunathan, et al.:Efficiency, Starting Torque and Prevention of "Run-Away" with Wells Self-Rectifying Turbines, The 2nd International Symposium on Wave Energy Utilization, 1982

[7] T.Miyazaki, et al.:Investigation of the Wave Propagation Detection Method in Phase Control, Proceedings Symposium of the 1st Energy Utilization, November 1984, (Japanese)

Simulation of a Pneumatic Wave-Power Buoy with Phase Control

G. OLTEDAL

Division of Experimental Physics.
University of Trondheim
N-7034 Trondheim-NTH. Norway.

Summary

A time-domain mathematical model is presented for a heaving buoy containing an oscillating water column. The model is prepared for numerical computation of the heave motion of the buoy and its power production. The hydrodynamical parameters of the device are converted from an approximation of rigid-body model parameters to applied-pressure model quantities, and they are expressed as kernel functions of convolution integrals. The non-causality of the kernel functions representing excitation quantities is discussed.

Introduction

The wave power team at the Norwegian Institute of Technology has for several years been working on optimisation of the motion of wave power converters [1-2] and a pneumatic phasecontrolled point absorber has been designed [3]. This paper presents a mathematical model describing in the time domain the hydrodynamics of the buoy as well as its power take-off system. The hydrodynamical performance of the device is computed from an approximation to the applied pressure model [4].

The wave energy converter

The main body of the converter is a freely floating spherically shaped hull with an opening to the sea in its lower end [3]. To obtain a fixed reference, a strut is passing through the buoy to a universal joint on the sea bed. The heave motion of the buoy is phase controlled by clamping the hull to the strut during one or two intervals of each wave period. The water column enclosed by the buoy acts as a piston forcing air through a turbine. As indicated in fig.1, the air flow is rectified by non-return valves and runs from a high pressure chamber through a turbine to a low pressure chamber. To some extent the air chambers have a smoothing effect on the air flow.

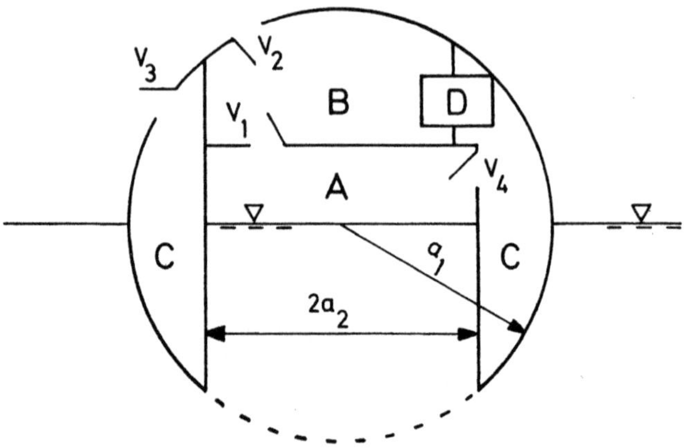

Fig.1. Schematic diagram of wave-power buoy with energy-extracting mechanism [3]. The hull is spherical with radius a_1 enclosing a cylindrical water column with radius a_2. The dashed line represents the inlet of the water column. A is the pump chamber, B and C are the high and the low pressure chamber respectively and D is the turbine. V_1, V_2, V_3 and V_4 are non-return valves rectifying the air flow.

Mathematical model

When the hull is free to move, its vertical acceleration $\dot{u} = \ddot{s}$ is computed from Newton's 2nd law. The equation of motion of the hull reads

$$m\dot{u} = \begin{cases} f_w + f_s + (p_A - P_0)A_2 & \text{(free buoy)} \\ 0 & \text{(fixed buoy)} \end{cases} \tag{1}$$

Here, m is the displacement of the hull, and $p_A - P_0$ is the dynamical pressure of the pump chamber with cross section $A_2 = \pi a_2^2$. The buoyance force f_s is non-linear due to the spherical hull, and is expressed as

$$f_s = -\rho g(A_1 - \pi\, s^2/3)s \tag{2}$$

where $A_1 = \pi a_1^2 - A_2$ is the water plane area of the hull, s its vertical position from equilibrium, ρ the density of water and g the acceleration of gravity. The total hydrodynamical wave force f_w includes excitation forces as well as reaction forces due to wave radiation. To allow for transient motion and non-linear power take-off computations are performed in the time domain. Hence, the wave force f_w is computed from convolution integrals as

$$f_w = \int_{-\infty}^{\infty} k_1(t')\eta(t-t')dt' - \int_0^{\infty} k_{11}(t')u(t-t')dt' - m_{11}(\infty)\dot{u}(t)$$

$$- \int_0^{\infty} k_{12}(t)p_A(t-t')dt' \qquad (3)$$

The first term represents the excitation force exerted by an incident wave with surface elevation η. The second convolution integral and the acceleration term describe the radiation force due to waves generated by the oscillating hull, and $m_{11}(\infty)$ is the infinite-frequency added mass of the hull. The last convolution integral represents the force originating from waves generated by the oscillating pump chamber pressure p_A. It should be noted that the excitation force is not a causal function [5], whereas the radiation forces are [6]. The kernels $k_1(t)$, $k_{11}(t)$ and $k_{12}(t)$ will be discussed later.

The rate of change of pump chamber volume V is given by

$$\dot{V} = A_2 u_1 - q_w \qquad (4)$$

where q_w is the total hydrodynamical volume flux expressed as

$$q_w = \int_{-\infty}^{\infty} k_2(t')\eta(t-t')dt' + \int_0^{\infty} k_{12}(t')u(t-t')dt'$$

$$- \int_0^{\infty} k_{22}(t')p_A(t-t')dt' \qquad (5)$$

The first integral is the excitation volume flux, the second and third integral represent the volume flux induced by radiated waves due to the oscillating hull and to the oscillating pump chamber pressure, respectively. From a known reciprocity relation [7-9] for the radiation problem in the frequency domain it follows that the kernels $k_{12}(t)$ of eq (4) and of eq (5) are identical.

The next step is to model the power take-off mechanism as indicated in fig.1. The first ingredient is the principle of conservation of energy which states that the rate of accumulation of internal energy plus the energy flux through valves out of an air chamber is balanced by the energy flux through valves into that chamber plus the mechanical power and the heat flow supplied by the surroundings. If the air in a chamber is assumed to behave like an ideal gas with uniform density and temperature, and if the power losses of the valves are neglected, a power balance for each of the three chambers A, B and C (fig.1) is established. For the state variables p (pressure), T (temperature), V (volume) and $\dot{Q}$ (heat flow) we use, subscript A for the pump chamber, B for the high pressure chamber, C for the low pressure chamber and 0 for the outer atmosphere. The principle of conservation of energy then reads for the pump chamber

$$\frac{\dot{p}_A V + p_A \dot{V}}{\gamma - 1} + \dot{m}_1 C_P T_A = \dot{m}_4 C_P T_C - p_A \dot{V} + \dot{Q}_A \tag{6}$$

where γ is the adiabatic constant, C_p the constant pressure heat capacity of air and $\dot{m}_n$ the mass flow through valve no. n (fig.1). Similar expressions are established for the high pressure chamber

$$\frac{\dot{p}_B V_B}{\gamma-1} + \dot{m}_t C_P T_B = \dot{m}_2 C_P T_0 + \dot{m}_1 C_P T_A + \dot{Q}_B \tag{7}$$

and for the low pressure chamber

$$\frac{\dot{P}_C V_C}{\gamma-1} + (\dot{m}_4+\dot{m}_3)\,C_P T_C = \dot{m}_t C_P T_t + \dot{Q}_C \tag{8}$$

where T_t is the outlet temperature from the turbine, and $\dot{m}_t$ is the mass flow through it. The heat flow into a chamber is computed from the thermal conductivity of its boundaries and the corresponding temperature differences across them.

The pneumatic power P_t extracted by the turbine equals the entalphy difference across it giving

$$\dot{m}_t = \frac{P_t}{C_P(T_B-T_t)} \tag{9}$$

The turbine performance is known empirically and it is expressed as a function of the pressure drop across the turbine, of the overall turbine efficiency η_t and of the speed of rotation. The inertia of the turbine is neglected and the turbine speed is assumed adjusted for optimum efficiency at any time. The outlet temperature is given by the pressure ratio and the mechanical efficiency η_m

$$T_t = T_B[1 - \frac{\eta_t}{\eta_m}\,(1 - \left(\frac{P_C}{P_B}\right)^{1-1/\gamma})] \tag{10}$$

In each of the chambers a mass balance is obtained by stating that the rate of accumulation of mass within a chamber equals the difference between mass flow into and mass flow out of that volume. The temperatures of the chambers are computed from the ideal gas law.

The last ingredient of the model is the operation of the non-return valves which are assumed to have negligible inertia. When the pressures of two adjacent chambers become equal, the connecting valve opens and it stays wide-open until a closing pressure occurs.

Time-domain hydrodynamical parameters

From frequency-domain forces and volume fluxes, time-domain quantities are computed by applying the inverse Fourier transform

$$f(t) = \mathcal{F}^{-1}\{F(\omega)\} = \frac{1}{2\pi}\int_{-\infty}^{\infty} F(\omega)e^{i\omega t}d\omega \tag{11}$$

provided the integral $\int_{-\infty}^{\infty}|F(\omega)|d\omega$ exists. Here, ω is the angular frequency, and i is the imaginary unit. $F(\omega)$ is a product of a hydrodynamical transfer function $H(\omega) = R(\omega) + iX(\omega)$ and an input function, which in our case is either the heave speed, the pump chamber pressure, or the surface elevation of the undisturbed incident wave. If $X(\infty) \neq 0$ existence of the inverse transform is achieved when applied to the difference $H(\omega) - iX(\infty)$. Setting $X_{11}(\omega) = \omega m_{11}(\omega)$ where m_{11} is the added mass of the hull, the radiation force on the hull due to its own oscillation may be expressed as

$$f_{11}(t) = -\mathcal{F}^{-1}\{[H_{11}(\omega) - i\omega m_{11}(\infty)]u(\omega)\}$$
$$-\mathcal{F}^{-1}\{i\omega m_{11}(\infty)u(\omega)\} \tag{12}$$

where $H_{11}(\omega)$ is the hull's radiation impedance, the real part of which is the radiation resistance $R_{11}(\omega)$. This equation (12) gives the second convolution integral and the acceleration term of eq (3). All the other transfer functions have zero high frequency limits, $H(\infty) = 0$ [10]. Hence, they contribute with only a convolution integral to the time-domain forces or volume fluxes. The integral kernels are computed from the inverse Fourier transform (eq. (11)) of a hydrodynamical transfer function with the high frequency limit subtracted, if necessary. If the transfer function is causal, the computation simplifies to a real transform which may be expressed as

$$k(t) = \frac{2}{\pi}\int_{0}^{\infty} R(\omega)\cos\omega t\, d\omega \tag{13}$$

The transfer functions describing the radiation properties are causal since the input is either a pressure or a speed which actually sets up the waves exerting the forces or volume fluxes. The transfer functions describing excitation quantities are, however, not causal since an incident wave actually hits the buoy and exerts an excitation force on the hull and an excitation volume flux in the pump chamber at an earlier moment than the moment at which the envisaged undisturbed incident wave would arrive at the origin of the system. Note that the envisaged wave elevation at the origin is used as the input function for the excitation transfer functions.

Numerical values of most of the hydrodynamical parameters for the applied pressure model are available [8], but only for a limited region of frequencies, and no numerical values at all are presented for the imaginary part of the radiation admittance of the pump chamber. Therefore a conversion of rigid-body model parameters is performed [10] by applying Newton's 2nd law to both models when the lower bounding surface of the water column is defined by the dashed line of fig.1. By assuming the two models to give the same vertical speed of the hull, the same pump chamber pressure and the same vertical speed u_2 of the water column ($u_2 = q_w/A_2$), the applied-pressure parameters are expressed as a function of the radiation impedance and the phase of the excitation force of the rigid body model and of geometrical quantities. The converted parameters contain information on the fundamental resonance frequency of the water column but not on its higher resonances. These parameters do not allow for spatial variation of the internal water surface as genuine applied-pressure parameters do. Hence, this approach gives an approximation to the applied-pressure parameters which is valid provided the wavelength is not too short.

It should be noted that the integral kernels of fig.2 include all information contained in both the real and imaginary parts of the corresponding frequency-domain parameters. For example, the radiation heave force kernel of fig. 2c carries the information contained in both the added damping and the added mass of the hull.

(2a)

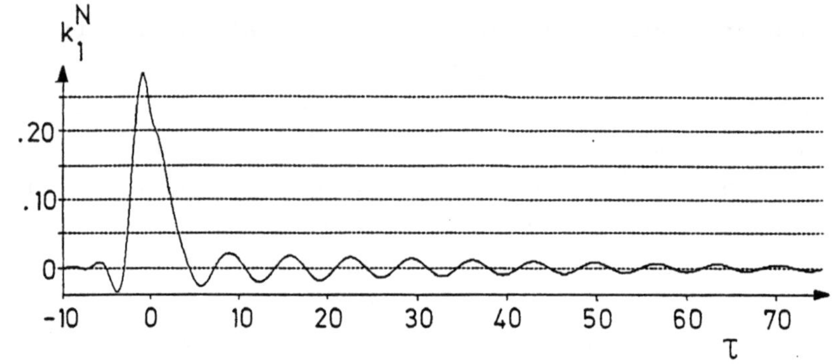
k_1^N
.20
.10
0
-10
0
10
20
30
40
50
60
70
τ

(2b)

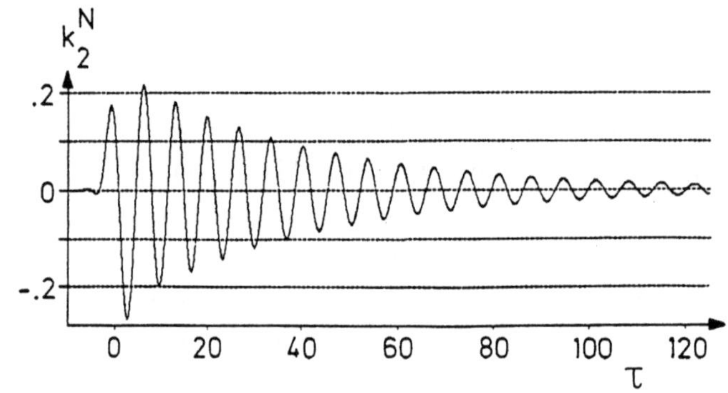
k_2^N
.2
0
-.2
0
20
40
60
80
100
120
τ

(2c)

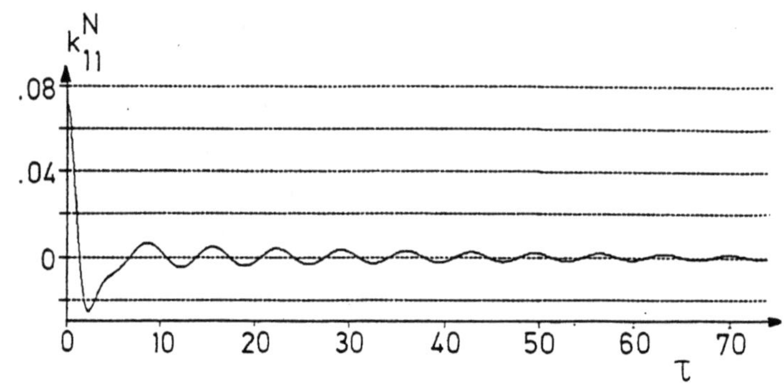
k_{11}^N
.08
.04
0
0
10
20
30
40
50
60
70
τ

(2d)

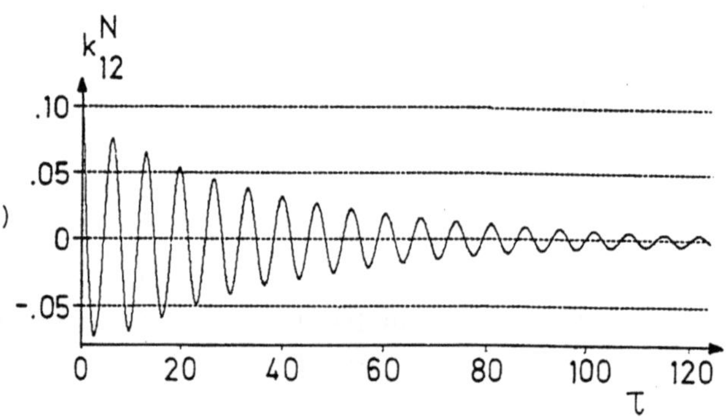
k_{12}^N
.10
.05
0
-.05
0
20
40
60
80
100
120
τ

(2e)

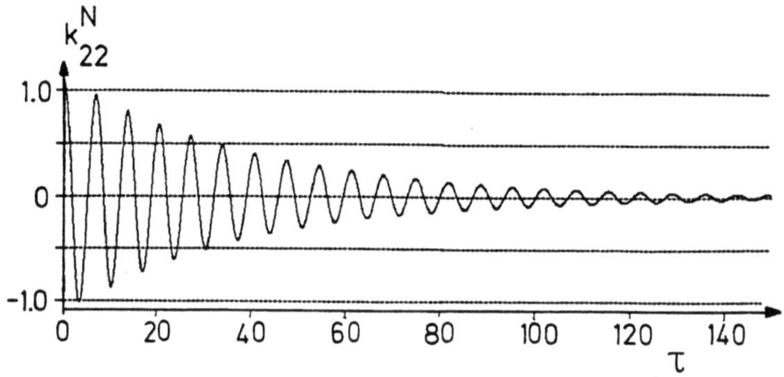

Fig.2. Normalised integral kernels versus non-dimensional time $\tau = \sqrt{g/a_1}\, t$ when $a_2/a_1 = 0.73$. 2a) is the excitation heave force kernel $k_1^N(\tau) = k_1(\tau)/\rho g A_1$. 2b) is the excitation volume flux kernel $k_2^N(\tau) = k_2(\tau)\sqrt{a_1/g}/A_2$. 2c) is the radiation heave force kernel $k_{11}^N(\tau) = k_{11}(\tau)/(\rho 2\pi a_1^3/3)$. 2d) is the hydrodynamical radiation coupling $k_{12}^N(\tau) = k_{12}(\tau)/A_2$. 2e) is the radiation volume flux $k_{22}^N(\tau) = k_{22}(\tau)\rho 2\pi a_1^3/(3A_2^2)$.

These integral kernels represent a rather strong memory of the past which is due to the inclusion of information of the resonance frequency of the water column in the applied-pressure parameters. This is different from a rigid body model of the water column where the information of the resonance frequency is contained in the equation of motion, leaving the integral kernels with a relatively short memory of the past. As an example is displayed in fig.3 the integral kernel for the radiation force on a heaving hemisphere. The kernel is computed by using the Fourier cosine transform (eq. 13) on the radiation resistance of the hemisphere [11].

The radiation impedance for the rigid body model with geometry as shown in fig. 1. is computed from empirical formulaes [12] and the phase angles of the excitation forces has been computed by Greenhow [13].

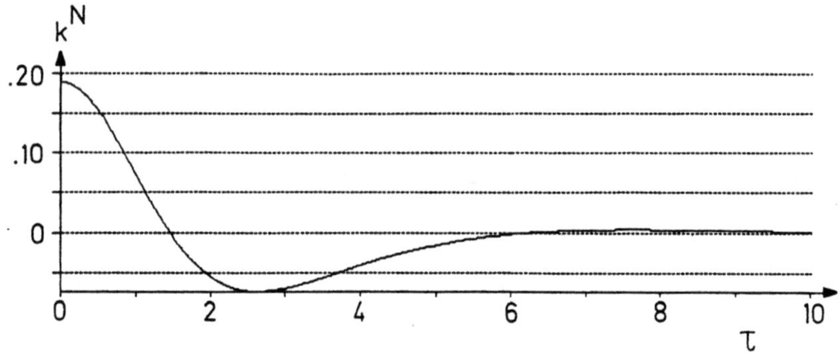

Fig. 3. Normalised integral kernel $k^N(\tau) = k(\tau)/(\rho 2\pi a^3/3)$ for the radiation force on a heaving hemispere of radius a versus non-dimensional time $\tau = \sqrt{g/a}\, t$.

Conclusion

A time-domain mathematical model of a pneumatic wave-power buoy is prepared for numerical computation of the heave motion of the buoy and its power production. The hydrodynamical parameters of the device, obtained from an approximation to the applied pressure model, are expressed as integral kernels of convolution integrals. The kernels describing excitation quantities are not causal, whereas those representing radiation properties are causal.

References

1. Budal, K. and Falnes, J.: A resonant point absorber of ocean wave power. Nature, 255 (1977) 478-479.

2. Falnes, J. and Budal, K.: Wave-power conversion by point absorbers. Norwegian Maritime Research. Vol.4 No.6 (1978) 1-11.

3. Budal, K., Falnes, J., Iversen, L.C., Lillebekken, P.M., Oltedal, G., Hals, T., Onshus, T. and Høy, A.S.: The Norwegian wave-power buoy project. Proceedings 2nd Int. Symp. on Wave Energy Utilization. Trondheim-NTH, Norway, 22-24 June 1982, 323-344.

4. Evans, D.V.: Wave-power absorption by systems of oscillating surface pressure distributions. Journal of Fluid Mechanics 114 (1982) 481-499.

5. Tick, L.J.: Differential equations with frequency-dependant coefficients. Technical Note, Journal of Ship Research, Vol. 3 No.3 (1959) 45-46.

6. Kotik, J. and Mangulis, V.: On the Kramers-Krønig relation for ship motions. Int.Ship Prog. Vol.9 No.97 (1962).

7. Falnes, J.: Unpublished lecture note from informal seminar, Trondheim, 15th April 1982.

8. Fernandes, A.C.: Analysis of an axisymmetric pneumatic buoy by reciprocity relations and a ring-source method. Ph.D. thesis, MIT, Des. 1983.

9. Falnes, J. and McIver, P.: Wave interaction with oscillating bodies and water columns. This symposium.

10. Oltedal, G.: Pneumatic wave-power conversion by a phase-controlled buoy. Dr.ing. thesis under preparation.

11. Havelock, T.: Waves due to a floating sphere making periodic heaving oscillations. Proc. Royal Soc., 231a (1955) 1-7.

12. Falnes, J.: Technical Note: Comments on "Added mass and damping of a sphere section in heave.", Applied Ocean Research, 6 (1984) 229-230.

13. Greenhow, M.J.L.: The hydrodynamic interactions of spherical wave-power devices in surface waves. "Power from Sea Waves", edited by B. Count, Academic Press 1980. 287-343.

Dynamic Response and Input Adaptive Maximum Energy Converting Control of an Ocean Wave Energy Converter

M. MASUBUCHI and R. KAWATANI

Department of Mechanical Engineering
Faculty of Engineering
Osaka University
Yamadaoka, Suita
Osaka, JAPAN

Summary

This paper presents an adaptive load control strategy for a floating body type wave energy converter. Two nonlinear control laws for large oscillating movement between floating body and connecting link are applied and compared. Good control response assures maximum energy conversion and stable operation for high incident wave height.

1. Introduction

A number of wave energy extraction methods have been devised since so-called energy crisis in 1973[1] and among them a method using the oscillating motion of floating bodies represented by Salter's Duck[2] can be cited as one of the most promising systems. A system devised by one of the authors[3] belongs to this category and its feature is to use the oscillating motions of several floating bodies. The merits of this system are: 1) energy converting efficiency is high over wide frequency range, 2) to extract energy, no direct fixed point is necessary, only mooring is needed. This system has been confirmed to have an excellent energy converting characteristic for regular waves by theoretical and experimental researches[4~6]. In the energy converter systems of this kind , the motions are generally converted into electrical energy by using secondary energy converter such as hydraulic turbines and electric generators etc.. To convert wave energy at the maximum efficiency, the amplitudes of oscillation for incident waves should be maintained at some values and this condition will mostly be satisfied at the design stage where the wave conditions are known at a given coastal region.

However, meteorological variations frequently give several effects such as wave height, wave spectrum and wave directions to ocean waves. Above all, wave height variation give large effects on the energy converter and in some cases destructive power. This paper proposes some control strategies to adaptively maximize the energy absorbing rate and to limit the amplitude of body oscillation in spite of the variation of incident wave height. Results of numerical simulations are also shown.

2. Wave Energy Converting System with Two Floating Bodies (WECS-2)

2.1 Basic characteristics[5]

By the simulation study of several systems, it has been proved that high energy converting efficiency over wide frequency range is obtained when large and small sized two floating bodies are combined as shown in Fig.2.1. This system is called WECS-2 in this paper.

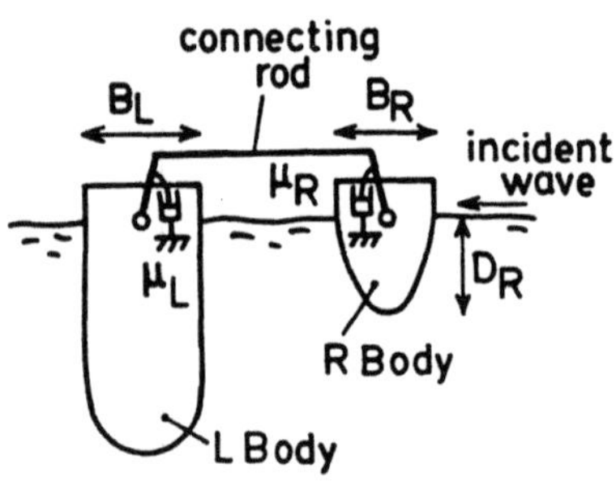

Fig.2.1 Wave energy converter system

Sinusoidal incident wave are transformed into relative motions between the link and floating bodies. Transformed mechanical energy is assumed to be absorbed in the two mechanical dampers which absorb energy proportional to the relative velocity.

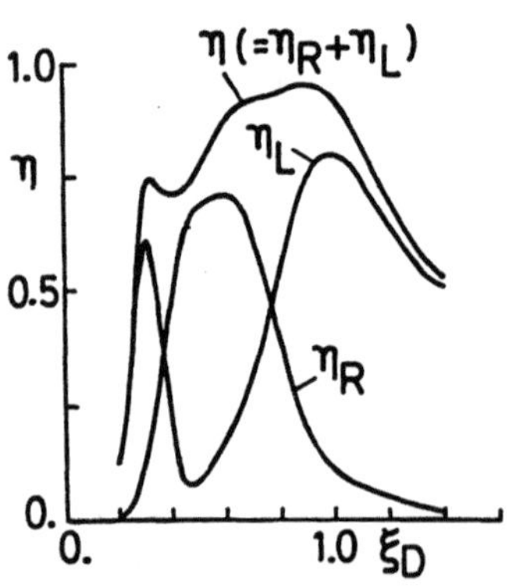

Fig.2.2 Energy converting efficiency

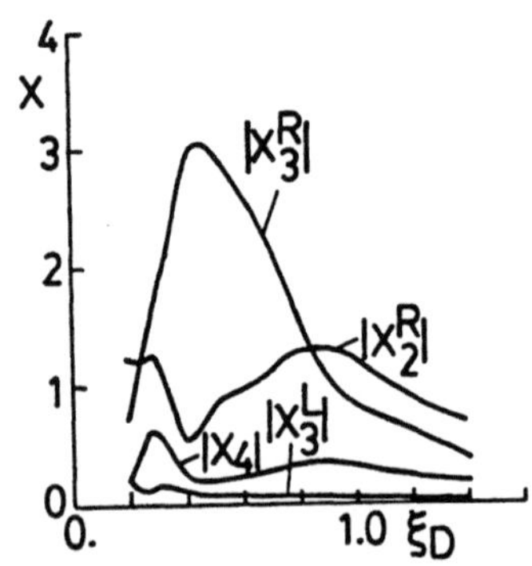

Fig.2.3 Amplitudes of R and L Body

Since several assumptions for deriving basic equations are left to the literature, only energy converting efficiency is described here. Figs.2.2 and 2.3 show energy converting efficiencies for sinusoidal incident wave and amplitudes of motion of two bodies, respectively. Floating body configurations follow Lewis Forms and the parameters are shown bellow.

for R Body : $(Ho,\sigma)_R=(0.5,0.75)$ $\mu_R'(=\mu_R/\rho\omega_0 B_R^4)=0.153$

for L Body : $(Ho,\sigma)_L=(0.25,0.95)$ $\mu_L'(=\mu_L/\rho\omega_0 B_L^4)=0.959$

and $l'(=l/B_R)=3.71$ **nondimensional length**

$n(=B_L/B_R)=1.5$ **width ratio**

$\xi_D=\omega^2 D_R/g$ **nondimensional frequency**

$\rho=$ **fluid density**

$\omega_0=$ **reference frequency which is taken to be 0.7.**

μ_R', μ_L' and l' are numerically searched to make the mean efficiency $\bar{\eta}$ be maximum for nondimensional frequency $\xi_D = 0.3 \sim 1.0$.

In Fig.2.2, η is the total efficiency, η_R and η_L each shows the efficiency at R and L Body, respectively. As seen from Fig.2.3, the motions of R Body predominate and the amplitude of L Body is small; this can be understood by the condition $n = 1.5$, that is, L Body operates as a kind of wave reflector. This is the reason why this system has high energy converting efficiency. When the incident wave height increases greatly, the amplitudes of heaving and rolling motions of R Body will increase so much that it becomes necessary to limit them for stable operation.

2.2 Means of control

This system is robust for the change of wave spectrum as explained above. To control the effect of incident wave height, it may be necessary to detect the wave height variation indirectly, for example, by measuring the relative rolling amplitude or velocity between the floating body and connecting link. Because, direct precise measuring of wave height variation on the sea surface will be in most cases impractical.

For the control variables which have the most direct effect on the energy conversion rate, two damper parameters μ_R and μ_L are used here.

3. Selection of Control Strategy

Since the above WECS-2 is a complicated variable parameter system, a simple fixed parameter mass-spring-dashpot system is considered to simplify the selection of the optimum types of control strategy.

Fig.3.1 shows the schematic diagram of this m-k-c system and the nondimensionalized governing equation for sinusoidal forcing input $P_0 \sin\omega t$ is

$$\frac{d^2x}{d\tau^2} + 2\zeta\frac{dx}{d\tau} + x = \frac{Po}{ky_o}\cos(\lambda t) \tag{3.1}$$

where

$$x = y/y_o,\ \tau = \omega_n t,\ \omega_n = \sqrt{k/m},\ \zeta = c/(2\sqrt{mk}),\ \lambda = \omega/\omega_n \tag{3.2}$$

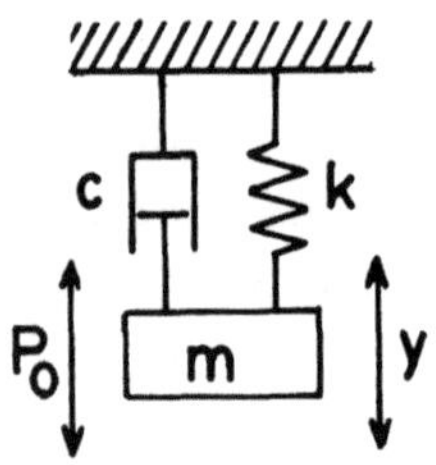

Fig.3.1 A simple mass-spring-damper system

y_0 is the upper limiting value of the amplitude of motion which is determined at the design stage, therefore, $x_{max} = 1.0$.

In this system, it is assumed that x is the controlled variable to be controlled within the limiting value x_{max} and damper parameter ζ is the manipulated variable. Fig.3.2(a),(b) show sinusoidal forcing and its response x , respectively.

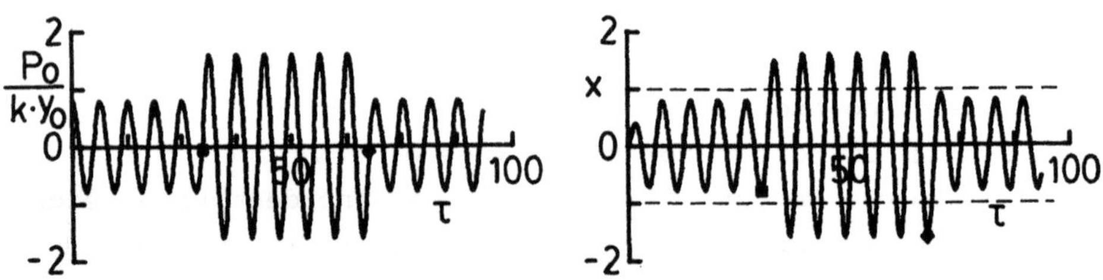

(a) (b)

Fig.3.2 Response for sinusoidal forcing

3.1 Control Law (I)

$$\zeta(\tau+\Delta T)=\zeta(\tau)+K_1(|x(\tau)|-|x(\tau-\Delta T)|)$$
$$\text{for } |x(\tau)|>x_{max} \quad \text{and} \quad |x(\tau)|>|x(\tau-\Delta T)| \qquad (3.3)$$

$$\zeta(\tau+\Delta T)=\zeta(\tau)-K_2(x_{max}-|x(\tau-\Delta T)|)$$
$$\text{for } |x(\tau)|<|x(\tau-\Delta T)|>|x(\tau-2\Delta T)| \quad \text{and} \quad |x(\tau-\Delta T)|<x_{max} \qquad (3.4)$$

where ΔT is a sampling time, K_1 and K_2 are gain constants.

Parameter ζ is controlled when $x(\tau)$ exceeds the limiting value x_{max} and it will become large when the difference $|x(\tau)|-|x(\tau-\Delta T)|$ corresponding to the exceeding speed becomes large. But, if ζ is set too large value, the subsequent response amplitudes are reduced below x_{max}, then the second control law is applied. That is, the damper parameter is reduced as described by equ.(3.4) until it reaches the minimum value ζ_0 which corresponds to the optimal value.

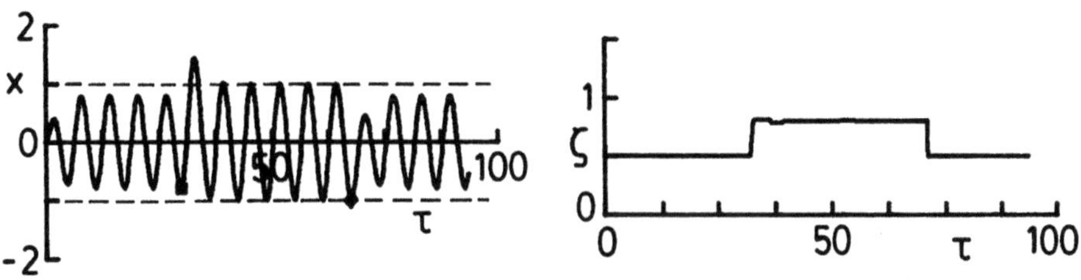

(a) (b)

Fig.3.3 Responses with control law (I)

Fig.3.3 shows the response of amplitude x and its manipulated variable ζ for gain constant $K_1 = K_2 = 0.7$ whose selection is shown later. Symbols ■ **and** ◆ correspond to the instant of step change of input amplitude in Fig.3.2(a). It is seen that good response is observed except for the first overshoot.

3.2 Control Law (II)

To get better control performance, consider state plane where velocity $\dot{x}(\tau)$ is also added besides $x(\tau)$. Upper limiting value is described by a circle with radius 1. Then, we get a new variable by taking the difference

$$z(\tau)=\sqrt{(x(\tau))^2 + (\dot{x}(\tau)/\lambda)^2} - 1 \qquad (3.5)$$

and control law (II) is defined in the following

$$\zeta(\tau+\Delta T)=\zeta(\tau)+K_3 z(\tau)+K_4(z(\tau)-z(\tau-\Delta T))$$

$$\zeta(\tau+\Delta T)=\zeta_o \quad \text{for} \quad \zeta(\tau+\Delta T)<\zeta_o \tag{3.6}$$

This control law corresponds to a digital PI control. Fig.3.4 shows the simulated responses for gain constants $K_3 = 0.6$, $K_4 = 10.0$. It is seen that the defect of control law (I) is completely solved and that the response x is excellent without any overshoot.

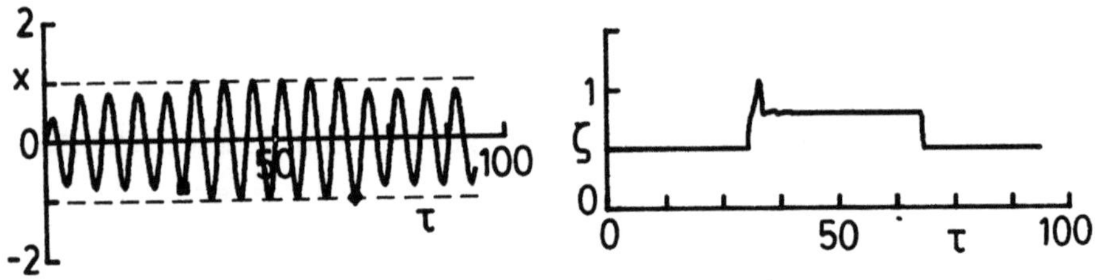

(a) (b)

Fig.3.4 Responses with control law (II)

In using these control laws, the control performance is so influenced by the gain constant that its optimal selection is very interesting. The ITAE criterion which is well known for its selectivity and which is widely used for control system design[7)] is adopted for two performance criteria S_E and S_D. Namely

$$S_E=\int_{\tau_o}^{\infty}(\tau-\tau_o)|x-x_\infty|d\tau \quad , \quad S_D=\int_{\tau_o}^{\infty}(\tau-\tau_o)|\zeta-\zeta_\infty|d\tau \tag{3.7}$$

where τ_0 is the instant of step change of forcing input, x_∞ and ζ_∞ are the reference values of x and ζ respectively.

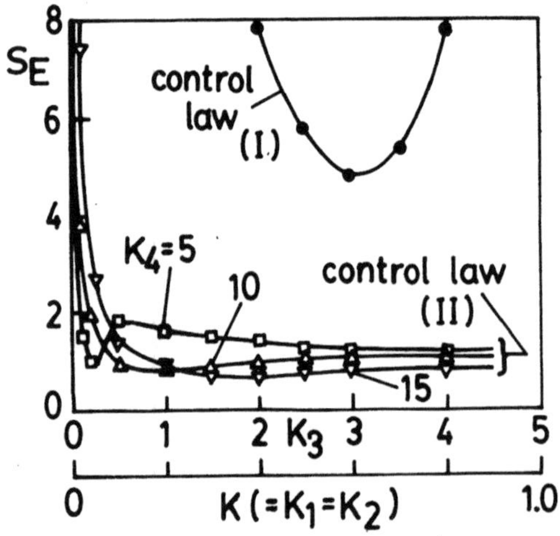

Fig.3.5 ITAE of x, S_E vs gain constant

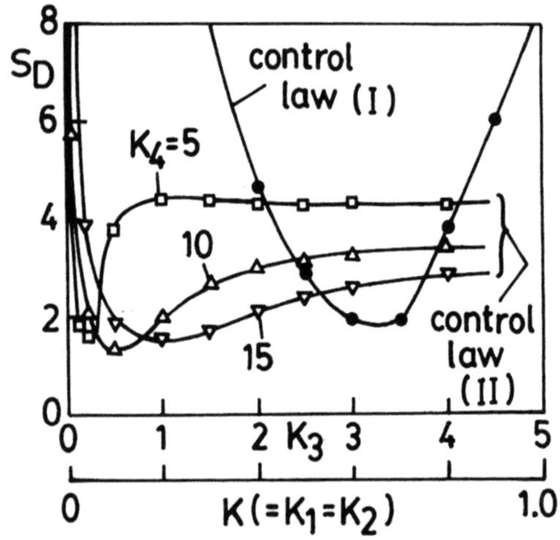

Fig.3.6 ITAE of ζ, S_D vs gain constant

Fig.3.5,3.6 show the values of S_E and S_D for various values of gain constant. For control law (I), $K_1 = K_2 = K$. In Figs.3.3,3.4, these optimal values $K_i (i=1,\cdots,4)$ are used.

4. Adaptive control of WECS-2

Two control laws discussed in the former section are applied to WECS-2. Fig.4.1 shows the schematic diagram of the control system in which the system enclosed by a dashed line is only considered in this section.

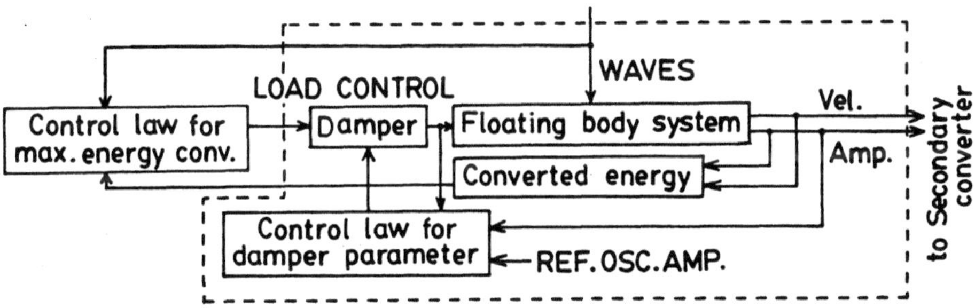

Fig.4.1 Total control system

Basic equations of motions of two floating bodies connected by a link whose mass is neglected for simplicity are given as

$$\underline{M}\cdot\ddot{\underline{x}} + \underline{D}\cdot\dot{\underline{x}} + \underline{K}\cdot\underline{x} = \underline{f}\cdot Ae^{i\omega t} \qquad \underline{x}=[x_1^R\ x_2^R\ x_3^R\ x_3^L\ x_4]^T \tag{4.1}$$

where $x^R_{1,2,3}$ = amplitude of swaying, heaving and rolling motions of R Body, respectively

x_3^L = amplitude of rolling motion of L Body

x_4 = amplitude of rolling motion of link

$\underline{M}$ = matrix of mass terms including added mass

$\underline{D}$ = matrix of viscous terms consisting of wave damping coefficients and damper parameters μ_R and μ_L

$\underline{K}$ = matrix of restoring force terms due to buoyancy

$\underline{f}$ = forcing term due to incident waves

A = amplitude of incident wave

Putting $\underline{x}_1 = \underline{x}$, $\underline{x}_2 = \dot{\underline{x}}_1$ in equ.(4.1), nondimensionalizing equ.(4.3) by dividing with body width B_R of R Body and by using $\tau = \omega t$, $\bar{A} = A/B_R$, we get

$$\dot{\underline{X}} = \underline{A}\cdot\underline{X} + \underline{B}\cdot\bar{A}e^{i\tau} \tag{4.2}$$

By numerically solving this equation by using Runge Kutta Method under appropriate initial conditions, the responses of the floating bodies for sinusoidal waves will be known. The object of control is to reduce within upper permissible values the relative rolling amplitudes $x_3^R - x_4$ and $x_3^L - x_4$ of R and L Bodies, respectively, which increase with the increase of amplitude $\bar{A}$ by manipulating damper parameters μ_R and μ_L in matrix $\underline{D}$.

First consider the system in Sec.2.1. Sinusoidal waves are supposed to be incident at $\xi_D = 0.7$. The amplitude of incident wave is assumed to vary stepwise from the first value $\bar{A} = 0.2$ to some large value $\bar{A} = 0.5$ as in the former section.

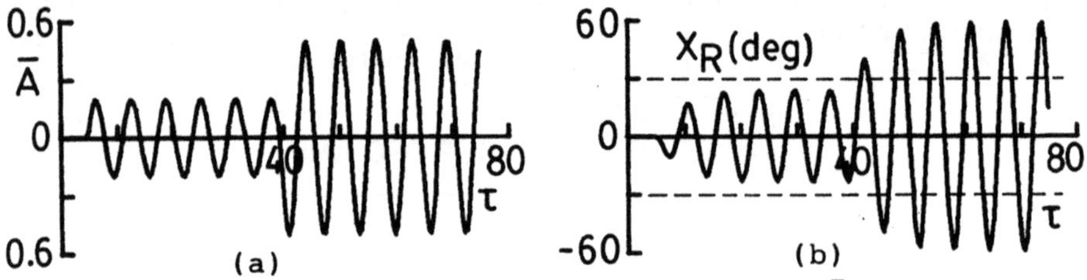

Fig.4.2 Response without control action for $\bar{A}$=0.2 to 0.5

Fig.4.2 shows the responses of the relative rolling motion for incident wave height variation when the system is at optimal state but has no control action. It is seen that fairly large amplitude response will be caused. As for the upper limiting value of oscillating amplitude, 30 deg. is taken as an example of permissible upper value, which is indicated in figure (b) by two broken lines. It is clearly shown that for large amplitude $\bar{A} = 0.5$ the control of damper parameters becomes necessary.

Then, the control laws (I) and (II) are applied to the example of Fig.4.2 in which cases x is replaced by $x_3^R - x_4$ or $(x_3^L - x_4)$ and ζ is by μ_R (or μ_L) in equs.(3.4) and (3.6). The results for control law (I) are shown in Fig.4.3 and those for (II) in Fig.4.4.

The responses for both control laws rapidly approach the upper limiting values just as is seen in the former Section, particularly, it is found that the control law (II) is superior. It is confirmed that for other wave frequencies the

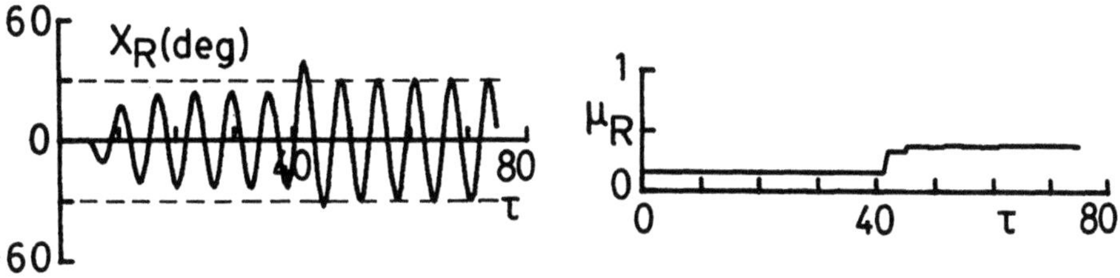

(a) (b)

Fig.4.3 Control response with control law (I)

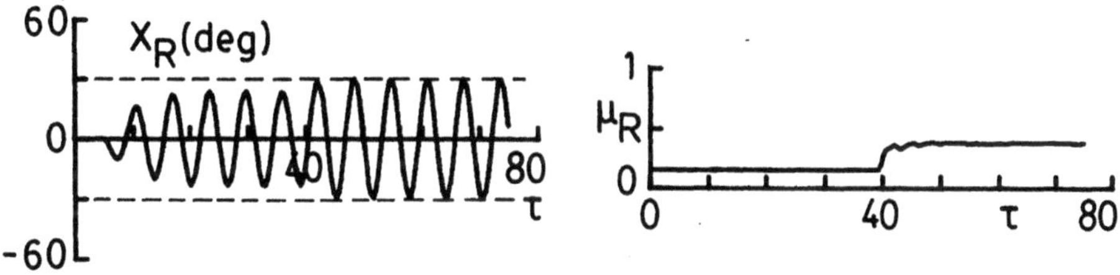

(a) (b)

Fig.4.4 Control response with control law (II)

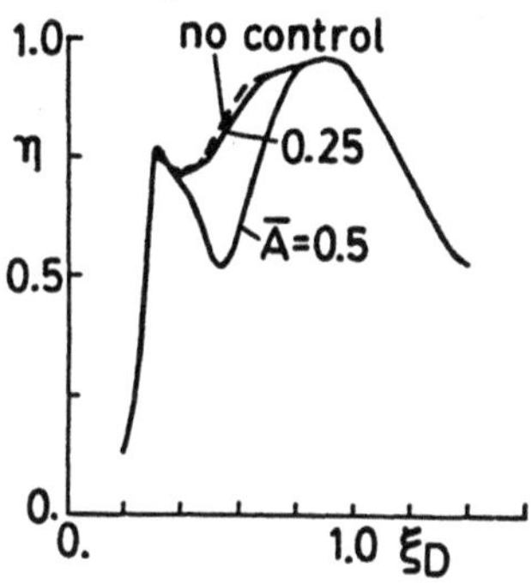

Fig.4.5 Energy conversion efficiencies η for excessive wave height

results of control are all excellent though the optimal gains are different.

Fig.4.5 shows how the conversion efficiency is reduced when above mentioned control laws are applied. Efficiency reduction of η_R caused by the increase of μ_R from the optimal value is noticed.

5. Conclusions

An adaptive control system is formulated for a wave energy converter utilizing the oscillating motions of floating bodies and control responses are discussed for varying amplitude of incident waves. The results are summarized as follows:

1) Two nonlinear control laws proposed in this paper are first applied to a simple fixed parameter m-k-c system in order to estimate the ability to reduce the oscillating amplitude within some fixed limits. The responses are favorable.
2) In WECS-2 which has large size L Body, the motion of R Body

is dominant. It has been found by numerical simulation that our control system has quite excellent responses when the proposed two control laws are applied control damper parameters for R Body.

In this paper, the incident waves have been restricted only to sinusoidal waves, however, a numerical simulation has been tried in the system of Fig.3.1 for an irregular waves assuming that the waves have Bretschneider-Mitsuyasu type spectrum and it has been confirmed that a better response is obtained for control law (II) than for (I).

It is supposed that if WECS-2 is controlled to have good transient response for regular waves and if each transient response is favarable for each component wave, the system will show corresponding good response practically for irregular waves.

Acknowledgement
The authors are grateful to Dr. H. Kanoh, Assistant Professor of Osaka University , for his valuable comments and thank to Mr. K. Fujita, student of our University, for his many efforts in numerical calculations. Thanks are also to Grant-in-Aid for Scientific Research of the Ministry of Education, Japan.

Literature

1) M.M.Leishman and G.Scobie (ed.), The Development of Wave Power-A Techno Economic Study, N.E.L., Scotland, 1976

2) S.H.Salter, Wave Power, Nature, 1974, p.249

3) UK Patent No.2, 073, 824; US Patent No.4, 474, 578

4) R.Kawatani and M.Masubuchi, Theoretical Analysis of Wave Energy Converter Consisting of Three Floating Bodies, Bull. of JSME, 26, 214, 1983, pp.667-674

5) M.Masubuchi and R.Kawatani, Frequency Response Analysis of an Ocean Wave Energy Converter, Trans. ASME, J-DSMC, 104, 1, 1983, pp.30-38

6) R.Kawatani and M.Masubuchi, Experimental Study on the Dynamic Characteristics of Wave Energy Converter having Two Floating Bodies (in Japanese), Trans. Japan Soc. of Mech. Engineers (in printing)

7) D.Graham, R.C.Lathrop, The Synthesis of Optimum Transient Response, Criteria and Standard Forms, Trans. AIEE, 1953

A Comparative Evaluation of Numerical Methods in Free-Surface Hydrodynamics

Ronald W. Yeung

Department of Naval Architecture & Offshore Engineering
University of California, Berkeley, CA 94720, USA

Abstract

An expository review is given on three major types of numerical methods commonly used in free-surface and body interaction problems of the deterministic type. The general principles governing boundary-fitted coordinates are described. Formulations commonly used for finite-element type methods are explained. Boundary-integral equation methods of various forms are discussed. Special attention is paid to features of each type of methods that are efficient and potentially attractive and also to features that are known to be problematic. Various current research activities in time-dependent type solutions are briefly summarized. A novel idea involving the new concept of "shell functions" in unsteady problems is proposed. Once properly calculated, the "shell coefficients" can be used for a large number of methods and for arbitrary body geometry.

1. Background

Similar to many other branches of engineering sciences, the subject of free-surface and body interaction is characterized by a set of boundary-value problems that is well-established, but defies simple analytical treatment except for the simplest body geometry. In most problems of practical interest, the need for numerical solutions exists. It is the nature and means of such numerical approximations of the exact solutions that we will consider here.

We shall confine our discussion to problems with bodies that have no mean forward speed, and to methods that have the flexibility or the potential to treat geometry of an arbitrary type. The present survey is not meant to be exhaustive, but does cover some of the recent efforts in the field since the more comprehensive review by Yeung (1982a). Related articles on "methods for free-surface flows" by Wehausen (1977), on "three-dimensional ship and platform motions" by Newman (1983), and on "steady nonlinear bow flows" by Baba (1981) are fine additional references. Two recent texts (Susbielles & Bratu, 1981; Mei, 1983) provide much of the background on formulation and other types of analytical and numerical approximations.

The development of solution methods for wave-body problems has been an eventful one. In the realm of linear problems,

Ursell (1949) was the first to apply the technique of multipole expansions, which subsequently formed the basis of many interesting semi-analytical works for two dimensional bodies using conformal mapping (Tasai, 1959; Porter, 1960) and for hemispherical geometry (Havelock, 1955; Wang, 1966; Hulme, 1982).

About the same time as Ursell's well-known work, John (1950) examined very carefully an integral-equation formulation based on a free-surface Green function. His work was to have lasting influence in the next 30 years. Such an integral-equation formulation, though powerful by its own right, could not be easily implemented until the advent of high-speed computers. The work of Hess & Smith (1960) in incompressible aerodynamics paved the way for such an implementation. Their discretization procedure and treatment of singular integrals have formed the backbone of much of the present-day computing. Solutions for two-dimensional problems with a free surface using integral-equation methods were first published independently by Lebreton & Margnac (1966) and Frank (1967). W.D. Kim's (1965) work, though limited only to ellipsoidal geometry, stood out as one of the few three-dimensional contributions in the 60's.

The early 70's marked the arrival of a number of non-traditional formulations: finite-elements methods (Chenault, 1970; Bai, 1972), boundary-element method based on 'simple source' (Yeung, 1973), and finite-difference methods (Chan & Hirt, 1974), all with the attractive feature of by-passing the need to compute the relatively complicated Green function, and all with fine potential for growth. Most, if not all, of these then new methods relied on an analytical knowledge of the behavior of the solution beyond the numerical truncation boundary. Several hybrid formulations were later developed to couple an interior domain of intensive numerical compuations with an external domain based on analytical representations. Meanwhile, steady progress was being made towards the implementation of Green-function methods for practical three-dimensional problems (Garrison, 1974; Michelsen & Faltinsen, 1974). With few exceptions, all methods were directed towards obtaining solutions in the frequency domain.

A renewed interest in integral-equation formulation occurred towards the late 70's. This might have been in part driven by a need to resolve the difficulties associated with "irregular frequencies" encountered in Green-function methods and in methods coupling finite-element and integral-equation formulations (Lenoir & Jami, 1978). The renewed interest was also motivated by the very successful Lagrangian and integral-equation procedure proposed by Longuet-Higgins & Cokelet (1976) for treating nonlinear problems. A parallel development in boundary-fitted coordinates techniques for finite-difference methods also began to receive some attention.

Research interest in the 80's has been focussed on time-dependent problems of both linear and nonlinear types. This is particularly timely in view of the computational need of research work in wave-energy extraction and in fluid-structure-soil interaction problems.

In order to establish a unifying framework for discussing he numerical techniques mentioned, we now recall the basic equa- ions that are of interest to this audience of wave-energy orkers. Let Ω denote a body of fluid which is bounded by a free urface ∂F on top, by a solid bottom surface ∂B, and by a body urface ∂H. On various occasions, it would be helpful to intro- uce a fictitious mathematical boundary Σ which represents either n artificial truncation of the computational domain or some nclosing surface at large distance for the purpose of analysis. he union of these surfaces that bounds Ω will be denoted by $\partial\Omega$. et the Oxz plane be a horizontal plane that coincides with the alm-water surface, with the y-axis pointing upward. Let $=Y(x,z,t)$ be the free-surface elevation at time t and (x,y,z,t) be the velocity potential that describes the flow. The ypical initial-boundary-value problem of interest is defined by he following equations:

$$\nabla^2\phi(x,y,z,t) = 0\ , \qquad \text{for } (x,y,z) \text{ in } \Omega, \qquad (1.1)$$

$$D_tY(x,z,t) = \phi_y\ , \qquad D_t\phi = \frac{1}{2}|\nabla\phi|^2 - gY \quad \text{on } \partial F, \qquad (1.2a,b)$$

$$\phi_n = f \quad \text{on } \partial H, \qquad \phi_n = 0 \quad \text{on } \partial B. \qquad (1.3a,b)$$

ere g is the gravitational constant and f a prescribed function. D_t denotes the material derivative with respect to time in a ixed frame of reference and the subscript n denotes normal erivative with respect to an exterior normal to Ω. Thus, the ypical nature of such hydrodynamic problems consists of a simple ield equation, Neumann type boundary conditions on solid urfaces, and a relatively complicated nonlinear 'double- ondition' on an unknown free surface. The linearized version of 1.2a,b) is given by the classical free-surface condition on the lane y=0 (denoted henceforth by ∂F_o):

$$\phi_{tt}(x,y=0,z,t) + g\phi_y = 0\ . \qquad (1.4)$$

t is understood that a unique solution of ϕ requires the addi- ional imposition of some appropriate initial conditions associ- ted with the fluid and free-surface motions at $t=0^-$ (see e.g. eung, 1982b).

If the motion attains a time-harmonic type steady-state, the inearized condition of (1.4) reduces to :

$$-\nu\phi(x,0,z) + \phi_y = 0\ , \qquad \nu=\omega^2/g\ , \qquad (1.5)$$

here ω is the angular frequency of the periodic motion. We note ere that, in the interest of notational brevity, we omit the ime argument in ϕ to indicate that it represents the time- omplex spatial portion of the original potential. Thus, in 1.5), the total potential occurs in the form of $\phi(x,y,z)e^{-i\omega t}$, $i=\sqrt{(-1)}$. Radiation conditions, which state that wave-like disturbances of the solution are of an outgoing kind, must be imposed if these time-harmonic problems are to have unique solutions.

In the next three sections (§2 - §4), we will describe some of the principles that govern the formulation and implementation of the methods of finite-differences, finite-elements, and boundary-integral equations. The intent is to provide the reader with some flavor of many of the methods currently in use.. Additional details can be found in many of the references cited. §5 describes some of the latest developments in transient or time-dependent solutions. In particular, a new idea called 'shell functions' is proposed. This can provide a universal link for a variety of hybrid or local methods that relies on the availability of a time-dependent open-boundary condition. In §6, conclusions are drawn and future directions are discussed.

2. Boundary-Fitted Coordinates

In the last decade, the development of grid-generation techniques has progressed at a rapid rate. This may be due to the activities of two somewhat competing groups: the finite-difference enthusiasts who strive to overcome the difficulties of applying classical differencing schemes to gemoetry of arbitrary shapes and the finite-element workers who desire to have the mesh-generation process and grid-control process reasonably automated. Recent reviews of the first were given by Thompson et al. (1982) and of the second by Thacker (1980). Since both types of methods are field-oriented solution methods, it is understandable that an efficient treatment of the mesh process is an important consideration. However, in the case of finite differencing, the grid generation process serves two purposes: a) it produces a set of curvilinear coordinates that is specially adapted to the geometry in question, b) it provides a crucial 'numerical transformation' that allows differencing schemes to be applied in, geometrically speaking, a much simpler computational domain. Because of the enormous amount of recent activities in this area together with its derivatives, it is worthwhile to give a brief exposition on the motivation of the development, the underlying principles, and the advantages and disadvantages of such numerical techniques.

Figure 1 illustrates what one typically encounters when a regular Cartesian grid is overlaid onto a physical domain having curved boundaries that do not intersect with the grid points. A typical interior node located away from the boundary corresponds to a 'regular star'. The Laplacian operator of Eqn. (1.1) at node (i,j), for instance, will have the following difference form to second-order accuracy in terms grid-spacing h:

$$\nabla^2 f = \frac{1}{h^2}\left\{ f_{i+1,j}+f_{i-1,j}-4f_{i,j}+f_{i,j+1}+f_{i,j-1} \right\} + O(h^2) \qquad (2.1)$$

However, a boundary-interior node involving its four neighboring nodes, with two lying not on the regular grid as shown in Fig. 1b, will at best have the following representation for the

Laplacian operator:

$$\nabla^2 f = \frac{2}{h^2}\left\{ \frac{1}{1+\alpha}\left(f_{i+1,j}+\frac{1}{\alpha}f_{i-1,j}\right) + \frac{1}{1+\beta}\left(f_{i,j+1}+\frac{1}{\beta}f_{i,j-1}\right) - \left(\frac{1}{\alpha}+\frac{1}{\beta}\right)f_{ij}\right\}$$

$$+ O([1-\alpha)h) + O([1-\beta]h) \qquad (2.2)$$

Note that although the "irregular stars" enable the boundary values on $\partial\Omega$ be assigned (for a Dirichlet type condition) at the actual boundary locations, a loss of accuracy of O(h) results. If one were to use regular stars near $\partial\Omega$ and assign values to the nearest nodes, one would have been, in essence, solving a geometrically different problem, i.e. one with jagged corners. For a Neumann type condition, more complications arise and it is not uncommon to use double interpolation at these boundaries to estimate normal derivatives. While higher-order schemes involving more points could be a remedy in accuracy, their implementations often result in rather cumbersome algorithms. Accuracy loss due to increase in truncation errors can lead to physically unacceptable solutions (see e.g. Radder, 1979). This is an important consideration in finite-difference method.

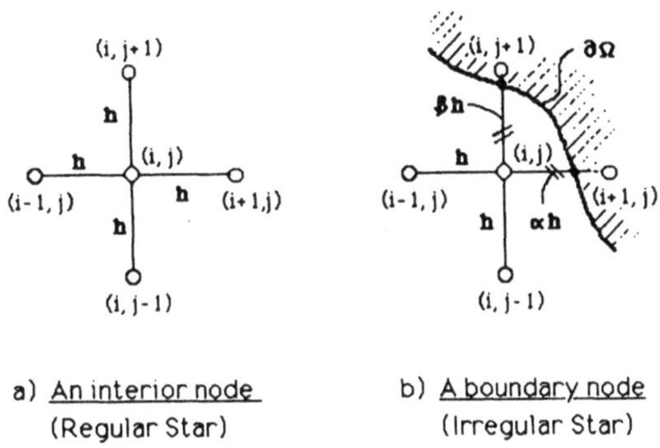

Figure 1. Regular star versus irregular star in finite differencing.

As an illustration of the method of boundary-fitted coordinates, consider solving a two-dimensional problem not in the original physical (x,y) domain but in a new, or so-called "logical" domain of mappped variables ($\xi.\eta$). The boundary curves in the physical plane will be transformed to coordinate lines of the mapped plane. This is conceptually not very different from analytical approaches such as cylindrical or parabolic coordinate systems. However, it is the use of numerical transformation that makes the idea general and powerful. Winslow (1967) was apparently the first to carry out this type of approach during a study of electro-magnetics. Figure 2 (taken from Thames et al., 1977) exemplifies well the use of this idea for a fluid-dynamics type application. "O-type" curves encircling the body map into constant η lines, whereas "C-type" curves eminating from the body map into constant ξ lines. The cuts AB and CD open up the flow region into a periodic-type computational domain.

To examine the effects of the mapping, let the transformation be:

$$\xi = \xi(x,y), \qquad \eta = \eta(x,y) \;. \tag{2.3}$$

While a variety of transformations exists, one popular choice is an elliptic type mapping function, which yields a smooth set of grid lines in the x-y plane:

$$\nabla^2\xi = P(x,y), \qquad \nabla^2\eta = Q(x,y) \tag{2.4}$$

The functions P and Q can be used to control the local density of the grid and to adapt to solution gradient as necessary. Since it was suggested earlier that ξ and η, rather than x and y, should be treated as independent variables, it is necessary to work with the inverse transformation, $x=x(\xi,\eta)$ and $y=y(\xi,\eta)$, instead of (2.3). Generally speaking, such an inverse exists if no two coordinate lines of the same family cross each other. Using results from Implicit Function Theorem, one can arrive at the following relations between the derivatives:

$$\xi_x = y_\eta/J, \quad \eta_x = -y_\xi/J; \qquad \xi_y = -x_\eta/J, \quad \eta_y = x_\xi/J \tag{2.5}$$

where the Jacobian J is given by $x_\xi y_\eta - x_\eta y_\xi$. Further, the differential operator with respect to the x and y variables transforms to:

$$\partial_x = (y_\eta/J)\partial_\xi - (y_\xi/J)\partial_\eta \;, \qquad \partial_y = -(x_\eta/J)\partial_\xi + (x_\xi/J)\partial_\eta \tag{2.6}$$

Using these relations in (2.4), the (inverse-)mapping functions can be shown to satisfy the following quasi-linear elliptic equations in the Cartesian coordinates (ξ,η):

$$\tilde{\nabla}^2_{\xi,\eta}x(\xi,\eta) = x_\xi P + x_\eta Q \;, \quad \tilde{\nabla}^2_{\xi,\eta}y(\xi,\eta) = y_\xi P + y_\eta Q \;, \tag{2.7a,b}$$

where

$$\tilde{\nabla}^2_{\xi,\eta} \equiv [(x_\eta^2+y_\eta^2)\partial^2_{\xi\xi} - 2(x_\xi x_\eta + y_\xi y_\eta)\partial^2_{\xi\eta} + (x_\xi^2+y_\xi^2)\partial^2_{\eta\eta}]/J^2 \tag{2.8}$$

The pair of mapping functions in (2.7) can be solved simultaneously with Dirichlet conditions, where the boundary coordinates (x,y) are supplied at constant values of ξ or of η. This is normally carried out by relaxation methods after standard differencing techniques are introduced. It is of interest to note that the boundary condition for x and y can be reformulated to generate a family of η lines normal to the body boundary, which facilitates the implementation of a Neumann type condition in the physical plane.

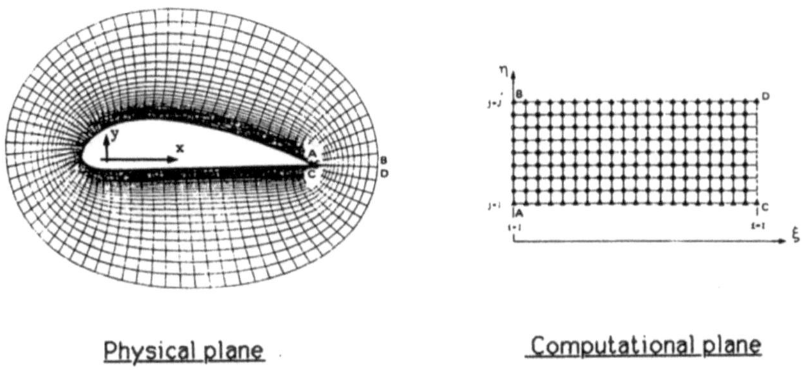

Figure 2. Boundary-fitted coordinates for an airfoil (Thames, et al., 1977). The functions P and Q were taken as decaying exponentials to attract lines towards the body.

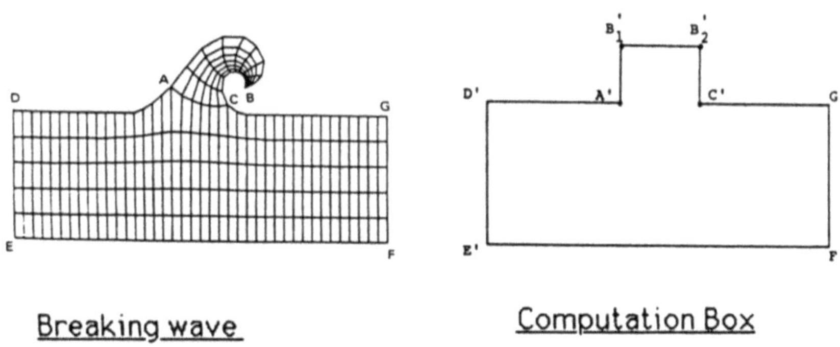

Figure 3. Boundary-fitted coordinates for a breaking wave (Ghia et al. 1981). A special solution-initialization scheme was required to obtain non-intersecting grid lines.

It would appear that the mapping must be first carried out before the flow solution is sought. However, if the field equation for the physical variables is also elliptic as in (1.1), namely, being given by:

$$\tilde{\nabla}^2_{\xi,\eta}\, \phi(\xi,\eta) = 0 \,, \qquad (2.9)$$

which is now written in terms of the mapped variables, the left-hand sides of (2.9) and (2.7) are identical. Thus the coordinate transformation and the physical solution may be solved concurrently in the same " sweep", with of course the proper adjustment made to accomodate the boundary conditions.

The power of this method in handling arbitrary geometry is apparent. Once the general alogrithm is developed, a change of geometry corresponds to a mere change of boundary coordinates. A more elaborate application of this procedure was carried out by Ghia et al. (1981) in the context of free surface flows (Figure 3). Unlike a Lagrangian-oriented method, the present requires a fairly complicated algorithm to handle a 'reentrant structure' such as that of a breaking wave.

A discussion of boundary-fitted coordinates method in two dimensions can give the misleading impression of the simplicity of these numerical techniques. The complexity associated with a three-dimensional transformation needs to be pointed out. In order to gain a proper perspective, one should first recall that the three-dimensional Jacobian J (2.5) will now have 6 terms, each consisting of a product of three derivatives, which is still reasonable. It follows then the three-dimensional version of (2.6) will have 6 terms each, which subsequently generates a 72-term(!) equation for a conceptually simple three-term Laplacian operator. The chances for coding error should not be overlooked. It is an area that can very well exploit the help of symbolic manipulations. Some work in that direction has been suggested by Steinberg & Roache (1985). We conclude by commenting that we have not considered in this short exposition the possibility of time-dependent boundary that changes with the flow solution. A discussion of some of the difficulties in implementation can be found in Haussling (1982).

3. Finite-Element Variational Forms

Finite-element methods form the mainstay of computational tools in structural and stress analysis. They have also been successfully applied to a variety of field-oriented problems. A good indicator of the directions of the recent activities in fluid related problems can be found in the biennial sequel of Gallagher et al.(1984) . The focus of the present discussion is on wave-body interaction problems.

This class of methods is known for its desirable characteristics of being able to handle curved boundaries. The approximating functions in the field are usually of a polynomial type, which are simple to work with. Figure 4, for instance, shows an 8-node isoparametric element that is widely used. This element can be mapped into a local (ξ,η) square, in which the calculation of the element's contribution to the field and boundary properties can be made. The unknown function is normally represented by a set of nodal values and a set of predefined 'shape functions'. These are chosen to satisfy certain continuity requirements across the elements. The requirements depend on the differential order of the field and

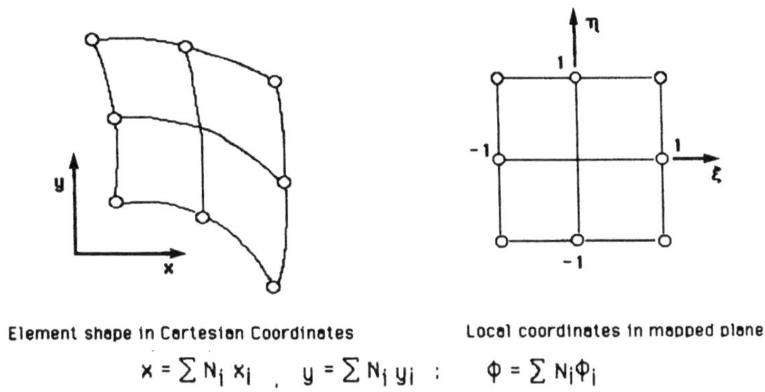

Figure 4. A (two-dimensional) 8-node isoparametric element. The shape functions N_i maps the curved edges to straight lines in the local coördinates. ϕ is represented by the same shape functions.

the boundary conditions. The determination of the nodal unknowns relies on a global or integral criterion which will be discussed in more detail momentarily. A variety of choices of element type exists. Discussions on these choices as well as their effects on solution can be found in a number of texts (e.g. Zienkiewicz, 1970; Carey & Oden, 1983).

The development of finite-element methods for external free-surface flows was handicapped , during its early stages, by the absence of a precise closure condition. Such a condition is necessary in order to truncate numerical computations at some finite boundary Σ . For the time-harmonic problems defined by the linear equations (1.1), (1.3), and (1.5), the Sommerfeld radiation condition (1949) was known long ago:

$$\sqrt{R}(\phi_R - i\kappa\phi) \to 0 , \qquad \text{as } R \to \infty . \tag{3.1}$$

Here R is a polar radius in the horizontal plane Oxz, and κ is the wave number given by the solution of the transcendental equation: $\nu\kappa = \tanh \kappa h$, h being the water depth (see Wehausen & Laitone, 1962). It was in the early 70's that Chenault (1970) and Bai (1972) demonstrated that useful results can be obtained by applying (3.1) at a finite Σ . A condition such as (3.1) is inefficient since it is only asymptotically correct, which necessitates an undesirably large Σ. The turning point appeared to be when two basically similar variational principles were obtained independently by Bai & Yeung (1974) and Chen & Mei (1974). These principles permit an "exact" coupling of finite-element calculations in Ω with any exterior solution outside of Σ. These formulations led to the so-called "localized finite-element" or "hybrid-element" methods. Finite-element methods can be implemented without the use of variational principles. Their availability in this instance provides a symmetric form that reduces both storage requirement and computational efforts. In providing a short derivation of the variational principle below, we hope to demonstrate some of the other forms that may be used.

Suppose ϕ represents an admissible set of functions to be used for constructing the solution of the time-harmonic problems defined in §1 but subject to some additional matching conditions on Σ. This boundary-value problem, in very general terms, may be summarized below as:

$$\nabla^2\phi(x,y,z)e^{-i\omega t} = 0. \tag{3.2}$$

$$\phi_n + \beta\phi = V_n \qquad \text{on } \partial F_o \cup \partial B \cup \partial H \tag{3.3}$$

$$\phi_n - \phi_n^e = 0\ , \quad \& \quad \phi - \phi^e = 0\ , \qquad \text{on } \Sigma\ . \tag{3.4a,b}$$

where ϕ^e is an exterior solution defined in Ω^e (outside of Σ) to be determined simultaneously with ϕ itself. It is clear that (1.3a,b), (1.5), and even (3.1) all fall into the general category of (3.3) by special choices of the constant β and the function V_n. Thus, for instance, $\beta=-\nu$ and $V_n=0$ on ∂F_o. (3.4a,b) represents an enforcement of the continuity of the solution and its normal derivative across Σ.

Given a set of 'trial functions' to be used to represent ϕ and ϕ^e, we may consider applying the method of weighted residuals in the following manner:

$$\int_\Omega \nabla^2\phi\ w_1 d\Omega - \lambda_1\int_{\partial\Omega}(\phi_n+\beta\phi-V_n)\ w_2 d\partial\Omega - \lambda_2\int_\Sigma(\phi_n-\phi_n^e)\ w_3 d\partial\Omega$$

$$+ \lambda_3\int_\Sigma(\phi-\phi^e)\ w_4 d\partial\Omega = 0\ , \tag{3.5}$$

where w_i, $i=1,\ldots,4$ are weighting functions and λ_i are constants. This is the integral constraint alluded to earlier. It provides a logical procedure of minimizing the error of such a trial. If the weighting or test functions w_i are chosen to be the linear independent set that forms the basis of the trial functions, the procedure of (3.5) is known as Galerkin's method. Let us take w_1, w_2, and w_3 as $\delta\phi$, a variation of ϕ in Ω and $\partial\Omega$, w_4 as $\delta\phi_n^e$, a variation of the normal derivative of ϕ^e on Σ, and all λ_i's as unity. This turns out to be a 'judicious choice'. It results in a variational principle that has a symmetric form:

$$\delta J(\phi,\phi^e) = \delta\left\{\int_\Omega \frac{1}{2}(\nabla\phi)^2\ d\Omega + \int_{\partial F_o\cup\partial H\cup\partial B}\left(\frac{\beta}{2}\phi^2 - V_n\phi\right)\ d\partial\Omega\right.$$

$$\left. - \int_\Sigma \phi_n^e(\phi - \phi^e/2)\ d\partial\Omega\right\} = 0\ . \tag{3.6}$$

The first two terms of the functional J are classical. It s the third term that enables one to fully couple the interior nd exterior solutions. In arriving at (3.6), it is necessary to ssume that ϕ^e is a solution of (3.1-3.3), thus satisfying the adiation condition explicitly. Aside from that, no restrictions re necessary. (3.6) is derived at present with the conditions 3.2-3.4) as a starting point. Conversely, one can verify that a inimization of the functional J with respect to ϕ and ϕ^e will ield (3.2) as the Euler Lagrange equations with (3.3) and 3.4a,b) satisfied as natural boundary conditions. This form was irst given in Bai & Yeung (1974) and the corresponding version nder the long-wave slowly-varying bottom approximations was erived by Chen & Mei (1974). Other pseudo-variational forms hat can accomodate the conditions stated are available (see, .g. Zienkiewicz et al. 1977).

Figure 5 shows how typically Σ may be chosen to facilitate he representation of the exterior solution. In fact, we may roceed by letting

$$\phi = \sum_{i=1}^{N} \phi_i N_i \quad , \tag{3.7}$$

here ϕ_i are the nodal values in the Ω, and N_i the shape unctions in the usual finite-element sense, and

$$\phi^e = \sum_{i=1}^{L} a_i \Psi_i^e \quad , \tag{3.8}$$

n the exterior domain Ω^e, where a_i are unknowns and Ψ_i^e is a set f independent exterior solutions. The result of such a iscretization process used in conjunction with (3.6) is a banded atrix shown in Figure 6. The bandwidth can be minimized if the odes are ordered in a layer by layer fashion around Σ. Here $\tilde{\phi}_i$ enotes the M unknowns on Σ, with the half-bandwidth equal oughly to M. Bai & Yeung (1974) chose a set of orthogonal eigen unctions for Ψ_i^e as shown in Fig. 5a. It had the advantage of educing the exterior submatrix to a diagonal form. This ubmatrix can be quickly "compacted" to give a new matrix $\{\tilde{A}\}$ as hown in Fig. 6, which does not involve the exterior unknowns a_i. The matrix $\{\tilde{A}\}$ is predominantly real and symmetric except or the small block $\{F\}$, which is complex and symmetric. The igen series is a very efficient representation. In the deep-ater limit, it becomes an integral representation. The number f elements increases rapidly with water depth, and a few workers ave used a truncation boundary as shown in Figure 5b, instead f 5a. It is noteworthy that, contrary to what some workers had uggested, the original formulation (3.6) was <u>not</u> restricted to he use of eigen-functions even though eigen functions were used s a demonstration.

The truncation boundary Σ can be any arbitrary shape if an ntegral-equation representation for ϕ^e is introduced. There has een a considerable amount of activities in coupling finite-lements to integral-equation formulation. More background on

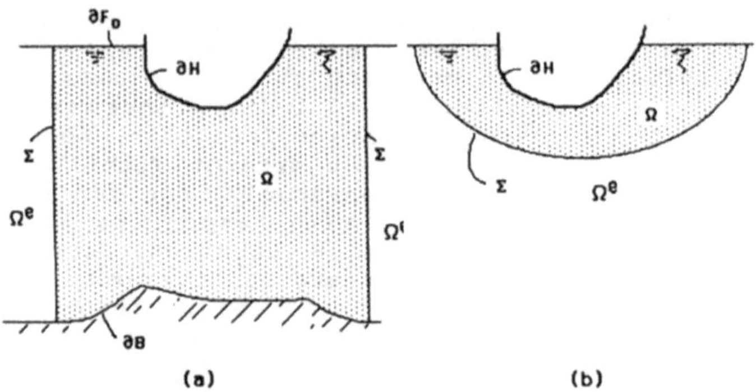

Figure 5. Two common choices of Σ in hybrid-type methods. (a) A vertical surface(s) is convenient for eigen-functions representation; (b) An arbitrarily shaped surface relies on integral-equation representation.

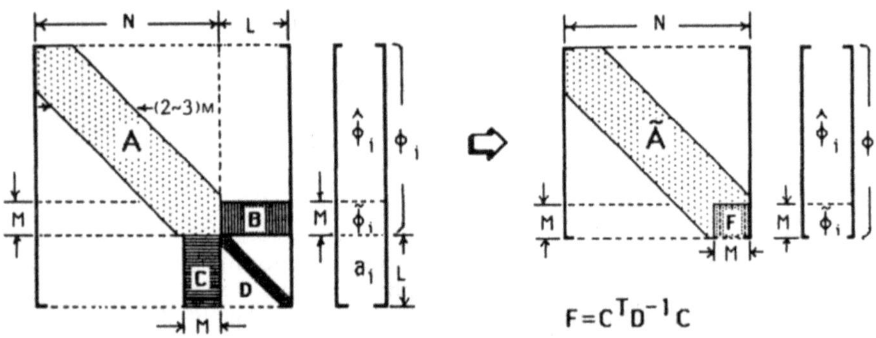

Figure 6. Structure of finite-element matrix. Formulation is based on (3.6) with eigen-function expansion.

integral-equation method as a stand-alone method will be given in the next section. Here it suffices to mention that the spatially singular solution is normally constructed to satisfy all of the homogeneous conditions, including (3.1). The first effort of such a coupling scheme was apparently due to Berkhoff (1972), who employed a source distribution on Σ in shallow-water problems. Seto and Yamamoto (1975) carried out a similar treatment in deep water. There are basically two ways that one may proceed using (3.6). The first is to keep the last term in (3.6), and then simply introduce the approporiate singular solution in its integrated or discretized form for Ψ_i^e in (3.8). Such an approach will preserve the symmetric form but will involve double integrals of the singular solutions. The second approach is to simply consider the effects of Σ as an additional term like the second term on the right-hand side of (3.6). The unknown V_n on Σ

an now be related to the unknown ϕ on the same boundary by the ntegral-equation formulation (Zienkiewicz et al. 1977). This is omputationally simpler but requires a little more book-keeping ork for the unknowns on Σ . The original symmetry property is ow lost.

The eigen-series method has been implemented by Seto (1976), ue et al. (1978), Eatock Taylor & Zietsman (1981), among others. lethods that couple finite-elements with boundary-integral quations have been examined and evaluated by Lenoir & Jami 1978) and Zietsman (1984), both methods based on a pseudo-ariational formulation. A short survey of the convergence and niqueness theorems available for the type of the formulations liscussed has been given by Euvrard et al.(1981).

The original motivation of introducing finite elements echniques in these wave problems was to avoid the evaluation of omplicated wave kernels associated with integral-equation ethods. The finite-element boundary-integral-equation coupling ethods bring back precisely the integrals that one was trying to void at the start. One may ask, why bother to use finite-lements at all? Several reasons have been offered: 1) The inite-element region can well serve as a transition zone that leforms a geometry of higher complexity into a geometrically impler Σ where the singular kernels can be more conveniently alculated; 2) The finite-element approximation is less ulnerable to inaccuracies when the interior region has multiple odies the are in close proximity, and singularity sheets must be lealt with more carefully when they are close to each other; 3) f the singularity distribution is assumed to be on ∂H, the lesired values of ϕ and ϕ_n on Σ are evaluated away from the ingularities, thus more amenable to numerical treatment.

The biggest disadvantage of these 'FEM'-'BIE' coupling ethods is that they bring back all of the unpleasant features of irregular frequencies' associated with the use of the integral-quation formulation (described in §4). Although such requencies may be detuned to a certain extent by choosing the ppropriate size of Σ, they cannot be completely eliminated using he standard integral-equation formulation. Jami & Polyzakis 1981) has proposed a modified boundary condition on Σ of the ollowing form that will remedy the situation:

$$(\partial_n + \mu)\phi \,\Big|_{\Sigma} \;=\; (\partial_n + \mu) \int_{\partial H} (\phi G_\nu - G V_n)\, d\partial\Omega \;, \quad \mathrm{Im}(\mu) \neq 0 \tag{3.9}$$

here μ is an arbitrary constant subject to $\mathrm{Im}(\mu)\neq 0$ and G is the ingular source (or Green) function. The representation on the ight-hand side is related to an augmented integral equation roposed earlier by Burton & Miller (1971) for Helmholtz' equa-ion. A linear combination of ϕ and ϕ_n on ∂H can yield an irre-ular-frequency free integral equation, which is the property eing exploited in (3.9); of course, at the expense of an increa-e in computational efforts.

4. Boundary-Integral Equations

The traditional approach to solving hydrodynamic problems of the type described here is based on integral-equation formulation. This was well outlined in the classic paper of John (1950). Since the field equation is homogeneous and Green functions are available for the linearized conditions (1.3b), (1.4) or (1.5), this is a natural path to take. In comparison with methods discussed in §2 and §3, this is an approach more superior in memory requirement and coding convenience. Indeed, the book-keeping work associated with an integral-equation method is only minimal.

The implementation of classical formulation based on Green functions is not entirely free of complications. There are two outstanding difficulties: 1) the standard formulation for time-harmonic problems breaks down at a discrete set of frequencies for free-surface piercing bodies, and 2) a non-trivial amount of numerical computations exists for the Green functions and its derivatives.

Most of the formulations in use are either a direct or indirect result of Green's Theorem. In terms of a Green function G that satisfies the radiation condition (3.1) and homogeneous conditions (1.3b) and (1.5), the time-harmonic problems, for instance, can be shown to reduce to solving the following integral equation:

$$\frac{1}{2}\phi(P) + \int_{\partial H} (\phi-\bar{\phi})G_\nu d\partial\Omega_Q = \int_{\partial H} (f-\bar{\phi}_\nu)Gd\partial\Omega_Q \quad , \quad P \in \Omega_\cap\partial H \tag{4.1}$$

Here ν, when occurring as a subscript, denotes normal derivative with respect to the integration variable Q on $\partial\Omega$. G is assumed to have a singularity of the form $1/(4\pi r)$, $r \equiv |PQ|$, and a harmonic part that satisfies the homogeneous boundary conditions stated earlier. $\bar{\phi}$ and $\bar{\phi}_\nu$ in (4.1) are associated with an "artificial domain" $\bar{\Omega}$ interior to ∂H that one would normally conjure to have no physical relevance to the external problem in Ω. In fact, if one assumes $\bar{\phi}$ is trivial for a homogeneous $\bar{\phi}_\nu$, the so-called Green's mixed distribution results:

$$\frac{1}{2}\phi(P) + \int_{\partial H} \phi(Q)G_\nu d\partial\Omega = \int_{\partial H} fGd\partial\Omega \; , \qquad P\in\partial H \tag{4.2}$$

This is a Fredholm integral equation of the second kind. Other representations involving either just source (G) or dipole (G_ν) distribution alone follow from other fictitious conditions imposed on $\bar{\phi}$ and $\bar{\phi}_\nu$. More details may be found in Yeung (1982a). (4.2) is also referred to as the "direct" formulation whereas those involving an intermediate unknown such as source or dipole strength are referred to as "indirect".

Uniqueness of the solution of $\phi(P)$ in (4.2) cannot be

stablished when the Fredholm determinant associated with the eft-hand side of (4.2) or its like (in the case of other epresentations) vanishes. This occurs at a set of discrete requencies $\nu=\omega^2/g$ asscociated with the free-surface condition 1.5). It may be interpreted as the presence of certain resonant or eigen modes" in the fictitious domain $\bar{\Omega}$ if $\bar{\Omega}\cap\partial F_o \neq 0$. he existence of such "irregular" frequencies was first proven by ohn (1950); in fact they are sometimes appropriately referred to s "John's frequencies". Figure 7 shows some typical omputations in which the phenomenon is prominently exhibited.

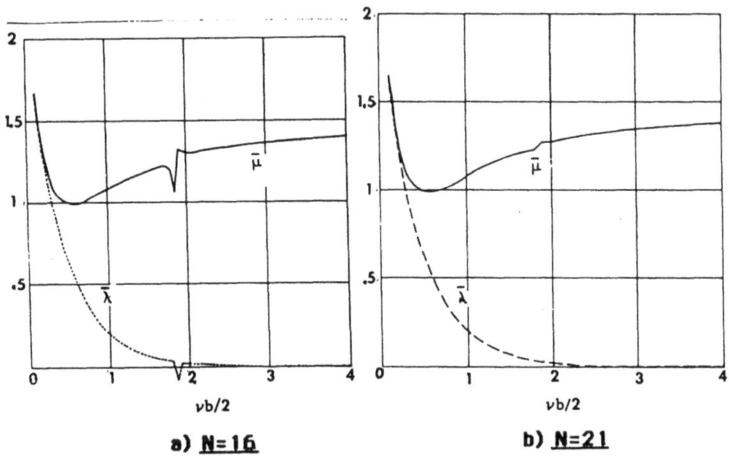

Figure 7. Added mass and damping of a two-dimensional rectangular cylinder (Frank, 1967). Behavior at irregular frequency is shown for two different values of N, the number of grid points on the half body.

Several methods have been found successful in eliminating rregular frequencies in Green-function formulation. Among them, he following are noteworthy: Kobus (1976) proposed solving an ver-specified problem in $\bar{\Omega}$. Ogilvie and Shin (1978) suggested sing an additional point singularity at the origin. This idea as further analyzed by Ursell (1980). Webster (1978) found it ffective to discard the free-wave term in high-frequency alculations. Exploiting a technique used by Waterman (1969), artin (1981) solved Equation (4.2) with P taken in $\bar{\Omega}$, in which ase, the first term on the left-hand side vanished. A set of null-field equations" was derived and was proven to be uniquely olvable. Martin's procedure requires the use of a bilinear form f G in terms of the variables P and Q. In the same spirit as leinman and Roach (1974), Kleinman has recently devised an augented integral equation based on forming a complex linear comination of (4.2) and its normal derivative. This yields:

$$\frac{1}{2}\phi + (1 + \mu\partial_n)\int_{\partial H} \phi G_\nu \, d\partial\Omega = (1 + \mu\partial_n)\int_{\partial H} fG \, d\partial\Omega - f/2 \tag{4.3}$$

ith the restriction that $\mathrm{Im}(\mu) \neq 0$. The integral operator on the left-hand side of (4.3) has the interesting property that the homogeneous solution is always trivial. Kleinman provided a uniqueness proof but existence was not established at the time. A related formulation in the finite-element methods was alluded to earlier.

Analytical expressions of free-surface Green functions in three dimensions consist typically of single and double integrals of wave functions. Efficient means of evaluating these oscillatory integrals have always been a subject of interest (Kim, 1965; Hogben & Standing, 1974; Hearn, 1977; Noblesse, 1982; Newman, 1984). The expressions can be written in a variety of forms, and depending on the ratio of the horizontal distance to vertical distance from the source point, as well as their absolute magnitude, series expansions, rational polynomial approximations, or integral representations may be used. A certain degree of optimization is necessary. A survey by Newman (1983) discussed some of the procedures. Accurate calculations of Green functions have always been a laborious and non-trivial task. These computational difficulties are gradually overcome and they do not pose as big an obstacle as they did in the 70's.

The integral equation (4.2) is analytically exact. In practice, a numerical approximation in the values of ϕ is sought with an approximation of the boundary geometry ∂H. We can formalize the procedure by applying the method of weighted residuals to (4.2) Thus, in the same spirit as that discussed in (3.5), one may consider weighting the error in (4.2) by writing:

$$\int_{\partial H} w(P)\left\{ \frac{1}{2}\phi(P) + \int_{\partial H} \phi(Q)G_\nu d\partial\Omega_Q \right\} d\partial\Omega_P = \int_{\partial H} w(P)\int_{\partial H} f(Q)G\, d\partial\Omega_Q d\partial\Omega_P \tag{4.4}$$

where w(P) is again a weighting function. Consider now the simplest form of approximation of $\phi(P)$ and f(P):

$$\phi(P) \cong \sum_{i=1}^{N} \phi_i N_i \;, \qquad f(P) \cong \sum_{i=1}^{N} f_i N_i \;, \qquad N_i = \begin{cases} 1 & P \in \partial H_i \\ 0 & P \notin \partial H_i \end{cases} \tag{4.5}$$

where ϕ_i and f_i are discretization variables representing ϕ and f. (4.5) are basically 'pulse' functions within each boundary element ∂H_i. If the weighting function is chosen as a delta function: viz., $w(P) = \delta(P-P_i)$, $P_i \varepsilon \partial H_i$, $i=1,\ldots,N$, (4.4) becomes the method of point collocation. The linear algebraic system under the present approximation of ϕ is given by:

$$\frac{1}{2}\phi_i + \sum_{j=1}^{N} \left(\int_{\partial H_j} G_\nu \, d\partial\Omega\right)\phi_j = \sum_{j=1}^{N} \left(\int_{\partial H_j} f_j G \, d\partial\Omega\right), \quad i=1,..,N \quad (4.6)$$

essence, (4.2) is satisfied at a finite number of points on ∂H d a linear system for ϕ_j, j=1,...,N, results once the coefficients in the parentheses of (4.6) are calculated.

On the other hand, if w(P) is chosen as the set of N_i that ans ϕ_i , the so-called Galerkin method, a more complicated proximate form results:

$$\frac{1}{2}\phi_i A_i + \sum_{j=1}^{N} \left(\int_{\partial H_i} d\partial\Omega \int_{\partial H_j} G_\nu \, d\partial\Omega\right)\phi_j = \sum_{j=1}^{N} \left(\int_{\partial H_i} d\partial\Omega \int_{\partial H_j} f_j G \, d\partial\Omega\right)$$

$$i=1,..,N \quad (4.7)$$

ere A_i is the surface area of the i-th ∂H_i. (4.2) is now tisfied in an integrated sense over each ∂H_i.

The Galerkin form of integral-equation method has been carried it by Sayer (1980), Nihous & Webster (1983), and Barnell (1984) two dimensions. It is generally believed that for the same lue of N it gives better results than the collocation method. scretization errors of (4.6) and (4.7) have been studied carelly by Brite et al.(1985) for three-dimensional bodies. Very w works have addressed this systematically. Figure 8 shows a ble of results, reproduced from Brite et al., for a floating heroidal body. It shows the rather surprising finding that the llocation method actually outperformed the Galerkin method. wever, Brite et al. suggested that the Galerkin method may ill be preferred because it has better stability near irregular equencies. This conclusion must be viewed with some reervation. The coefficients in front of the unknowns in (4.7) volve double quadrature compared with those of (4.6). Since e computations of these are the primary cost contributors, a pical 9-point quadrature formula used in (4.7) would require ughly the same effort as that of the collocation method for 3N knowns; or perhaps 2N unknowns if auxilliary effects associated th the increase in matrix solution time is accounted for. An crease in N is known to improve the solution behavior near the regular frequencies (see Fig. 7). Such an increase also ovides a better discretized definition of ∂H. Evidence seems suggest that point collocation methods are more accurate than e Galerkin method at the same computation efforts. Computation st is a more equitable basis for comparison.

One very interesting feature of the Galerkin method, wever, is that the mixed-distribution formulation and sourceistribution formulation yield the same discretized matrix form. e paper of Brite et al. also suggests using Richardson's rmula to extrapolate the limiting behavior of the results if e solutions for two or more values of N are available.

	Added Mass				Damping			
	"C"	"G"	"C"	"G"	"C"	"G"	"C"	"G"
$\nu L/2$	N=16		N=64		N=16		N=64	
0.08	.0742	.0870	.0178	.0216	.0090	.0083	.0023	.0020
0.16	.0753	.0876	.0180	.0217	.0169	.0155	.0043	.0037
0.32	.0726	.0843	.0173	.0209	.0295	.0268	.0075.	.0064
0.80	.0533	.0648	.0123	.0162	.0486	.0428	.0122	.0101
1.60	.0239	.0365	.0047	.0094	.0486	.0405	.0121	.0094
3.20	.0110	.0232	.0013	.0064	.0216	.0129	.0052	.0025
4.80	.0145	.0253	.0022	.0071	.0143	.0034	.0032	.0003
6.40	.0170	.0279	.0028	.0077	.0076	-.0047	.0016	-.0015
8.00	.0202	.0311	.0040	.0084	.0050	-.0083	.0008	-.0024
16.00	-.0558	.0493	-.0050	.0029	.0520	-.0215	-.0003	-.0071
24.00	.0362	.0826	.0027	.0174	.0275	-.0309	-.0039	-.0038

"C" = collocation method, "G" =Galerkin method.

Figure 8. Discretization errors of the added mass and damping of a spheroidal body obtained by a three-dimensional panel program (Brite, et al., 1985). Errors are compared with the "benchmark results" based on spheroidal-panel method. First irregular frequency occurs at $\nu L/2 \cong 15$.

An alternative integral-equation formulation to the Green function method was proposed by Yeung (1973) (see also Bai & Yeung, 1974). It utilizes a 'simple (non-wavelike) source', i.e. $G = 1/(4\pi r)$. The resulting integral equation is given by:

$$2\pi\phi(P) + \int_{\partial F_0 \cup \partial B \cup \partial H} [\phi\partial_\nu + \beta\phi](\frac{1}{r})d\partial\Omega + \int_{\Sigma} [\phi^e\partial_\nu - \phi^e_\nu](\frac{1}{r})d\partial\Omega = \int_{\partial H} f(\frac{1}{r})d\partial\Omega\ , \quad \text{for } P \in \partial\Omega. \tag{4.8}$$

where β takes the same definition as that defined by (3.3). The unknowns ϕ^e and ϕ^e_ν on Σ can be determined by collocation after introducing the eigen expansion of (3.8) (Bai & Yeung, 1974) or by a relation of ϕ^e and ϕ^e_ν (Harten & Efrony, 1978) on Σ after introducing the appropriate analytical expression for ϕ^e in the exterior region Ω^e. This methodology has been implemented by Yeung (1975) for geometry involving different asymptotic water depths, by Harten & Efrony (1978) for multiple bodies, by Ferdinande & Kritis (1980) for axi-symmetric shapes, and by Nestegard & Sclavounos (1984) using a multipole expansion for ϕ^e. More recent works based on a hybrid approach in the time domain are described in the next section. The number of discretized unknowns based on this formulation is considerably larger than those based on Green functions. However, the kernel is simpler to evaluate and more receptive to far-field approximation. Further, numerical experience has indicated that it is not plagued by the presence of irregular frequencies. Angell et al. (1983) have recently provided a proof that this is indeed the case for a wave-like condition of (3.1) applied on Σ.

Higher-order approximations of the boundary-geometry and of the unknown functions are active areas of research in the field of boundary-element method. This exploits a finite-element type local representation of the boundary surface and unknowns. Hess & Smith have developed most of the expressions for a plane quadri-lateral element or panel. In ship hydrodynamics, expressions for triangular elements with a linear variation in the unknown function have been given by Webster (1975) and Yeung (1973). A spline technique for two-dimensional rectilinear element was described in Yeung & Bouger (1979). A 'tent function' repesentation was used by Nihous & Webster (1983).

5. Time-dependent Problems

Most of the discussions in earlier sections were confined to solutions in the frequency domain. Formulation for treating non-linear problems using a systematic perturbation procedure was discussed in an extensive review by Ogilvie (1983). Under such a procedure, higher-order problems have structures similar to the linear ones although complications associated with inhomogeneous boundary conditions on the free surface normally arise. There are, however, many occasions that a direct time-domain solution would be desirable, particularly in devising control strategy related to energy extraction in an unsteady wave train. If the free-surface condition (1.4) is linearized, the time-domain solution and frequency domain solution are related. However, there are certain features in the time-domain problem that permit more expedient solution treatment than those of the frequency domain.

A fully nonlinear procedure for calculating transient waves of a periodic type has been developed by Longuet-Higgins & Cokelet (1976). A Lagrangian type approach is employed to describe the evolution of the potential and position of the free surface:

$$D_t\vec{x} = \nabla\phi \quad , \qquad D_t\phi = \frac{1}{2}|\nabla\phi|^2 - gY \quad , \qquad \text{on } \partial F \qquad (5.1a,b)$$

These are basically the same conditions given earlier by (1.2a,b). The potential on any two-dimensional profile $\partial\Omega$ is now related by Green's theorem based on a simple source:

$$\pi\phi(P,t) + \int_{\partial F_\cup \partial H} [\phi\partial_\nu - \phi_\nu] \ln(1/r) \, d\partial\Omega = 0 \; , \quad P\epsilon\partial\Omega \; . \qquad (5.2)$$

which can be used as an integral equation of the first kind to solve for ϕ_ν for a given ϕ on ∂F , and as an integral equation of the second kind to solve for ϕ for a given ϕ_ν on ∂H. (5.1) provides the basis for the time advancement of ∂F and of ϕ on ∂F. The effects of a periodic boundary Σ was accounted for in the case of Lounget-Higgins and Cokelet by a conformal mapping. This can also be carried out by using a spatially periodic Green function. (5.1) and (5.2) form a simple yet powerful means of studying nonlinear wave-body interaction. Vinje & Brevig (1981) offered an improved version of (5.2) based on Cauchy's theorem. By considering the potential and stream function as unknown on ∂H (or ∂B) and ∂F respectively, the integral equation remains the second kind for both types of boundaries. Greenhow et al. (1982) applied the procedure to study the capsizing of a Salter's duck. Figure 9 illustrates some of the capabilities of the procedure in handling breaking waves. Lin et al. (1984) developed an algorithm to study the fluid motions in front of a wavemaker. A related integral equation based on a vortex-sheet formulation was also given by Baker et al. (1983). Attempts to generalize the procedure to genuine external flows have been met with limited success (see e.g. Isaacson, 1982). Most of the difficulties can be overcome by the development of a reliable nonlinear open boundary condition on Σ. A short discussion of the ones that are currently in use may be found in Yeung (1982a). The success of such a development may well depend on experience to be gained in using linear time-depenedent theory and on knowledge contributing to the understanding of nonlinear free-surface flow.

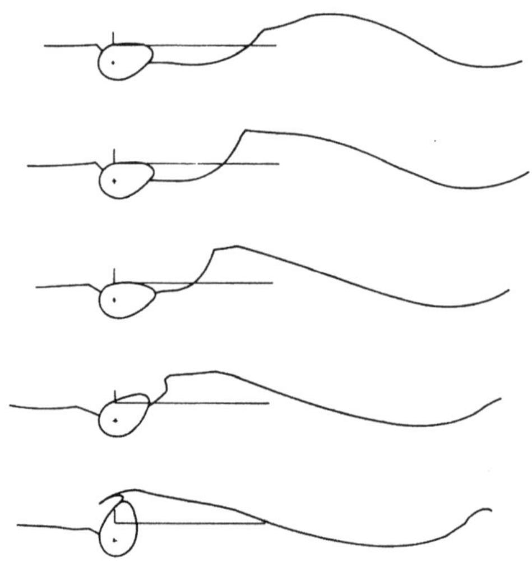

Figure 9. Time history of the capsizing of a Salter's duck by breaking waves. Solution obtained by using the integral-equation method of Vinje & Brevig (1981). Figure taken from Greenhow et al. (1982).

Within the context of linear theory, a time-dependent Green function satisfying (1.4) and two homogeneous initial conditions on the free surface was first derived by Finkelstein (1957). This may be conveniently written as:

$$G(P,Q,t) = (\frac{1}{r} - \frac{1}{r_1})/4\pi + H(P,Q,t)$$

$$\equiv G^\circ(P,Q) + H(P,Q,t) \qquad (5.3)$$

here P and Q denote field and source variables respectively and is the time elapsed after the Green function has been brought nto existence. The term G° represents the source and its negtive image above y=0. It corresponds to the high-frequency imit of the condition (1.4) in frequency-domain analysis. H epresents free-surface memory effects and is sometimes referred o as the retardation function. For a fluid whose motion is iniially at rest, the following integral equation for $\phi(P,t)$ on ∂H as given by Yeung (1982b):

$$\frac{1}{2}\phi(P,t) + \int_{\partial H} \phi(Q,t)G_\nu^\circ(P,Q)d\partial\Omega_Q = \int_{\partial H} f(Q,t)G^\circ d\partial\Omega +$$

$$+ \int_0^t d\tau \int_{\partial H} d\partial\Omega[\phi(Q,\tau)H_{\tau\nu}(P,Q,t-\tau)-f(Q,\tau)H_\tau] , \quad P\varepsilon\partial H. \qquad (5.4)$$

ree-surface memory effects, which occur in the form of a conolution integral, only modify the right-hand side of the ntegral equation since H has the property of $H_\tau(P,Q,0)=0$. The ernel associated with the left-hand side of (5.4) is time inependent and needs to be inverted just once. The preponderant omputation efforts are associated with the time and spatial inegrals of H_τ and $H_{\tau\nu}$, particularly at large time argument.

Solutions of (5.4) or its variants have been sought by Daoud 1976), Adachi & Ohmatsu (1979), Guo(1981), Yeung & Kim (1981), eung (1982), and Newman (1985). In calculating the unsteady reponse of a two-dimensional body, subject to a unit step input: .e. $\phi_n(P,t)=u(t-0+)$ for P on ∂H, with u(t) being the heaviside tep function, Adachi & Ohmatsu noticed that the solution based n a source formulation oscillated violently but not the one ased on the mixed distribution (5.4). The unwarranted oscillaions have a period very close to the first irregular frequency f the interior problem in $\bar{\Omega}$. Newman (1985) observed similar scillations even by using (5.4) for an axisymmetric body in eave. In obtaining the transient response of a freely floating ylinder, Yeung (1982) solved for the boundary condition f on ∂H imultaneously with the rigid-body dynamics instead of using the sual step-response technique. No such oscillations were bserved.

The occurrence of such an anomalous oscillatory behavior in he time domain seems peculiar. The kernel of (5.4) is one ssociated with the fluid motion of the body and its image above =0 in an infinite fluid (for the case of heave motion) and its redholm determinant does not vanish except for very shallow-draft geometry. In order to understand the origin of such a behavior, the original algorithm of Yeung (1982b) was run at a number of combinations of space and time discretization. As in the original paper, the evaluation of the last integral in (5.4) at time $t=t_k$, and at collocation point P_i, was carried out by the approximation:

$$I(P_i) \cong \sum_{m=1}^{k} \sum_{j=1}^{N} \left(\phi_j^m \gamma_{ij}^{k,m} - f_j^m \lambda_{ij}^{k,m} \right) \tag{5.5a}$$

where

$$\begin{matrix} \gamma_{ij}^{k,m} \\ \lambda_{ij}^{k,m} \end{matrix} \equiv \int_{t_{m-1}}^{t_m} d\tau \int_{\partial H_j} d\partial\Omega_Q \begin{matrix} H_{\tau\nu} \\ H_\tau \end{matrix} (P_i, Q, t_k-\tau) \quad . \tag{5.5b}$$

The γ_{ij}'s and λ_{ij}'s were integrated analytically and evaluated to 6-digits accuracy to avoid inaccuracies due to large-time oscillations in the integrand. Approximations occur only in using step representations of ϕ and f by ϕ_j^m and f_j^m based on the midpoint approximation.

Figure 10a shows the step response of the memory portion of the force acting on a circular cylinder due to a unit step in the vertical velocity. The time step in (5.5b) is taken as constant, step size being denoted by $DT \equiv \Delta t/\sqrt{(g/a)}$. Results are given for various combinations of N and time steps. N was chosen to be excessively small so as to induce the said oscillations. Curves with the same N are offset vertically by a value of 0.1 for clarity of display. Apparently, equation (5.4) is susceptible to instability caused by space and time errors in the solution. Such errors could induce, in a manner similar to (4.1), a form of initial disturbances on the free surface which oscillates at the irregular frequencies (primarily only the first mode) of the interior adjoint homogeneous problem. Newman (1985) pointed out that an increase in N will reduce the amplitude of the oscillations. It appears that both time-step and N are interrelated. In a typical situation, a reduction of Δt or an increase in N can reduce the undesirable behavior. Figure 10b is a plot of two solutions, each having the same number of calculations of γ_{ij}'s and λ_{ij}'s, viz. N^2k=constant. The two curves practically overlay each other. Interestingly, the accuracy of the solution depends on how much total effort one 'invests' in the γ_{ij}'s and λ_{ij}'s. Fortunately, the unwarranted oscillations are removable by more precise time and space grid.

An important property of the γ_{ij}'s and λ_{ij}'s must not go unnoticed since it greatly affects the economy and feasibility of such computations. It was first pointed out in Yeung (1982b). The occurrence of the sequence γ_{ij}'s and λ_{ij}'s as k changes is illustrated in Figure 11, where m is associated with the dummy time variable τ and k with the time t when the solution is desired. For any given (i, j), the "pyramid" structure of Figure 11 suggests a total evaluation of $N^2(k+1)k$ such "memory" coefficients. However, if the time grid is constant, it is not dif-

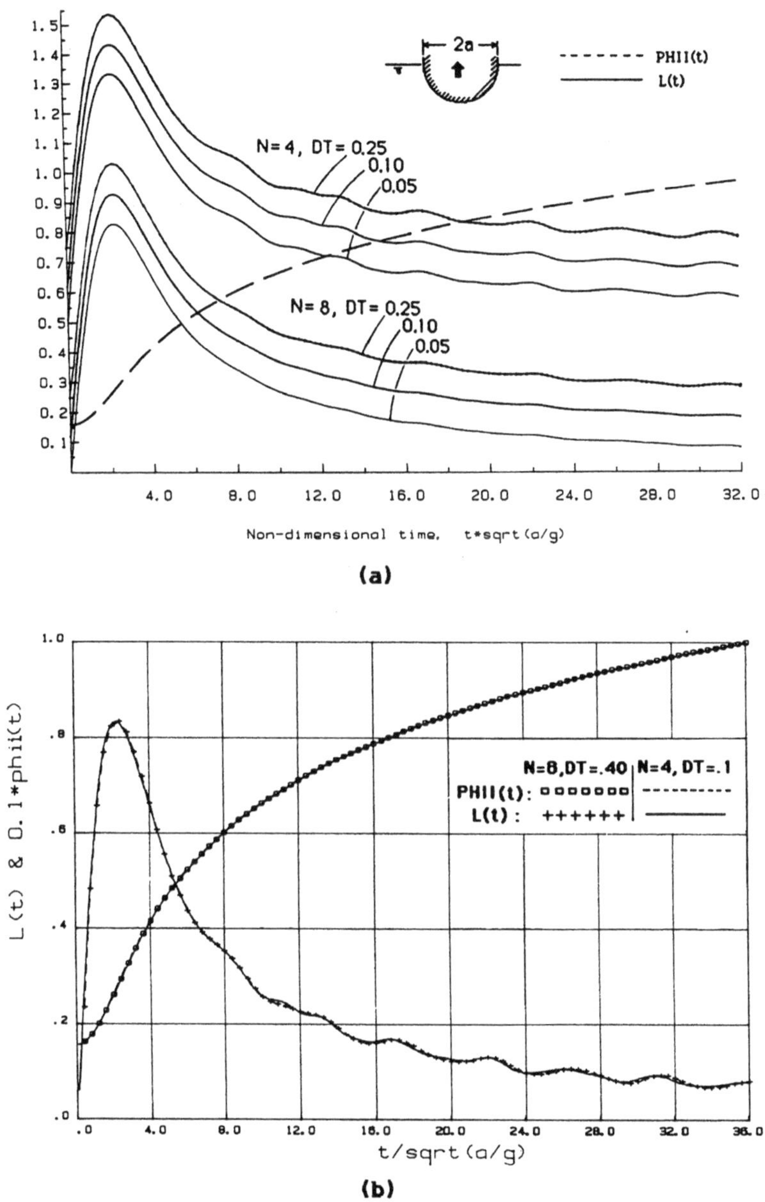

Figure 10. Effects of spatial and time grid on the step heave-response of a surface-piercing circular cylinder. The memory-effects force function L(t) is nondimensionalized by $\rho g a^2$. The total hydrodynamic force is given by $\mu(\infty)$ x(instantaneous acceleration/g) + (convolution of L(t) with past history of body acceleration), $\mu(\infty)$ being the infinite-freqeuncy non-dimensional added mass. PHII(t) denotes the integral of $[\phi(Q,t)n_y(Q)]$ over the cylinder. The first theoretical 'irregular period' is 1.348.

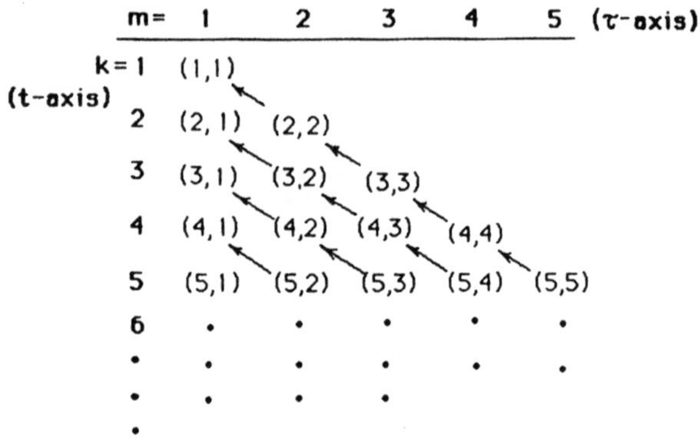

Figure 11. Recursive structure of the pyramid of memory coefficients in time-dependent type calculations.

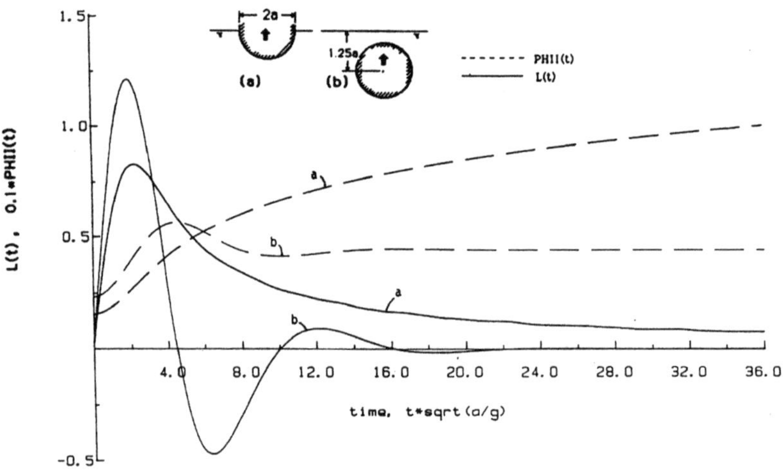

Figure 12. Step response of a source-like body versus that of a dipole-like body. Note the oscillatory behavior of a dipole-like body. Both bodies are given step-function type heave motion.

ficult to verify that

$$\gamma_{ij}^{k,m} = \gamma_{ij}^{k-1,m-1} , \quad \lambda_{ij}^{k,m} = \lambda_{ij}^{k-1,m-1} , \quad m\geq 2 . \qquad (5.6)$$

Thus, only the first element of each row is distinct. This implies immediately that the storage of these $2N^2k$ distinct numbers can save a factor of k/2 in computation time! This interesting property does not depend on the means of evaluating the memory coefficients (5.5b). However, it ceases to be true when variable time stepping is used.

The slow decay rate ($\sim t^{-1}$) shown in these repsonse curves makes it unattractive to apply Fourier transform techniques to recover frequency-domain results. An analytical knowledge of the tail of the response is needed to improve the Fourier resolution. Fortunately, the case shown here is just about the worst situation that one may be confronted with (three-dimensional heaving body in finite-depth water has the same property, see Yeung, 1981). In most of the other situations where the net flux from the body boundary condition f vanishes, a dipole type behavior results a much faster decay rate. For example, the step response of a submerged and heaving circular cylinder (vertical dipole) is shown and compared with that of a surface-piercing one in Figure 12. Note that the genuine oscillations of a dipole-like response should not be confused with the spurious numerical oscillations mentioned earlier.

Recently, Lee (1985) completed a thesis for the time-domain solution of a swaying axisymmetric structure based on a hybrid formulation that uses a 'simple-source' representation of (4.8) in the domain Ω. On the open boundary Σ, which was chosen as a vertical cylinder, (5.4) was used with ∂H replaced by Σ. Hence, one has:

$$\frac{1}{2}\phi^e(P,t) + \int_\Sigma \phi^e(Q,t)G^\circ_\nu(P,Q)\,d\partial\Omega_Q = \int_\Sigma \phi^e_\nu(Q,t)G^\circ\,d\partial\Omega + \int_0^t d\tau \int_\Sigma d\partial\Omega[\phi^e(Q,\tau)H_{\tau\nu}(P,Q,t-\tau)-\phi^e_\nu(Q,\tau)H_\tau] , \quad P\epsilon\Sigma. \tag{5.7}$$

Figure 13 illustrates that the surface Σ behaves like a cylindrical wavemaker whose surface motion flexes back and forth so that it is compatible with the solution in Ω and with the memory effects associated with the exterior domain Ω^e. In the time domain, a straight-forward Sommerfeld-type condition does not quite work out if the waves are dispersive and disparate wavelengths occur simultaneously. Lee (1985) showed that the matching condition of (5.7) was analytically exact for both wavelike and non-wavelike disturbances in linear thoeory. An additional approximation can be made in (5.7) to achieve only wave-type matching. Figure 14 shows a comparison of the step responses of a spherical body in sway motion, obtained from both the frequency domain and the time domain, the latter using the new matching condition of Eq. (5.7). Some sample calculations for a flexible body were also described by Lee.

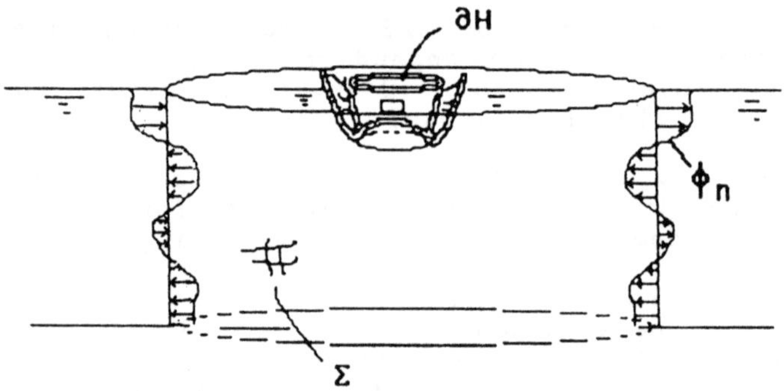

Figure 13. A flexible cylindrical wavemaker to represent time-dependent radiation, Eq. (5.7), (Lee, 1985).

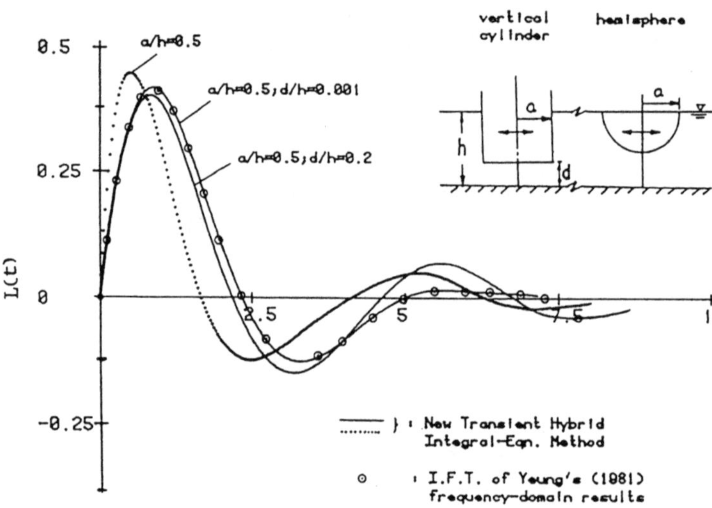

Figure 14. Transient sway response of a vetical cylinder and a sphere in water of finite depth obtained by a hybrid integral-equation method (Lee, 1985).

The Shell Functions - A New Idea

The successful use of the open-boundary condition (5.7) paves way for the following very useful idea for solving transient hydrodynamics problems. Since the effects of the body geometry enter only through ϕ^e and ϕ^e_ν in (5.7), the calculations of the functions γ_{ij}'s and λ_{ij}'s defined by (5.5), but now evaluated on Σ, are completely independent of the internal geometry. We shall name these "**shell functions**". For a given discretized form of Σ and given time grid, the "shell functions" or "**shell coefficients**" need to be calculated only once for all and then stored for later use. At any given time t, (5.7) can be sed to obtain the relation:

$$\{\phi^e\} = \{\mathbf{A}\}^{-1}[\{\phi^e_n\} + \{B\}] , \qquad \text{for } \{\phi^e\}, \{\phi^e_n\} \text{ on } \Sigma. \tag{5.8}$$

here the matrix $\{\mathbf{A}\}^{-1}$ is evidently independent of time, and the hell coefficients are used only to update the vector {B}. The olution in the region Ω for any internal geometry can now be btained using (4.8), as in the case of Lee (1985), or any 'local ethods' such as finite-element and finite-differences, for that atter.

We have just introduced a set of universal free-surface funtions, the shell functions, which can be used for all methods, ll internal geometry, and within certain limitations, a large ariation of water depths. The concept has the unique feature hat the shell needs to be generated only once and can be ubsequently used for all purposes. Further, if we take dvantage of the recursion form (5.6), only $2\tilde{N}^2k$ distinct numbers eed to be kept, $\tilde{N}$ being the number of discretized points on Σ. ince Σ can be chosen at will, efforts of calculating the shell unctions can be minimized by taking advantage of simplifications ue to the symmetry of Σ or other useful geometric properties. or a surface Σ formed by a revolution about a vertical (y) xis, the combination of 32 circumferential and 8 depthwise ectors yields a total of 204,800 shell coefficients for k=100. hese, when properly stored and retrieved, can be used to perform ny three-dimensional time-dependent calculations in Ω with a reatively trivial amount of computational effort. With this echnique, all linearized three-dimensional unsteady calculations eems to be within easy reach of the present-day super-micro or ven micro computers.

. Conclusions

It is clear that the development of numerical methods for requency-domain solutions in both two and three dimensions has eached a state of maturity. A multitude of integral-equation ethods and hybrid-type finite-element methods has already been mplemented for production-type calculations. We foresee an inreased emphasis on numerical accuracy and efficiency. With hose factors under control, extensions of these methods to igher-order theories based on perturbation analysis can be expected.

The difficulties occurring at irregular frequencies in reen-function methods and in finite-element methods that are oupled with Green-functions will prevent these methods from beoming truly versatile for some time. New formulations, owever, are available for overcoming these hurdles. Finitelement methods and simple-source integral-equation methods that xploit the eigen-function matching approach have now been proven o be free of irregular frequencies. They are viable and ompetitive computational alternatives. The finite-depth eigen eries is known to converge slowly in deeper water. In simulating infinite water-depth calculations, a trade-off exists between the lowest frequency desired and the number of terms to be used in the series.

There has been a bit of a myth that finite-difference formulations are inflexible for arbitrary body geometry. The extensive development in boundary-fitted coordinate methods in the last decade has basically overcome this deficiency. The potential in much of that area remains untapped. Finite difference and finite-element techniques will be especially useful when the field equation is inhomogeneous, which will be the case if viscous effects need to be considered.

Methods for time-dependent type solutions are being developed gradually. When properly implemented, the Green function approach in the time domain is not affected by irregular frequencies as severely as the corresponding formulation in the frequency domain.

A new time-dependent radiation (or open) boundary condition for free-surface flows has been validated and presented in this paper. This condition is effective for bodies of all shapes and is expected to be applicable to many methods using a localized type formulation. In particular, it can be used in conjunction with a novel concept known as "shell functions" which was introduced in §5. Being independent of the method of solution, these shell functions need to be evaluated only once, and can be used repeatedly for any body geometry in time-dependent type calculations.

Acknowledgement

Preparation of this paper as well as portions of the research reported in §4 and §5 have been supported by the Office of Naval Research under contract N00014-84-K0026. I am indebted to my friend Nathan Cheung whose generosity has enabled me to generate a decent final print-out of this paper.

References

Adachi, H., Ohmatsu, S. 1979. On the influence of irregular frequencies in the integral equation solutions to the time-dependent free surface problems. J. Soc. Naval Arch. Japan, 146:127-35.

Angell, T. S. , Hsiao, G. S. , Kleinman, R. E. 1983. An integral equation for the floating body problem. Proc. Int. Workshop Ship & Platform Motions, Berkeley, Calif. 443-49.

Bai, K. J. 1972. A variational method in potential flows with a free surface. Univ. of Calif., Berkeley, College of Engrg. Rep. NA 72-2.

Bai K. J., Yeung, R. W. 1974. Numerical solution to free-surface flow problems. Proc. 10th Symp. Naval Hydrodyn. Cambridge, Mass. 609-33.

Baba, E. 1981. Some free-surface phenomena around ships to be challenged by numerical analysis. Proc. 3rd Int. Conf. Numer. Ship Hydrodyn. 9-26.

Baker, G. R., Meiron, D. I., Orszag, S. A. 1981. Application of a generalized vortex method to nonlinear free surface flows. Proc. 3rd Int. Conf. Numer. Ship Hydrodyn. 179-92.

Barnell, A. 1984. Boundary-integral methods for solving potential flows. M.S. Thesis, Dept. Ocean Engrg. Mass Inst. of Technology.

Berkhoff, J. C. W. 1972. Computation of combined refraction - diffraction. Proc 13th Coastal Engrg. Conf., Vancouver, Britsih columbia, 471-90.

Brite, S. R., Newman, J. N., Sclavounos, P. D. 1985. A new generation of panel programs for radiation-diffraction problems. BOSS '85, Delft, Holland.

Burton, A. J., Miller, G. F. 1971. The application of integral equation methods to the numerical solution of some exterior boundary value problems. Proc. R. Soc. Ser. A., 201-10.

Chapman, R. B. 1981. Time-domain method for computaing forces and moments acting on a three-dimensional surface-piercing ship hull with forward speed. Proc. 3rd Int. Conf. Numer. Ship Hydrodyn. 237-48.

Chan, R. K. C., Hirt, C. W. 1974. Two-dimensional calculations of the motion of floating bodies. 10th Symp. Naval Hydrodyn, Cambridge, Mass. 667-82.

Carey F. C., Oden, J. T. 1983. Finite elements: a second course. New Jersey: Prentice Hall.

Chen, H. S., Mei, C. C. 1974. Oscillation and wave forces in a man-made harbor in the open sea. Proc. 10th Symp. Naval Hydrodyn. Cambridge, Mass. 573-96.

Chenault, W. 1970. Motion of a ship at the free surface. MS thesis. Naval Postgraduate School, Monterey, Calif.

Daoud, N. 1975. Potential flow near to a fine ship's bow. Dept. Naval Arch & Marine Engrg. Univ. Michigan, Rept. 177.

Eatock Taylor, R., Zietsman. J. A comparison of localized finite element formulations for two-dimensional wave diffraction and radiation problems. Int. J. Numer. Meth. Engrg. 17:1335-84.

Euvrard D., Jami, A., Lenoir, M., Martin, D. 1981) Recent progress towards optimal coupling between finite-elements and dingularity distribution procedures. Proc. 3rd Int. Conf. Numer. Ship Hydrodyn. 193-208.

Ferdinande, V. , Kritis, B. G. 1980. An economical method of determining added mass and damping coefficients of axisymmetric floating bodies in oscillatory heaving motions. Int. Shipbldg. Prog. 27:(313) 231-40.

Finkelstein, A. B. 1957. The initial problem for transient water waves. Comm Pure Appl. Math. 10:511-22.

Flokstra C., Berkhoff, J. C. W. 1977. Propagation of short waves over a shoal. Delft Hydraulic Lab. Rept. W 154-V (in Dutch).

Frank, W. 1967. Oscillation of cylinders in or below the free surface of deep fluids. Naval Ship Res. Dev. Center. Rept. 2375.

Gallagher, R. H., Oden, J. T., Zienkiewicz, Kawai, T., Kawahara, M. 1984, ed. Finite elemets in fluids, vol 5.

Ghia, U., Shin, C. T., Ghia, K. N. 1981. Analysis of a breaking free-surface wave using boundary-fitted coordinate for regions including reentrant boundaries. Proc. 3rd Int. Conf. Numer. Ship Hydrodyn., Paris, France, 133-46.

Greenhow, M., Vinje, T., Brevig, P., Taylor, J. 1982. A theoretical and experimental study of the capsize of Salter's duck in extreme waves. J. Fluid Mech. 118:259-75.

Guo, C-L. S. 1981. Nonlinear theory of ship maneuvering. J Ship. Res. 21-43.

Harten, A., Efrony, S. 1978. A partition technique for the solution of potential flow problems by integral equation methods. J. Comp. Phys. 27:71-87.

Haussling, H. J. 1982. Solution of nonlinear water wave problems using boundary-fitted coordinate systems. Numerical Grid Generation, ed. J. F. Thompson, North-Holland Publ., 385-408.

Havelock, T. 1955. Waves due to a floating sphere making periodic heaving oscillations. Proc. Roy. Soc. London, Ser. A. 231:602-8

Hearn, G. E. 1977. Alternative methods of evaluating Green's function in three-dimensional ship-wave problems. J. Ship Res. 21:89-93.

Hess, J. L., Smith, A. M. O. 1967. Calculation of potential flows about arbitrary bodies. Prog. in Aero. Sciences. 8:1-137. See also Hess, J. L. 1960, Douglas Aircraft Co. Rept. ES29812.

Hogben, N., Standing, R. G. 1974. Wave loads on large bodies. Proc. Int. Symp. Dyn. Marine Vehicles & Struct., Univ. College, London, 258-77.

Hulme, A. 1982. The wave forces acting on a floating hemisphere undergoing forced periodic oscillations. J. Fluid Mech. 121:443-63.

Isaacson M. de St. Q. 1982. Nonlinear-wave effects on fixed and floating bodies. J. Fluid Mech. 120:267-81.

John, F. On the motion of floating bodies, II, simple harmonic motions. Comm. Pure Appl. Math. 3:45-101.

Jami, A., Polyzakis, M. 1981. A finite element solution of diffraction problems in unbounded domain. Comput. Meth. Appl. Mech. & Engrg. 29:1-18.

Kleinman R. E. 1982. On the mathematical theory of the motion of floating bodies. D.W. Taylor Naval Ship Res. & Dev. Ctr., Rept. 82/074.

Kleinman, R. E., Roach. G. F. 1974. Boundary-integral equations for the three-dimensional Helmholtz equation. SIAM Review, 16:214-36.

Kim, W. D. 1965. Oscillations of a rigid body in a free surface. J. Fluid Mech. 21:427-51.

Kobus, J. M. 1976. Application de la methode des singularites au probleme des flotteurs cylindriques soumis a des oscillations harmoniques forcees de faible amplitude. These (Docteur-Ingeniere). Univ. Nantes, France.

Lee, M. Y. 1985. Unsteady fluid-structure interaction in water of finite depth. Ph.D. thesis, Dept of Naval Arch. & Offsh. Engrg., Univ. Calif., Berkeley.

Lebreton, J. C., Margnac, M. A. 1966. Traitement sur ordinateur de quelques problems concernant l'action de la houle sur corps flottants en theorie bidimensionelle. Bull. Centre Res. D'essais de Chatu.

Lin, W. M., Newman, J. N., Yue, D. K-P. 1984. Nonlinear forced motions of floating bodies. Proc. 15th Symp. Naval Hydrodyn., 33-47.

Longuet-Higgins M. S., Cokelet, E. D. 1976. The deformation of steep surface waves on water. I. A numerical method of computation. Proc. R. Soc. London Ser. A 350:1-26.

Lenoir, M. , Jami, A. 1978. A variational formulation for exterior problems in linear hydrodynamics. Comp. Meth. Appl Mech Engrg. 16:341-59.

Martin, P. A. 1981. On the null-field equations for water-
ıve radiation problems. J. Fluid Mech. 113:315-332.
Mei, C. C. 1983. The applied dynamics of ocean surface waves.
Wiley & Sons.
Newman, J. N. 1983. Three-dimensional wave interaction with
ıips and platforms. Proc Int. Workshop Ship & Platform Motion,
rkeley, Calif., 418-42.
Newman, J. N. 1984. An expansion of the oscillatory source
tential. Appl. Ocean Res. 6:116-7.
Newman, J. N. 1985. Transient axisymmetric motion of a float-
ıg cylinder. J. Fluid Mech. to appear.
Nestegard, A. , Sclavounos, P. D. 1984. A numerical solution
two-dimensional deep-water wave-body problems. J. Ship Res.
Nihous, G., Webster, W. C. 1983. An analysis of a wave-power
sorber. Proc Int. Workshop Ship and Platforms. Berkeley, Calif.
2-162.
Noblesse, F. 1982. The Green function in the theory of radia-
on and diffraction of regular waves. J. Engrg. Math. 16:137-69.
Ogilvie, T. F. 1983. Second-order hydrodynamic effects on
ean platforms. Proc. Int. Workshop Ship & Platform Motions.
rkeley, Calif. 205-65.
Ogilvie, T. F., Shin, Y. S. 1978. Integral-equation solutions
r time-dependent free-surface problems. J. Soc. Naval
chitects, Japan. 143:41-51.
Porter, W. R. 1960. Pressure distributions, added mass, and
ımping coefficients in a free surface. Univ of Calif., Inst.
ıgrg. Res. Rept. 8216.
Radder A. C. 1979. On the parabolic equation method for
ıter-wave propagation. J. Fluid Mech. 95:159-76.
Sayer, P. 1980. An integral-equation method for determining
ıe fluid motion due to a cylinder heaving on water of finite
pth. Proc. Roy. Soc. Ser A. 372:93-110.
Seto, H., Yamamoto, Y. 1975. finite element analysis of
ırface wave problem by a method of superpostion. Proc. 1st Int.
nf. Numer. Ship Hydrodyn., Gaithersburg, Md. 49-70.
Seto H. 1978. Fundamental studies on steady ship wave
oblems by the finite element method (the fourth report). J.
c. Naval Arch Japan, 144:(12)88-95 (in Japanese).
Susbielles, G., Bratu C. 1981. Vagues et Ouvrages Petroliers
Mer. Editions Technip, Paris.
Steinberg, S., Roache, P. J. 1985. Symbolic manipulations and
mputational fluid dynamics. J. Comp. Phys. 57:251-84.
Tasai, F. 1959. Damping force and added mass of ships heaving
ıd pitching, Rept. Res. Inst. Kyushu Univ., 7:131-52.
Thames F. C., Thompson, J. F., Mastin, C. W., Walker, R. L.
77. Numerical solutions for viscous and potential flow about
bitrary two-dimensional bodies using body-fitted coordinate
stem. J. Comp. Phys. 24:245-68.
Thacker, W. C. 1980. A brief review of techniques for
nerating irregular computational grids. Int. J. Numer. Meth.
ıgrg. 15:1335-42.
Thompson, J. F., Warsi, Z. U. A., Mastin, C. W. 1982. Boun-
ıry-fitted coordinate systems for numerical solution of partial
fferential equations - a review. J. Comp. Phys. 47:1-108.
Ursell, F. 1949. On the heaving motion of a circular cylinder
the free surface of a fluid. Quart. J. Mech. Appl. Math.
218-31.

Ursell, F. 1981. Irregular frequencies and the motion of floating bodies. J. Fluid Mech. 105:143-46.

Vinje, T., Brevig, P. 1981. Numerical simulation of breaking waves. Adv. Water Res. 4:77-82.

Wang, S. 1966. The hydrodynamic forces and pressure distributions for an oscillating sphere in a fluid of finite depth. Ph.D. dissertation, Dept. of Naval Arch. Marine Engrg., Mass. Inst. of Tech.

Waterman, P. C. 1969. New formulation of acoustic scattering. J. acoustic soc. Am. 45;1417-29.

Webster, W. C. 1975. The flow about arbitrary three-dimensional body. J. Ship Res. 19:206-18.

Webster, W. C. 1978. Computation of the hydrodynamic forces induced by general vibration of cylinders. J. Ship Res. 23:9-19.

Wehausen, J. V. 1977. Methods for boundary-value problems in free-surface flows. D. W. Taylor Naval Ship Res. & Dev. Ctr. Rept. 4622.

Wehausen, J. V., Laitone, E. V. 1960. Surface Waves. Handbuch der Physik 9:446-778. Berlin:Springer Verlag.

Winslow, A. M. 1966. Numerical solution of the quasilinear Poisson equation in non-uniform triangular mesh. J. Comp. Phys. 1:149-72.

Yeung, R. W. 1973. A singularity-distribution method for free-surface flow problems with an oscillating body. Univ. of Calif., Berkeley, Calif. College Engrg. Rept. NA73-6.

Yeung, R. W. 1975. A hybrid integral-equation method for time-harmonic free-surface flow. Proc. 1st Intl Conf. Numer Ship Hydrodyn., 581-608.

Yeung, R. W. 1982a. Numerical Methods in Free-surface flows. Ann. Rev. Fluid Mech. 14:395-442.

Yeung, R. W. 1982b. The transient heaving motion of floating cylinders. J. Engrg. Math. 16, 97-119.

Yeung, R. W., Bouger, Y. C. 1979. A hybrid integral-equation method for steady two-dimensional ship waves. Int. J. Numer. Meth. Engrg. 14:317-36.

Yeung, R. W., Kim, S. H. 1981. Radiation forces on ships with forward speed. Proc. 3rd Int. Conf. Numer. Ship Hydrodyn. Paris, France. 499-516.

Yue, D. K-P., Chen, H. DS., Mei, C. C. 1978. A hybrid element method for diffraction of water waves by three-dimensional bodies. Int. J. Numer. Meth. Engrg. 12:245-66.

Zienkiewicz, O. C. 1970. The finite element method in engineering science. London: McGraw Hill.

Zienkiewicz, O. C., Kelly, D. W., Bettess, P. 1977. The coupling of the finite element method and boundary solution procedures. Int. J. Numer. Meth. Engrg. 11:355-75.

Zietsman, J.F.W. 1984. The coupled finite element and boundary integral analysis of ocean wave loading: a versatile tool. Comp. Meth. Appl. Mech. Engrg. 44:153-76.

Survival of Surface-Piercing Wave Energy Devices in Extreme Waves

M. GREENHOW

Department of Mathematics
University of Manchester
Manchester

Abstract

This paper extends and exploits a numerical method of calculation of the development of a wave approaching a fixed or moving body. The theory is strictly two-dimensional which implies that the body length is much greater than its breadth or depth, and that it lies parallel to the wave crests. Many wave energy devices of the terminator type, either mounted upon the sea bed or on rigid or flexible spines, satisfy these requirements and therefore the present method can be used to study their behaviour in extreme waves. This paper considers the general problem of forces and pressures on a circular spine section when exposed to a large wave, and also considers two specific examples, the "Clam" and an Oscillating Water Column (O.W.C.).

Introduction

In recent years there has been a lot of theoretical and experimental activity in evaluating the performance of devices designed to extract energy from ocean waves. The devices divide up into three broad categories: attenuators, which are long devices lying along the wave direction; terminators, which are long devices lying parallel to the wave crests, and point absorbers. For details of device design see Cottrill (1981). Much mathematical work has been directed at ascertaining the efficiency of power extraction (or absorption length in the case of point absorbers) and extensive use has been made of linearised floating body theory (see eg the review by Evans (1981)). In early work the power extraction mechanism was modelled by springs and dampers but more realistic power take - off can be simulated in the time domain (see e.g. Greenhow, Rosen and Reed (1984) and other papers in this conference.) Terminators have special appeal since many features may be explored with two-dimensional theories or testing in a narrow wave tank, and it is probably for this reason that design of these devices is more advanced than the other device types. It is worth mentioning that in all cases theoretical and experimental device efficiences are uniformly high for all devices currently considered, and so does not give a rational basis for prefering one device above the others. Here engineering considerations must come into play: how much will the device cost, can it be made reliable without servicing for long periods of time and can the device survive extreme waves (20m crest to trough) and of extreme steepness which could cause breaking near or on the device. This paper aims to provide a partial answer to the last question.

A possible method of looking at extreme waves is simply to scale up the effects of small waves calculated from linearised theory. This approach is highly doubtful in view of the assumptions of neglecting quadratic terms in the free surface kinematic and dynamic conditions and applying boundary conditions on the equilibrium positions of both the body and the free surface. For example results given later show that allowing the cylinder to move freely reduces the wave forces considerably. Another objection is that even this simplified theory is not that easy to apply: an alternative empirical approach is to apply Morison's equation (see e.g. Newman (1977)):

$$F = (m + \rho\forall)\dot{U} + \tfrac{1}{2}\rho\ell^2 C_D U/U/ \qquad (1)$$

Here it is assumed that the body dimension ℓ is small compared to the wavelength λ. U is the velocity of the fluid relative to the body centroid, $\dot{U}$ is the relative acceleration, μ is the added mass, ρ = density and $\forall$ = volume of displaced fluid. C_D is a drag coefficient caused by viscous forces and the second term is omitted here for comparison with the inviscid theory below. Use of Morison's equation with the acceleration $\dot{U}$ given by linearised wave theory for a wave of amplitude 10m acting upon a cylinder of diameter 10m centred at the still water level gives a maximum force of 440 KN/m in both horizontal and vertical directions. Whilst this force is of the correct order, it is seen that as the wave engulfs the cylinder much higher forces are experienced.

It should also be noted that linearised theory usually deals with steady state periodic waves whereas wave breaking is highly transient. Thus while steady state theories give a limit on wave steepness ($a/\lambda \sim 0.14$), non-steady waves may be much steeper leading to overturning. The jets arising from overturning are thought to be especially dangerous, giving slamming loads of short duration. Possible approaches are numerical (see Vinje and Brevig (1981c)) or analytical, assuming self-similar expansion of the impact of a wedge of water onto a flat plate and ignoring gravity (see Cumberbatch (1960)). In the present paper, ejected jets do not strike the body before calculations are terminated for other reasons.

Perhaps the most rational way of proceeding is through extensive model testing coupled with wave statistics to ascertain the probability of exceedence of a certain value of wave loading (see Ochi and Tsai (1984)). Such knowledge would lead to an economical "death rate" for devices by wave action (almost certainly not equal to zero!). In the wave energy context, extensive testing of a Salter's Duck has been carried out at Edinburgh and is described by Greenhow et al (1982). In that paper the numerical method of Vinje and Brevig (1981a) was also used, giving good agreement with the experiments. This method, which is modified and applied in the present paper, is based upon earlier work by Longuet-Higgins and Cokelet (1976) except that bodies may be introduced either below the free surface (see Brevig et al (1981)) or at the free surface (see Vinje and Brevig (1981b)). Both that paper and Greenhow et al give details of the mathematical formulation and the numerical methods used, so the present paper is confined to an outline of the method only, and a discussion of recent progress towards treating the intersection of the body and free surface.

2. Outline of the Method

Under the usual assumptions of homogeneous, irrotational and incompressible fluid we may admit the use of a complex velocity potential

$$\beta(z,t) = \phi(x,y,t) + i\psi(x,y,t) \tag{2}$$

where $z = x+iy$, ϕ is the velocity potential and ψ is the stream function. Since ϕ and ψ satisfy Cauchy-Riemann equations, β is analytic and hence Cauchy's theorem holds:

$$\oint_C \frac{\phi + i\psi}{z - z_0}\,dz = 0 \tag{3}$$

for z_0 outside any closed contour C within the fluid. Here we take C to be the free surface, the wetted body surface, the bottom and two vertical end boundaries situated a long way from the body. Because of equation 3 the two-dimensional problem reduces to a one-dimensional one where the entire fluid motion is specified by values of β on the boundary C alone.

On the free surface ϕ is specified initially and calculated for later times by Bernoulli's equation:

$$\frac{D\phi}{Dt} = \tfrac{1}{2}ww^* - gy \tag{4}$$

Here the new value of ϕ is given at the new position of a marked fluid particle which is followed in time (Lagrangian description). The fluid velocity w^* and hence the new position of the marked particle on the free surface is given by

$$\frac{Dz}{Dt} = w^* = u + iv = \left(\frac{\partial\beta}{\partial z}\right)^* \tag{5}$$

On the bottom ψ is constant, while on the body ψ is given by the condition of matching normal fluid and body velocities. Thus:

$$\psi(x,y) = u_0(y - y_0) - v_0(x - x_0) \tag{6}$$

where (u_0,v_0) is the velocity of the centre of gravity of the body at (x_0,y_0).

On the vertical boundaries ψ and ϕ are both unknown but periodicity is assumed to hold. This reduces the number of unknowns in the final equations making the resulting matrix equation soluble.

On the boundary we know either ϕ, on part of the boundary designated by C_ϕ, or ψ on C_ψ and the problem reduces to calculating the unknown part of β from equation 3. Thus:

$$\alpha\psi(x_0,y_0,t) + \mathrm{Re} \int \frac{\phi + i\psi}{z - z_0}\,dz = 0 \tag{7}$$

for z_0 on C_ϕ and

$$\alpha\phi(x_0,y_0,t) + \mathrm{Re} \int \frac{\phi + i\psi}{z - z_0}\,dz = 0 \tag{8}$$

for z_0 on C_ψ. Here α is the angle between two tangents of C at z_0 (equal to π for any smooth part of C mathematically, but generally different from π in the numerical scheme.)

The principle valued contour integrals in equations 7 and 8 are not known explicitly since they include unknowns (either ϕ or ψ) at other points around the boundary. Discretising the contour into points z_k along C and applying either equation 6 or 7 for each of these points gives the required matrix equation. Typically in the calculation described here 84 points were used on the free surface, 20 on the body and 140 in total.

To solve for the pressures on the body points, and hence integrate to give the body forces and accelerations, requires knowledge of $\frac{\partial\phi}{\partial t}$ on the body, since:

$$\frac{-p}{\rho} = \frac{\partial\phi}{\partial t} + \tfrac{1}{2}ww^* + gy \tag{9}$$

On the free surface we have:

$$\frac{\partial\phi}{\partial t} = -\tfrac{1}{2}ww^* - gy \tag{10}$$

whilst on the body we have:

$$\frac{\partial\psi(x,y,t)}{t} = u_0 v - v_0 u + (y-y_0)a_{x_0} - (x-x_0)a_{y_0} \tag{11}$$

Here (u,v) is the fluid velocity at (x,y) and (a_{x_0},a_{y_0}) is the acceleration of the centre of gravity of the body which is unknown, but may be factored out of the equations. Thus equation 11 reduces to three separate problems:

$$\frac{\partial\psi}{\partial t} = \frac{\partial\psi_1}{\partial t} + a_{x_0}\frac{\partial\psi_2}{\partial t} + a_{y_0}\frac{\partial\psi_3}{\partial t} \tag{12}$$

Cauchy's theorem, equation 3, holds for each of the three composite problems with ϕ_1 on the free surface given by equation 10, and $\phi_2 = \phi_3 = 0$ on the free surface. The full non-linear problem is given by the superposition of these three (unphysical) potentials β_1, β_2 and β_3. Thus $\frac{\partial\phi}{\partial t}$ may be calculated on the body whence a matrix equation for a_{x_0} and a_{y_0} may be obtained and solved. Full details are given in Vinje and Brevig (1981b). The calculations described here typically took 7 seconds per time step on the CYBER 205 (although no vectorisation was used). The numerical scheme appears to be remarkably stable and no explicit smoothing was used in any calculation.

3. The Intersection Points

At the intersections of the free surface and body both the pressure condition at the free surface (equation 10) and the body condition (equation 6) hold and hence both ϕ and ψ are known here. In these calculations these points are therefore removed from the unknows in the matrix equation and the points are treated as free surface points for the purposes of time stepping. Obviously the fluid is not influenced by the unwetted shape of the body and some small adjustment is necessary to ensure that these points remain on the body also. This procedure, which is correct if the complex velocity potential at these intersection points remain regular (e.g. a standing wave against a vertical boundary is regular) appears to work well for singular cases also. Thus Lin, Newman and Yue (1984) have examined the impulsive start-up of a wavemaker for which the analytical solution is known:

$$\eta = \frac{-Ut}{\pi} \ell n[\tanh(\frac{\pi x}{4h})] + O(t^2) \qquad (13)$$

Here the impulsive velocity of the wavenumber is U, h = depth of water, t = time, x = distance measured horizontally from the wavemaker and η = vertical free surface displacement. This result, due to Peregrine (1972) has also been derived by Chwang (1983) and predicts infinite fluid displacement as $x \to 0$. Lin, Newman and Yue's solution gives a free surface profile very close to equation 13, except at $x = 0$. Experiments by Greenhow and Lin (1983) also validate equation 13 and the above numerical procedure, except near $x = 0$ where a jet is ejected.

Further work using this treatment of the intersection points by Greenhow (1985) on wedge entry into calm water give convincing results except for high entry speed when the pressure on the upper part of the wetted body surface can become negative. This is thought to be due to the numerical scheme surpressing jet formation. In this paper no problems of this type were encountered: during the engulfing of the cylinder by the wave the method also works well. Figure 1 shows the result of a preliminary run when a half-immersed cylinder, neutrally buoyant when totally submerged, is released in calm water.

4. Results

In all the runs described here initial conditions for the wave displacement and velocity potential on the free surface were given by linear theory, but linearly reduced in the range $\lambda/2 < |x| < \lambda$ to zero at $|x| = \lambda$. This initial condition lacks mathematical justification but does appear to be close to the wave profiles of transient plunging breakers in the Edinburgh narrow tank, and a similar procedure was used in Greenhow et al (1982). At full scale λ = 110m, depth = 60m and the wave amplitude = 10m giving a wave of extreme steepness which soon breaks.

Figure 2 shows the wave profiles when a fixed cylinder is placed at various positions whose centre coordinates are given. For simulations of a plunging breaker without bodies, the scheme may be continued unitl jet touch down in the preceeding trough when the theory ceases to be valid: mathematically the fluid has moved onto a different Riemann sheet. With a body present, this also happens when the body becomes totally submerged and the calculations must be discontinued. Other causes of breakdown are poor jet resolution, and local splashing close to the body (see e.g. figure 2, cylinder at (80, -2.5) on the downwave side.) The forces are given in figure 3 showing the horizontal force (dotted) and total vertical force including the hydrostatic componant (solid). For the fluid loading on the cylinder support for example, one should subtract the cylinder weight. For the moving cylinder calculations the cylinder was either half neutrally buoyant (weight = 385 KN/m) or $^3/4$ buoyant (weight = 578 KN/m). It is seen that as the wave engulf the cylinder very high forces are experienced although the caculations must be stopped before maximum forces are attained.

Figure 4 shows the circumferential pressure around the cylinder where θ is measured from the downward vertical from the centre and positive in the direction of wave propagation. Such information may be useful for stress analysis of the spine as well as deciding where to put weak points like holes, welds etc.

Figure 5 shows various runs with freely floating cylinders. Calculations may be carried further in these cases since the cylinder moves up to avoid submergence until a later time. Figure 6 gives the forces experienced in the various positions while figure 7 shows the effect of wave amplitude upon a cylinder at (80, 0.0). Comparison of forces from waves of amplitudes 5m and 10m shows an increase in the hydrodynamic forces of approximately 2.3, and thus simple scaling up of the wave forces (as predicted by Morison's equation) is likely to be too conservative. Again the calculations must be stopped before maximum forces are attained, but a significant decrease compared with the forces on a fixed cylinder is noted. Thus some flexibility in the spine is likely to be an advantage in reducing loads.

Figure 8 shows the circumferential pressures around the cylinder, again decreased from the fixed cylinder case.

Figure 9 shows the wave acting on a cross section of a clam spine with the air bag collapsed (i.e. the clam is not absorbing energy, but is in a survival mode.) Dimensions and shape of the clam were taken from Bellamy (1982). Figure 10 shows the pressure down the wave facing front of the clam and might be used to study bag collapse during the wave sequence. Forces are given in figure 11 and again the vertical force includes the hydrostatic component. They are considerably higher than for the circular spine, due to the larger clam size and lack of flexibility in the clam backbone which tends to hold any cross section fixed, as in these calculations.

Figure 12 shows the wave run against an O.W.C. with no induced water flow in the mouth. During the sequence the mouth is uncovered before the wave engulfs the device, when the pressure at the mouth rises very rapidly (see figure 13) giving the possibility of water ingestion into the turbines. The dimensions and shape of the O.W.C. are taken from Drew (1982). The forces on the device are very large (figure 14) despite the fact that the incident wave is of smaller amplitude (7m) as might be more realistic for shallower water depths (taken here to be 30m) needed for this bottom mounted device.

5. Conclusions

During early stages of wave breaking, the extreme wave loads and pressures on terminator type devices may be predicted by the present numerical method. Use of linearised water wave theory coupled with Morison's equation gives forces which are significantly less than those experienced when the body is engulfed by the wave. These large forces may be reduced by allowing the body to move freely, and it is likely that the same effect can be achieved by a flexible spine. The computations cannot predict the maximum forces which presumably occur after complete body submergence, and for which model testing is thought to be necessary.

6. References

1. Bellamy, N.W.: Development of the SEA Clam Wave Energy Converter. Proc.

Int. Symp. Wave Energy Util. (1982) Trondheim, Norway, pp.175-190.

2. Brevig, P.; Greenhow, M.; Vinje, T.: Extreme Wave Forces on Submerged Cylinders. Proc. B.H.R.A. 2nd Int. Symp. Wave and Tidal Energy (1981) Cambridge.

3. Cottrill, A.: Wave Energy: main U.K. contenders line up for the 1982 decision. Offshore Engineer Jan. 1981.

4. Cumberbatch, E.: The Impact of a Water Wedge on a Wall. J. Fluid Mech. Vol. 7 (1960) pp. 353-374.

5. Drew, S.: Recent Developments with the Vickers O.W.C. Devices. Proc. Int. Symp. Wave Energy Util. (1982) Trondheim, Norway pp.191-210.

6. Evans, D.V.: Power from Water Waves. Ann Rev. Fluid Mech. (1981) 13 pp.157-187.

7. Greenhow, M.; Vinje, T.; Brevig, P; Taylor, J.: A theoretical and experimental study of the capsize of Salters Duck in Extreme Waves. J. Fluid Mech. Vol. 118 (1982) pp. 221-239.

8. Greenhow, M.; Lin, W-M.: Nonlinear Free Surface Effects: Experiments and Theory. M.I.T. Rep. no. 83-19 (1983).

9. Greenhow, M.; Rosen J.; Reed, M.: Control Strategies for the Clam Wave Energy Device. App. Ocean Res. Vol. 6 No.4 (1984) pp.197-206.

10. Greenhow, M.: Water Entry of a Wedge. Unpublished results.

11. Lin,W-M.;Newman, J.N. and Yue, D.: Nonlinear Forced Motions of Floating Bodies. 15th Symp. Naval Hydro Hamburg, Germany (1984).

12. Longuet-Higgins, M.S. and Cokelet, E.D.: The Deformation of Steep Surface Waves on Water. I.A. Numerical Method of Computation. Proc. Roy. Soc. Lond A. Vol. 350 pp.1-26.

13. Newman, J.N.: Marine Hydrodynamics. M.I.T. Press (1977).

14. Ochi, M.K., Tsai C-H.: Prediction of Impact Pressures by Breaking Waves on Vertical Cylinders in Random Seas. Appl. Ocean Res. Vol. 6 No.3 July (1984) pp.157-165.

15. Vinje, T. and Brevig, P.: Nonlinear Ship Motions. Proc. 3rd Int. Conf. on Numerical Ship Hydrodynamics. Paris (1981a).

16. Vinje, T. and Brevig, P.: Nonlinear two-dimensional Ship Motions. S.I.S. Project rep. no. R-112.81 Trondheim, Norway, (1981b).

17. Vinje, T. and Brevig, P.: Numerical Calculation of Forces from Breaking Waves. Proc. Int. Symp. Hydrodynamics in Ocean Engineering (1981c). Trondheim, Norway, pp.547-566.

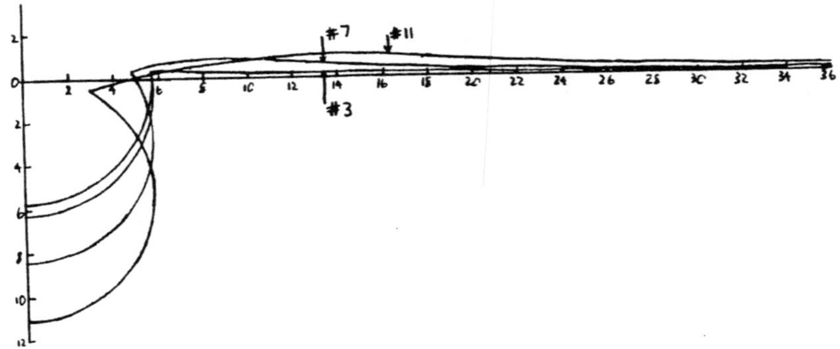

Figure 1 Cylinder Submergence

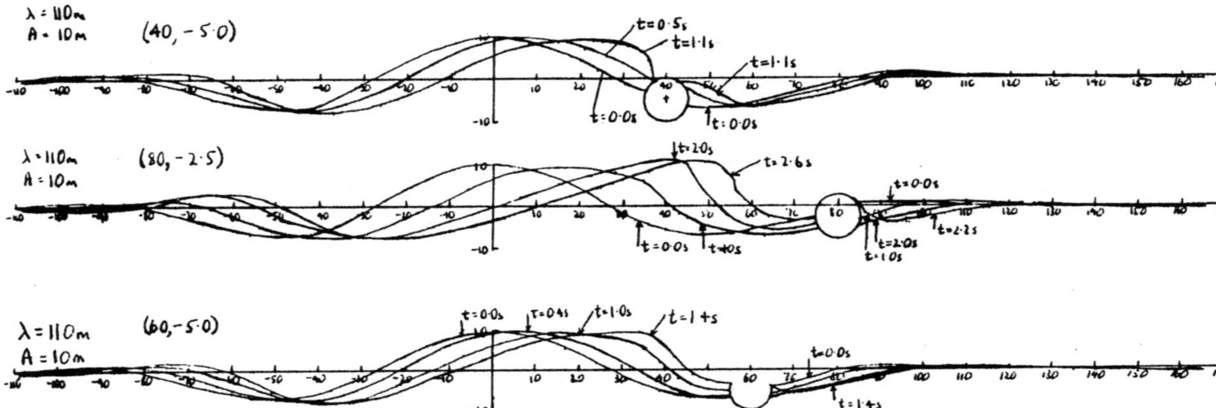

Figure 2 Development of a Breaking Wave Against a Fixed Cylinder.

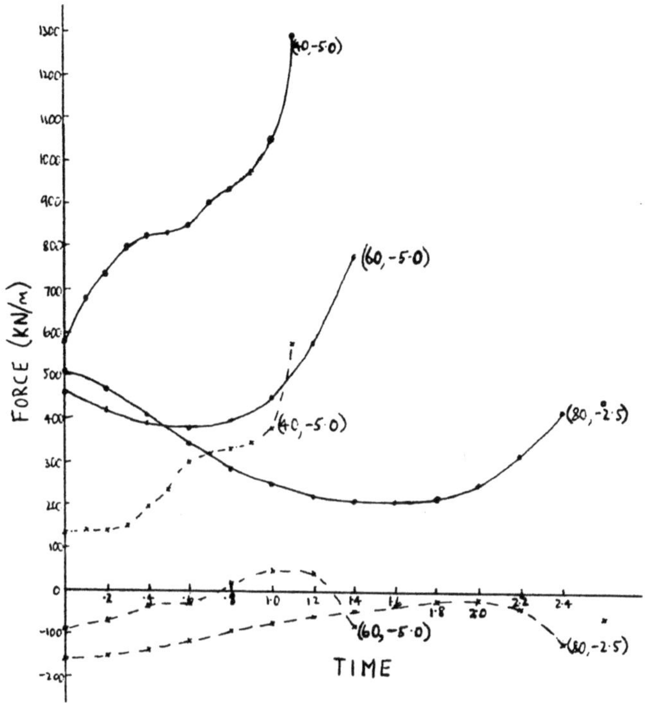

Figure 3 Wave Forces on a Fixed Cylinder
(-Vertical - - - Horizontal)

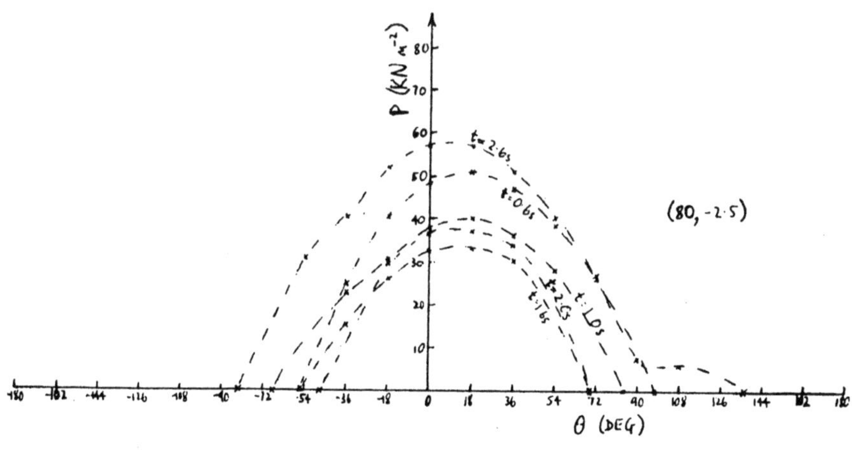

Figure 4

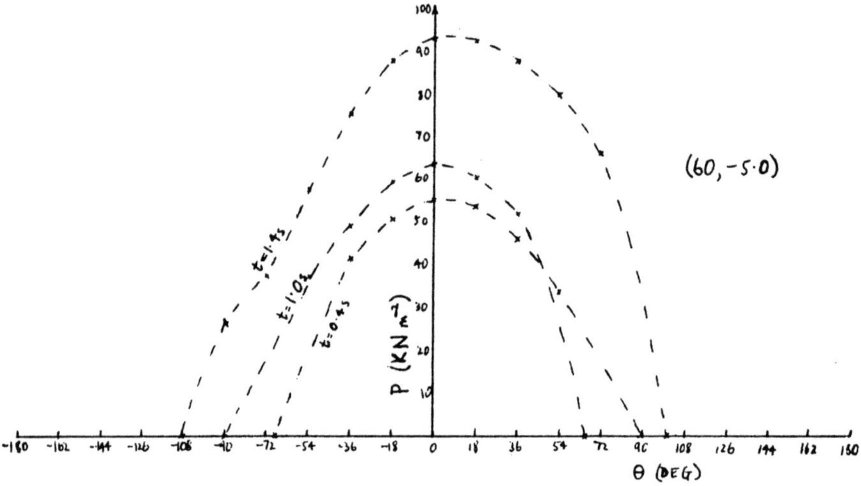

Figure 4

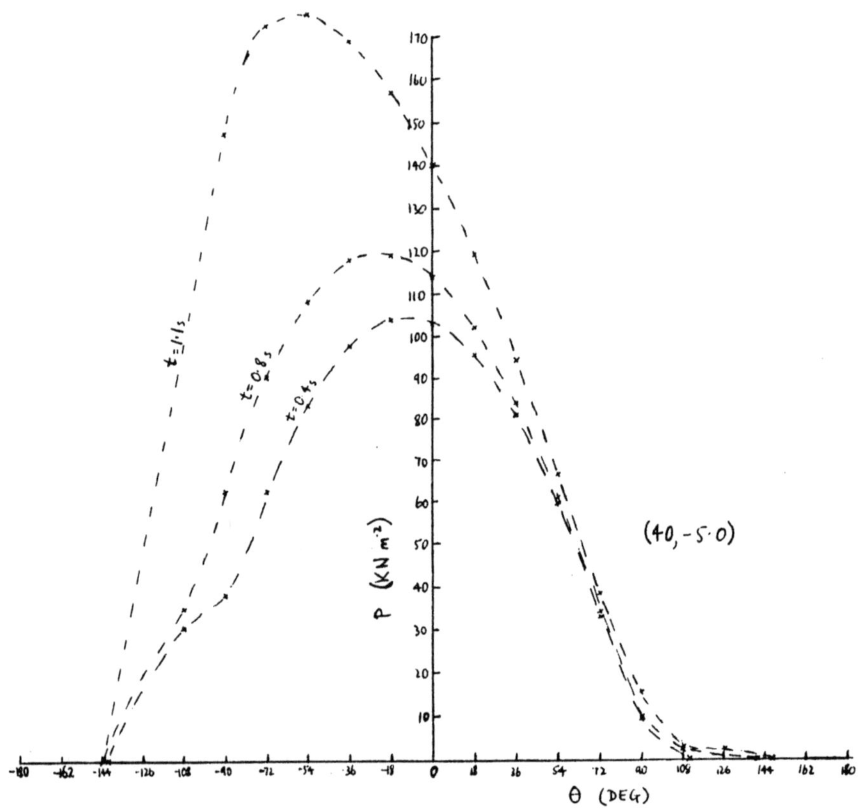

Figure 4 Circumferential Pressure around a Fixed Cylinder

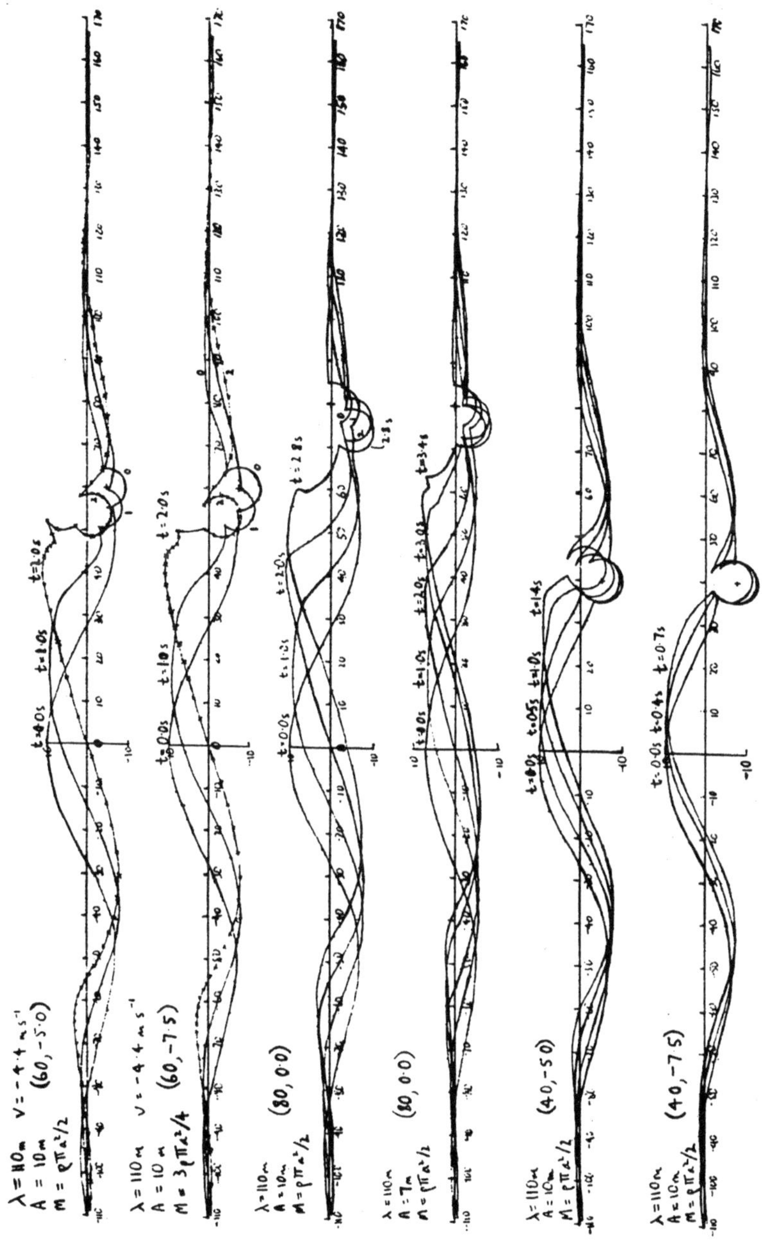
λ = 110m V = -4.4 m s⁻¹
A = 10m
M = ρπa²/2
(60, -5.0)
λ = 110m V = -4.4 m s⁻¹
A = 10m
M = 3ρπa²/4
(60, -7.5)
λ = 110m
A = 10m
M = ρπa²/2
(80, 0.0)
λ = 110m
A = 7m
M = ρπa²/2
(80, 0.0)
λ = 110m
A = 10m
M = ρπa²/2
(40, -5.0)
λ = 110m
A = 10m
M = ρπa²/2
(40, -7.5)

Figure 5

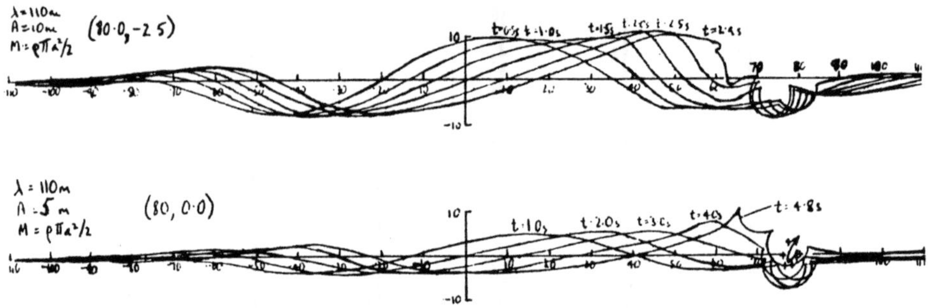

Figure 5 Development of a Breaking Wave against a Freely Floating Cylinder.

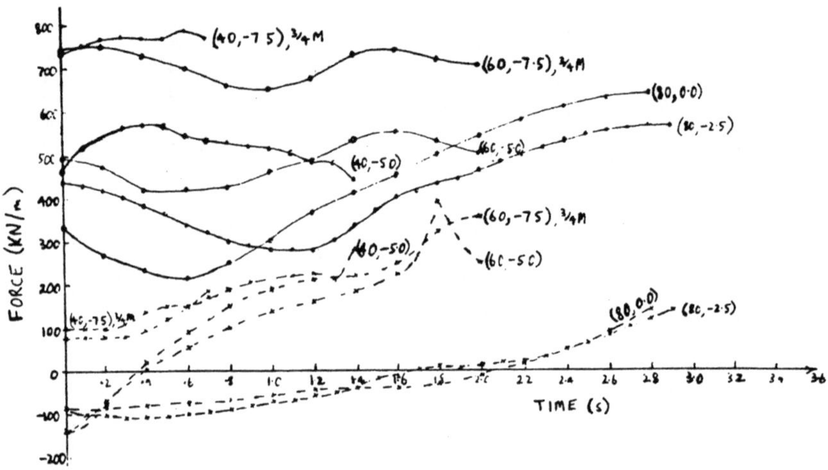

Figure 6 Wave Forces on a Freely Floating Cylinder (-Vertical ---Horizontal)

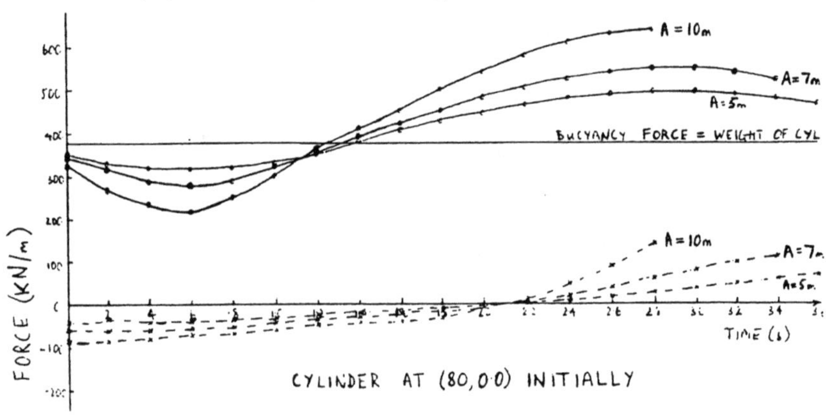

Figure 7 Wave Forces on a Freely Floating Cylinder - Wave Amplitude Variation (-Vertical --- Horizontal)

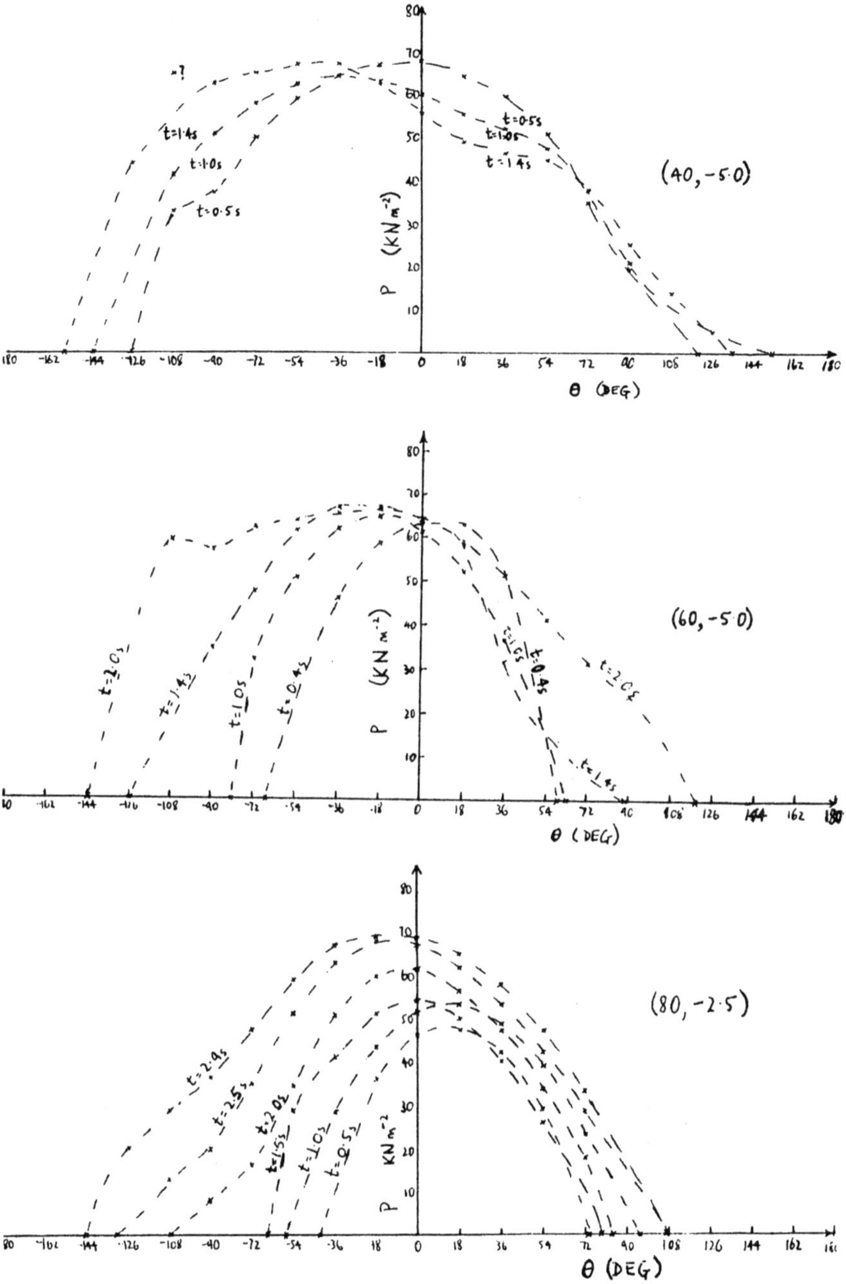

Figure 8 Circumferential Pressure around a Moving Cylinder.

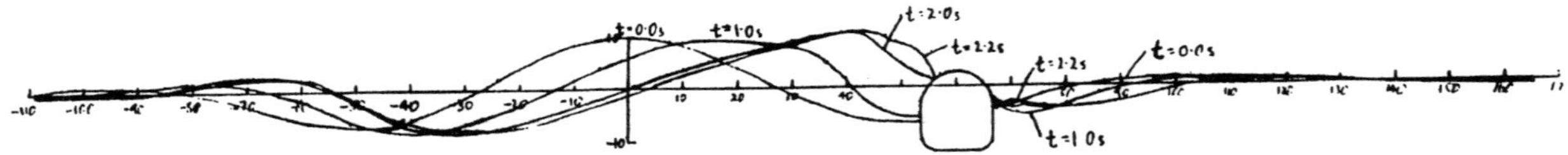

Figure 9 Development of a Breaking Wave Against a Clam Spine

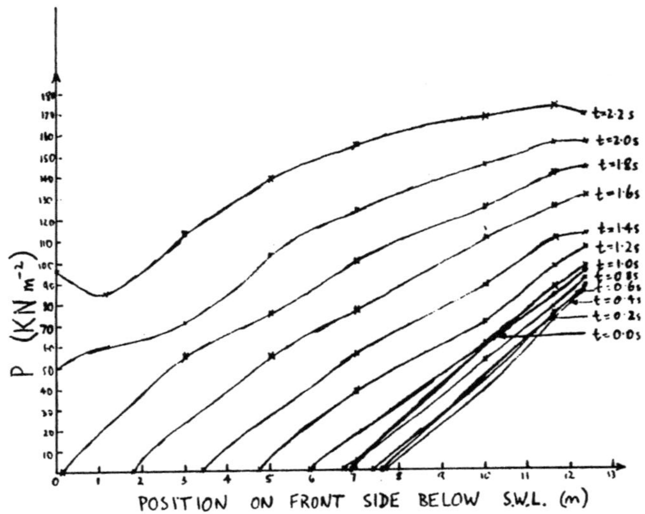

Figure 10 Pressures on the Front Face of the Clam.

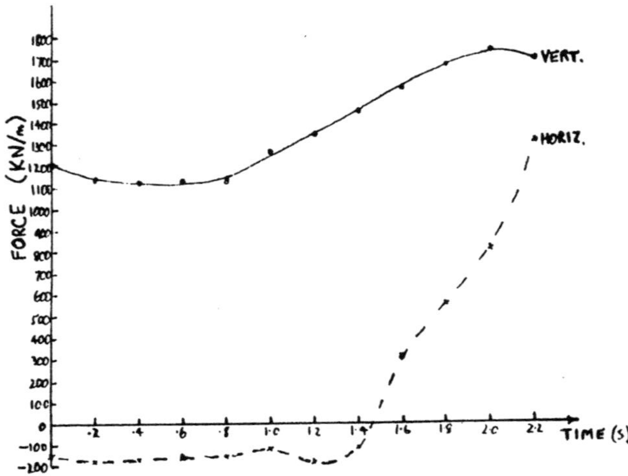

Figure 11 Wave Forces on the Clam (-Vertical --- Horizontal).

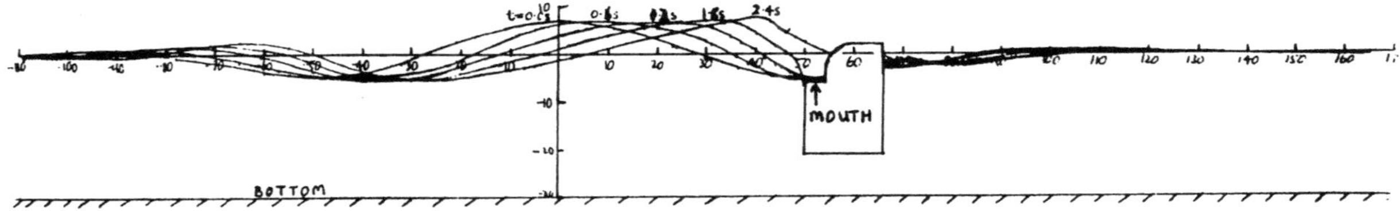

Figure 12 Development of a Breaking Wave Against an O.W.C.

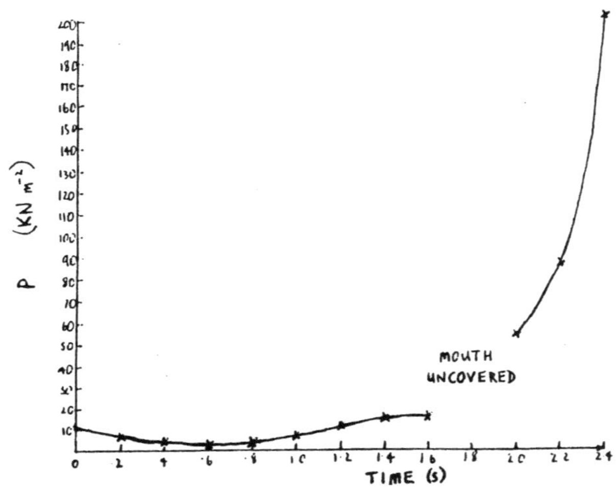

Figure 13 Pressure at the Mouth of the O.W.C.

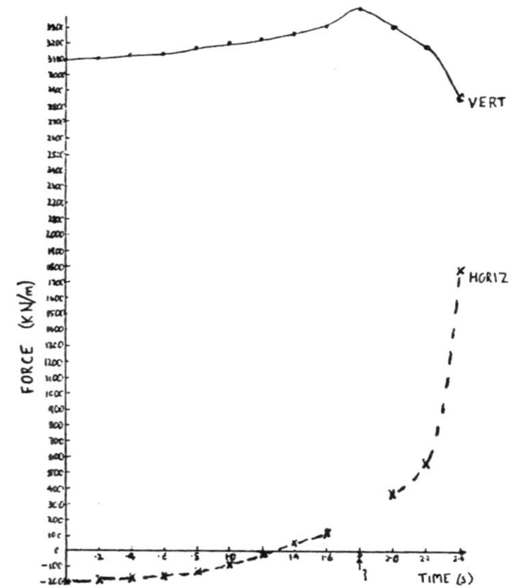

Figure 14 Wave Force on the O.W.C.
(-Vertical ---- Horizontal).

Progress on Flap-type Wave Absorbing Devices

ROBERT M. SCHER

John J. McMullen Associates
Arlington, Virginia

Summary

Numerical and experimental analyses of a class of wave-absorbing devices were performed over a four-year period at The University of Michigan. The particular type of device consists of one or two thin, vertical flaps, arranged to respond in sway or roll motion, pivoted about an axis in the plane of the flap. The general results of these studies, particularly with regard to engineering guidance for further prototypes, are reviewed here.

INTRODUCTION

It is in the nature of water waves that there are an almost infinite variety of device concepts that can be proposed to absorb energy from them. It is in the nature of people in general, and scientists and engineers in particular, to actually propose most of them at one time or another. So many different types have been developed over the past decade or so, with offshoots and refinements of each, that it has become customary to speak of second- and third-generation devices, even though the majority of the first generation have never actually been built.

The flap-type absorbing device, by comparison with many others, is too simple to be technically eye-catching. In essence, the device consists of a thin, vertical plate, usually (but not necessarily) at the free surface. The body may be arranged to respond either in sway (as on a tracked carriage), or in roll, pivoted about an axis which may be above or below the surface, as desired. The simplicity of the body shape, which suggests certain economies in fabrication, and also valuable simplifications in analysis, nonetheless carried with it certain disadvantages as well. Most importantly, from the outset, the flap-type device was compromised by the premise of "2-D Efficiency": the ratio of absorbed to incident wave power in a two-dimensional problem. This ratio, for a symmetric device oscillating in a single degree of freedom, was very early shown to be a maximum of one-half, at optimal tuning and damping. By contrast, devices that are non-symmetric, or that oscillate in two or more degrees of freedom, were shown to be capable of absorbing all of the incident wave power, at least within the context of linear theory. Thus, inevitably, the single-flap device was widely judged as "inferior" by a factor of two. The concept had the additional misfortune of being fostered in the United States, which has not been as devoted to wave-energy research as some other countries.

The preoccupation with maximum 2-D efficiency tended to obscure several key issues, of course. First, if a device absorbs power by damping (whether linearly or otherwise) the motion of a body, or even a part of the free surface itself, then the motion under certain conditions, including the conditions for maximum power absorption, may become large. This can be shown to be especially important when a small device is tuned and damped to generate a large power at resonance. Therefore, the obtainable absorption of a real device is often not limited by its "efficiency," but by its mechanical arrangements. Second, the concept of 2-D efficiency was misleading. It implied that more power can be developed by increasing the width of the device: true enough in normal wave encounter, but not true at all in oblique wave encounter. It also implied that there is a meaningful denominator in the fraction absorbed/incident power, which is also only true in a 2-D case. Third, maximum absorption in regular waves of a given frequency (and a given amplitude, too, if maximum responses of the device are considered) by no means ensures maximum time-averaged absorption in a random sea. Fourth, and perhaps most obvious, there is no point in incorporating design features that increase absorption (even the useable absorption), if the cost of the device is increased faster.

In any case, early thinking, dominated as it was by the consideration of 2-D efficiency, led to the idea of the twin-flap device, which, because of its two independent degrees of freedom, entered the promised land of 100 percent absorption. There were other perceived advantages, too: including the ability to obtain partial or complete cancellation of first-order mooring forces by adjusting the tuning and damping of the two flaps to give the required phasing of the responses. As it turned out, many of these advantages were more easily obtained on paper than in the laboratory, although the ability to reduce mooring forces was verified. It was also questioned whether the performance of the devices would differ in a three-dimensional treatment.

DEVELOPMENTS

During 1981-82, a series of model experiments were performed with a twin-flap device in the towing tank at The University of Michigan. In these experiments, reported in Ref. (1), measurements were made of absorbed power, incident wave, flap motion responses, and forces. All experiments were performed with the supporting structure rigidly fixed, in normal wave encounter. Both 2-D experiments (with the flaps spanning the entire tank) and finite-width experiments were conducted. The principal object of these experiments was to obtain a direct comparison with the 2-D analytical predictions of Srokosz and Evans (2). Generally, good agreement was obtained between prediction and experiment with regard to 2-D efficiency. Maximum 2-D efficiencies of 100 percent were confirmed. Further, the predicted behavior of efficiency versus wavenumber and as a function of external applied damping was generally matched by experimental results. Results for flap motion responses and forces showed good agreement, as well.

In the course of these experiments, it became apparent that the feasibility of the concept would be strongly influenced by its characteristics under conditions that could not be studied experimentally, due to the limitations of the towing tank and its wavemaker, namely: (1) Oblique wave encounter with finite device width; and (2) Finite depth water (depth on the order of flap draft.) Therefore, during 1983-84, a numerical method was developed to estimate the motions, forces, and power absorption of flap type devices (single or twin-flap), incorporating the 3-D linear hydrodynamic coefficients for wave encounter from any angle, in the presence of a bottom, Ref. (3).

Summarizing the method, the flaps are modelled as distributions of normal dipoles. These distributions are subdivided into discrete quadrilaterals, on each of which the unknown dipole density is assumed constant. The normal velocity at the centroid of each quadrilateral is then approximated as the summation of the contributions of each quadrilateral, including the local one. These velocity contributions are computed by taking the appropriate derivative of the finite water-depth dipole Green's function. Finally, by applying the normal-velocity body-boundary conditions on the flaps for the radiation and diffraction problems, the unknown dipole densities can be found and the corresponding pressure jumps calculated directly. Integrating these over the flaps, the required hydrodynamic coefficients (added mass, damping, exciting forces, and coupling forces for twin-flap devices,) are calculated. These coefficients, substituted into the coupled linear equations of motion for the flaps, together with values of the external (power takeoff) damping, intrinsic mass, and restoring forces, give the predicted motions, forces, and power absorption. For simplicity of computation, the flaps were assumed to be rectangular in form, and pivoted about an axis in the plane of the flap (this point, however, may be placed at a great distance, resulting in a numerically satisfactory approximation of pure sway motion). In the case of twin-flap devices, the flaps were further assumed to be identical in dimensions, and directly behind one another, rather than partially overlapping. Because of this simple geometry, and the opportunity to obtain a high order of symmetry in the calculation of the influence matrix, computational requirements were less troublesome than for more complex body shapes. The Green's function matrix requires calculation only once for each frequency, the effects of encounter angle being confined to the diffraction-problem boundary condition. Nevertheless, in spite of these simplifications, the 3-D computation required about 80 times the computer resources of the 2-D analytical program derived from Ref. (2). The derivation, testing, and exercise of the 3-D numerical method are described in excruciating detail in Ref. (3).

Concurrently with this numerical work, a second experimental device was tested in the Michigan towing tank. This was a twin-flap device with flap dimensions of approximately 1.2 m draft, 2.4 m width, designed to be tested with the supporting structure either rigidly mounted or free-floating. The preliminary results, reported in Ref. (4), showed that a free-floating arrangement resulted in some degradation of power absorption, a conclusion that is in line with several earlier experimental efforts. The results of these experiments were compared directly with the predictions of the 3-D numerical method. Agreement in absorbed power was fair to good, in the case of rigidly supported experiments.

DESIGN GUIDANCE FOR PROTOTYPE FLAP-TYPE ABSORBING DEVICES

The results of numerical and physical models, as well as a preliminary economic analysis (also reported in Ref. (3)), tend to support several broad conclusions regarding the design of a large-scale prototype device. A brief summary of the main conclusions is offered, in the following sections.

Fixed versus Floating Platforms

Current thinking favors fixed structures. The principal argument is the difficulty of designing secure mooring systems for a full-scale system, rather than the degradation of power absorbing characteristics. The decrease in power absorption due to ship-like responses of the supporting structure was in the neighborhood of 10 percent, as reported in Ref. (4). Such a loss might well be compensated by the higher average power density in a deeper water site, where a fixed supporting structure would not be practical. Nevertheless, the vulnerability and expense of a mooring system, combined with the substantial costs of underwater power transmission, tend to support an initial preference for a site in relatively shallow water. Such a choice will also permit a relatively simple supporting structure, based on the use of mass concrete, bottom soil characteristics permitting.

Roll versus Sway Geometry

After gaining some experience with a pure sway arrangement during the initial experiments of 1981-82, a roll arrangement was adopted for the later experimental apparatus of Ref. (4). This preference is purely practical, and has nothing to due with linear predictions of absorption capabilities. The predictions, in fact, show no difference in maximum absorption between the two options. However, a rolling device, pivoted about an axis near the bottom of the flap, offers several advantages in terms of design and construction, and in operation as well. These include the avoidance of alignment problems, reduction in structural forces and moments, and the ability to operate the flap at large angles of inclination, away from the free surface, thus shedding some exciting force, when desired.

Flap Width

To the extent that flap width remains small with respect to wavelength (surely valid for a flap that is assumed to respond as a rigid body), the maximum absorption cross-section obtainable at any frequency has been shown to be primarily a function of the wavelength, and NOT of the width of the device. In this respect, the situation is analogous to that of a long-wave radio antenna, in which the frontal area of the antenna has nothing whatever to do with the amount of energy that can be absorbed. A flap-type absorber operating in this wavelength regime is in effect a "directional point absorber." In the limit of long wavelength (the point-absorber limit) the maximum power absorption for a single-flap has been shown to be equal to the incident wave power contained in an extent of wavefront of λ/π. In fact, this point absorber limit is reached, for practical purposes, at a wavelength approximately 3 times the device width. At shorter wavelengths, of course, the flap-type device gradually becomes a terminator. In the limit of short wavelength, in normal encounter, the situation then reverts to the 2-D model, with maximum absorptions of one-half for a single-flap device, and unity for a twin-flap.

Directional effects, that is, the variation of absorption with angle of encounter, are also different in the point-absorber and terminator limits. In the former case, the limit of long wavelength, the directional dependency of the exciting force magnitude is as the cosine of the encounter angle. Since it can be shown that power absorption varies as the square of the exciting force, the power absorption of a directional point absorber must decline as the cosine squared. For shorter wavelengths, the exciting force has been shown to decline more rapidly, and for a flap width on the order of wavelength there must exist certain angles at which the exciting force vanishes, due to cancellation of the pressures at different locations along the flap. (Presumably, a real flap will cease to respond as a rigid body before this limiting situation can be reached, in full scale waves.)

For both of these reasons, maximum normal absorption and directionality, practical advantages arise in connection with short flaps. This is particularly true when the costs of the oscillating body are substantial with respect to the costs of power-conversion machinery and other items, which depends strongly on the level of technology envisioned for the power conversion system.

On the other hand, although the maximum power absorption is essentially independent of flap width (within the point absorber regime), the corresponding flap motion amplitude has been shown to vary approximately inversely with flap width. Therefore, the selection of an appropriate flap width cannot be made independent of the motion limits imposed by power takeoff machinery, and the incident wave amplitude.

Flap Draft

Similarly, in the linear problem, it can be shown that flap draft has no effect on maximum power absorption. However, the corresponding motion amplitude at maximum absorption varies approximately as the inverse square of the flap draft. In numerical runs, this result has appeared to hold in finite depth water as well as in deep water. For this reason, flap draft, even more strongly than flap width, is the primary design variable for matching the size of the absorbing body to the maximum stroke of the power conversion device. In addition, for a bottom-pivoted rolling flap in which restoring forces are generated by hydrostatics, flap draft is also the primary design variable for tuning the device to a given frequency. For geometrically similar flap sections, dimensional analysis shows that the tuned wave period varies approximately as the square-root of the flap draft. Some variation is introduced by the relative scaling of intrinsic inertia of the flap (which is primarily a matter of structural design) and the added inertia. By contrast, the effect of flap width on tuning is slight.

Since an increase in flap draft, apart from retuning the device to a longer wave period, always reduces the motion of the flap at maximum absorption, it can be argued that for a shallow-water site an appropriate initial choice of draft is made by placing the lower edge of the flap (and the pivot) as close to the bottom as possible, subject only to purely engineering details. It may also be mentioned here that the effect of pivot position was also studied numerically, and that it was found to have no effect on maximum power absorption. However, corresponding motion amplitudes occur with the pivot at the lower edge of the flap.

Tuning and Related Design Features

Since the flap draft represents the key design variable in tuning the flap, the selection of draft in effect "targets" some component of the wave spectrum. Presumably, the target frequency is in the neighborhood of the power-peak of the spectrum under consideration. In general, the higher-frequency spectra are crucial for design purposes: it is relatively difficult to obtain a resonance at high frequency, while low frequency tunings can always be obtained by the free use of ballast to increase the inertia and decrease the hydrostatic restoring moment. Accordingly, the following rules of thumb arise: (1) The flap should be kept light, and with a low center of gravity, so that the higher frequencies can be absorbed when necessary, without an unduly shallow flap draft. (2) Flap freeboard should be as small as possible, but the flap should not be submerged below the still water line: (this has been shown to increase the required flap motion at constant power absorption.)

Non-Resonant Conditions and Flap Dimensions

The arguments presented in the previous sections give a sense of the effects of flap dimensions on maximum absorption, corresponding motion amplitude, and resonant tuning. For non-resonant frequencies, however, it can be shown directly from the equations of motion that larger flaps are inherently less sensitive to tuning, because they are relatively more heavily damped. Accordingly, for a narrow-banded design spectrum (such as swell) a very small device can be tuned and damped to absorb in the point-absorber regime, and its acceptable dimensions will be governed primarily by the limits of motion response desired. By contrast, the more broad-banded the spectrum, the larger the device should be, ultimately constrained by water depth, directional effects, and the rigid-body assumption.

TWIN VERSUS SINGLE-FLAP

Twin flaps represent a more complicated situation in three dimensions than in two. In 2-D, the ability to absorb 100 percent of the incident wave implies the cancellation of outgoing wave components, both reflected and transmitted. In 3-D, with a finite width twin-flap device, it is obvious that the entire radiated wave pattern cannot be completely cancelled at any frequency, although it can be reduced. The effect of the twin-flap, then, is to recapture part of the energy that would otherwise be lost as a radiated wave. However, the condition for optimal tuning and damping of the twin-flap in 3-D is not as easily solved as in 2-D. The results of numerical runs have shown that a twin-flap device in 3-D can still absorb twice the power of a single-flap device of equivalent width. However, the optimum settings for damping and restoring force coefficients seem to be more sensitive to frequency for a twin-flap. As was found in Ref. (2) for the 2-D case, optimal absorption can involve negative restoring force coefficients, as well. Accordingly, there may be practical reasons why optimal settings cannot be obtained for a twin-flap device.

Importantly, if a 3-D twin-flap array can be shown to be limited to only twice as much power as a single flap (this has not yet been demonstrated as a proof, although the numerical results suggest it), then no real economic improvement is expected unless certain special conditions can be shown to apply. These conditions are as follows: (1) Increased power output with one of the flaps free of power takeoff (either passively oscillating or locked). It should be noted that optimal absorption for a twin-flap device in 2-D resulted from a passive component, but that this did not seem to be the case in 3-D. (2) Broader absorption responses than for single-flap devices, with consequent advantages in average absorption in a random sea. This has not been shown in general, and the improvement would have to approximate a doubling in order to make compelling economic sense. (3) Reduced directionality at a given absorption. This, also, has not been found in general. In conclusion, while there are certainly enough interesting possibilities to warrant further study, it cannot yet be stated that a twin-flap device is economically superior to a single-flap system of equivalent combined flap width.

THE NEXT STEP

Plans are presently under way to build and test a "mid-scale" working prototype, suitable for installation in a field site, possibly on Lake Superior. As presently envisioned, this device will have dimensions of approximately 5 m draft, and 10-20 m width. It will be bottom pivoted, with positive-displacement hydraulic power takeoff. It will probably be built in two separate units, both for ease of emplacement, and so that single and twin-flap configurations can be tested, if desired.

ACKNOWLEDGEMENT

The work reported here was supported by the Q-Corporation, Troy, Michigan.

REFERENCES

1. R.M. Scher, A.W. Troesch, and G. Zhou, "Experimental and Theoretical Evaluation of a Twin-Flap Wave Energy Absorbing Device," Ocean Engineering, Nov 1983.
2. M.A. Srokosz and D.V. Evans, "A Theory for Wave Power Absorption by Two Independently Oscillating Bodies," J Fluid Mech, Vol 90, 1979, pp 337-362.
3. R.M. Scher, A Study of Flap-Type Wave Absorbing Devices, PhD dissertation, The University of Michigan, 1984.
4. D. Bauermeister, "Evaluation of a Three-Dimensional Wave Energy Recovery Device," University of Michigan, Dept of Naval Arch & Marine Eng, Ship Hydro Lab Rept 388771, July 1984.

Three-Dimensional Analysis of Oscillating-Water-Column Terminator Devices in Obliquely Incident Waves

A. J. N. A. SARMENTO and A. F. de O. FALCÃO

Mechanical Engineering Department, Instituto Superior Técnico, Universidade Técnica de Lisboa, 1096 Lisboa Codex, Portugal

Summary

A three-dimensional analysis is developed and numerical results are presented concerning the wave-energy absorption by a terminator OWC device consisting of an infinite row of contiguous air-chambers of rectangular planar form, each chamber being equiped with an independent air-turbine which can be phase-controlled. The water depth is in general finite and the incoming wave is assumed monochromatic and inciding obliquely on the chamber row.

1. Introduction

The paper presents a three-dimensional analysis of the hydrodynamics of oscillating-water-column devices of terminator type in obliquely incident regular waves, and can be considered as an extension of the two-dimensional analysis of Sarmento & Falcão [1]. The wave-energy absorbing system consists basically of an infinite row of contiguous identical air-chambers of rectangular planar form. The front wall, as well as the side walls separating the chambers, are thin and submerged to a negligible depth. The back wall is also assumed thin and submerged to a small depth, as in a floating system, or alternately spans vertically the whole water depth, as in the case of a bottom-standing system like the NEL breakwater device. Each chamber is equiped with an air turbine (or equivalent device), all turbines being identical and operating independently. The water depth is constant and in general finite. The air is assumed incompressible, and the pressure drop through each turbine is taken proportional to the air flow rate. This,together with the usual assumption of linear water-wave theory, ensures that the air pressure in each chamber is a simple-harmonic function of time [2]. As in [1], the turbine proportionality coefficient is assumed in general to be complex, in order to model a turbine exhibiting a phase difference between pressure and flow rate.

2. The Wave Field

We take constant water depth h, and chose a system of co-ordinates (x,y,z), with the positive y-direction pointing vertically upwards and the origin at the undisturbed free-surface. The internal free-surfaces are bounded by $x=\pm a$, with walls at $z=\pm(q+\frac{1}{2})b$ $(q=0,1,2,\ldots)$ separating contiguous chambers. We assume here that the submerged part of the structure is of negligible draught. The incident-wave potential can be written as

$$\Phi_i(x,y,z,t) = \frac{igA}{\omega}\frac{\cosh[K(y+h)]}{\cosh Kh}\, e^{i(\omega t - Kx\cos\beta - Kz\sin\beta)}, \qquad (1)$$

where β $(0<\beta<\pi/2)$ is the angle of the incident wave crests with the z-direction (row-parallel direction), g is the acceleration of gravity, and K is the wavenumber which is known to be the positive root of $K\tanh Kh - \omega^2/g$.

We introduce a dimensionless complex proportionality coefficient for the turbines

$$C\,e^{i\theta} = \frac{\rho}{b}\sqrt{g/a}\,\frac{Q}{P} \qquad (C \text{ real positive, } |\theta|<\pi/2), \qquad (2)$$

where Q and P are the complex amplitudes of the air-turbine volume-flow rate and the chamber pressure oscillations, respectively, and ρ is the water density. The real part of Q is taken positive for outward flow.

The chamber width will be written in the form $b=(j/\ell)\lambda/\sin\beta$, where j and ℓ are relatively prime positive integers, and $\lambda=2\pi/K$ is the incident wavelength. By introducing this assumption, which does not restrict the generality of the analysis, it can easily be found that the chamber air-pressure becomes a discontinuous periodic function of the row-parallel co-ordinate z, with period $\ell b=j\lambda/\sin\beta$.

The surface-pressure distribution can now be written as

$$p(x,z)\,e^{i\omega t} = \Pi\,X(x)\,Z(z)\,e^{i\omega t}, \qquad (3)$$

where Π is in general complex, $X(x)=0$ for $|x|>a$ and $X(x)=1$ for $|x|<a$, and

$$Z(z) = e^{-iqbK\sin\beta} \quad \text{for} \quad (q-\tfrac{1}{2})b<z<(q+\tfrac{1}{2})b, \quad q=0,\pm1,\pm2,\ldots \qquad (4)$$

(Note that $\Pi\,Z(z)$ equals the complex pressure amplitude P that appears in (2).) By expanding Z(z) in a Fourier series, we find

$$Z(z) = \sum_{\substack{n=-\infty \\ n\neq 0}}^{\infty} a_n e^{-iK_n z} \quad (K_n = \frac{n}{j} K \sin\beta) , \tag{5}$$

where $a_n=(n\pi/\ell)^{-1}\sin(n\pi/\ell)$ if $(j-n)/\ell$ is an integer and $a_n=0$ otherwise. The potential, $\Psi(x,y,z,t) = \psi(x,y,z)e^{i\omega t}$, of the radiated wave due to the surface-pressure distribution (3), can be obtained from the general expression given by Wehausen & Laitone [3]

$$\psi(x,y,z) = -\frac{i\omega}{2\pi\rho g} \iint_{-\infty}^{\infty} p(\xi,\zeta)d\xi\, d\zeta \;⨍_0^{\infty} \frac{k\cos[k(y+h)]}{k\sinh kn - \alpha\cosh kh} J_0\left[k\sqrt{(x-\xi)^2+(z-\zeta)^2}\right]dk$$

$$-\frac{\omega m K}{2\rho g}\frac{\cosh[K(y+h)]}{\sinh Kh} \iint_{-\infty}^{\infty} p(\xi,\zeta)\; J_0\left[K\sqrt{(x-\xi)^2+(z-\zeta)^2}\right]d\xi\, d\zeta, \tag{6}$$

where J_0 is a Bessel function, $\alpha=\omega^2/g$, and $m = (1 + \alpha h\,\mathrm{cosech}^2 Kh)^{-1}$. Replacing $p(x,z)$ by its expression (3) with $Z(z)$ given by (4), and taking into account that J_0 is an even function and that [4]

$$J_0(\alpha\sqrt{x^2+z^2}) \cos\beta x\, dx = \begin{cases} 0 \quad \text{for} \quad 0<\alpha\leq\beta, \\ \dfrac{\cos(z\sqrt{\alpha^2-\beta^2}}{\sqrt{\alpha^2-\beta^2}} \quad \text{for} \quad 0<\beta<\alpha, \end{cases}$$

we find the following expression for the radiation potential

$$\psi(x,y,z) = \Pi \sum_{\substack{n=-\infty \\ n\neq 0}}^{\infty} a_n \psi_n , \tag{7}$$

where

$$\psi_n(x,y,z) = -\frac{2\omega}{\rho g} e^{-iK_n z}\left\{\frac{i}{\pi} ⨍_{|K_n|}^{\infty} \frac{k\cosh[k(y+h)]\,\sin(a\sqrt{k^2-K_n^2})\,\cos(x\sqrt{k^2-K_n^2})}{(k-K_n^2)(k\sinh kh - \alpha\cosh kh)} dk\right.$$

$$\left. + H(j-|n|\sin\beta)\; \frac{m\cosh[K(y+h)]\;\sin(Ka\cos\tau_n)\;\cos(Kx\cos\tau_n)}{K\cos^2\tau_n\;\sinh Kh}\right\}. \tag{8}$$

Here, H() is Heaviside's unit step function, and $\tau_n=\arcsin[(n/j)\sin\beta]$, $|\tau_n|<\pi/2$.

It is important to know the radiated wave field far away from the chamber row, and for that purpose we must determine the asymptotic value of ψ_n when $|x|\to\infty$, particularly the asymptotic principal value of the integral of (8).

We note that the integrand is finite everywhere except at k=K, and so by the Riemann-Lebesgue lemma the only contribution to the principal value of the integral, when $|x|\to\infty$, comes from the vicinity of that point if $K>|K_n|$, i.e. if $|n|\sin\beta<j$. The denominator of the first fraction of the integrand, which has a root at k=K, can be written as follows

$$k\sinh kh - \alpha\cosh kh = \frac{\sinh kh}{m\,f(k)}(k-K),$$

where f(k) is a continuous function with finite first derivative in $-\infty<k<\infty$, such that f(K)=1. Then, making use of the results [5]

$$\lim_{x\to\infty} \fint_{-\varepsilon}^{\varepsilon} g(t)\,\frac{\cos xt}{t}\,dt = 0 \qquad \text{and} \qquad \lim_{x\to\infty}\int_0^{\varepsilon} g(t)\,\frac{\sin xt}{t}\,dt = \frac{\pi}{2}\,g(0) \qquad (\varepsilon>0),$$

where g(t) is continuous and of bounded variation in the interval of integration, it can be found that

$$\psi_n(x\to\pm\infty,y,z) = \frac{igB_n}{\omega}\,\frac{\cosh[K(y+h)]}{\cosh Kh}\,e^{-iK(|x|\cos\tau_n + z\sin\tau_n)}, \tag{9}$$

where

$$\left.\begin{aligned} B_n &= \frac{2im}{\rho g}\,\frac{\sin(Ka\cos\tau_n)}{\cos^2\tau_n} \quad \text{if} \quad |n|\sin\beta < j, \\ B_n &= 0 \quad \text{if} \quad |n|\sin\beta > j. \end{aligned}\right\} \tag{10}$$

It can easily be seen that $a_n\psi_n(x\to\pm\infty,y,z)e^{i\omega t}$ is the potential of a wave travelling away from the chamber row, whose crests make an angle $\pm\tau_n$ with the z-axis. Its amplitude, equal to a_nB_n, is nonzero only for a finite set of values of n, which are those that satisfy simultaneously the conditions $|n|\sin\beta<j$ and $(j-n)/\ell$ equal to an integer (it being assumed that $Ka\cos\tau_n \neq \pi, 2\pi, 3\pi, \ldots$). These two conditions are always satisfied for n=j, in which case it is $\tau_j=\beta$. Besides, from those two conditions it can be found that there are no other nonzero values of the wave amplitude a_nB_n (i.e. for $n\neq0$, $n\neq j$) if, and only if, $(j/\ell)\sin\beta<1+\sin\beta$, which is equivalent to

$$\frac{b}{\lambda}(1+\sin\beta) < 1. \tag{11}$$

An identical condition is known to apply to an infinite row of points absorbers with spacing b (see Falnes & Budal [6]).

3. Energy Extraction

In order to obtain the expressions for the power extracted from the incident wave and the corresponding efficiency, and also to relate these quantities to the turbine coefficient, we must know the air-flow rate $q\,e^{i\omega t}$ displaced by the water free-surface inside the chamber. For the chamber $-b/2<z<b/2$, we have

$$q = q_i + \Pi \sum_{\substack{n=-\infty \\ n\neq 0}}^{\infty} a_n q_n , \tag{12}$$

where

$$\{q_i, q_n\} = \int_{-a}^{a} dx \int_{-b/2}^{b/2} \frac{\partial}{\partial y} \{\phi_i, \psi_n\}_{y=0}\, dz .$$

It can be found that

$$q_i = \frac{2iAb\omega}{K} \gamma_j \sin(Ka\cos\beta) , \tag{13}$$

$$q_n = \frac{\omega b}{\rho g K} \gamma_n (a_n + a_{-n}) \left[i\mu_n - H(j-|n|\sin\beta) \frac{2m\sin^2(Ka\cos\tau_n)}{\cos^3\tau_n} \right] , \tag{14}$$

where $\gamma_k = (k\pi/\ell)^{-1}\sin(k\pi/\ell)$ and

$$\mu_n = -\frac{4K}{\pi} \int_{|K_n|}^{\infty} \frac{k^2 \sin^2(a\sqrt{k^2-K_n^2})}{k-\alpha\coth kh} \frac{dk}{k^2-K_n^2} . \tag{15}$$

More important than the geometry considered so far is the case when there is a reflecting wall submerged vertically from the surface to the bottom at x=0 (breakwater geometry). The oscillating surface-pressure is then applied over the strip $-a<x<0$, and the wave field is restricted to the half-space $x<0$. In order to complete the field, we have to add the wave resulting from the reflection of the incident wave on the wall; its potential is given by (1) if the signs of x and z are changed. The following analysis and the numerical results refer to this case, except where otherwise stated.

Instead of (12), we have now

$$q = q_i + \frac{\Pi}{2} \sum_{\substack{n=-\infty \\ n\neq 0}}^{\infty} a_n q_n , \tag{16}$$

where q_i and q_n are still given by (13) and (15). The quantity q_i now accounts for the flow rate due to the incident and the reflected waves.

Taking into consideration that, for the chamber $-b/2<z<b/2$, the pressure drop and the air flow rate through the turbine are respectively $P=\Pi e^{i\omega t}$ and $Q=q e^{i\omega t}$, and recalling that P and Q are interrelated by (2), we obtain

$$\Pi = -\frac{2A\rho g\gamma_j \sin(Ka\cos\beta)}{[u+r+i(v+s)]\cos\beta}, \tag{17}$$

where

$$u = \sum_{n=1}^{\infty}(a_n+a_{-n})\gamma_n\mu_n, \qquad v = 2m\sum_{n=1}^{N}(a_n+a_{-n})\gamma_n \frac{\sin^2(Ka\cos\tau_n)}{\cos^3\tau_n}, \tag{18}$$

$$r = -\frac{K}{\omega}\sqrt{ga}\, C\sin\theta, \qquad s = \frac{K}{\omega}\sqrt{ga}\, C\cos\theta. \tag{19}$$

In (18), N is the largest integer smaller than $j/\sin\beta$.

The time-averaged power extracted by one chamber is $W=\frac{1}{2}\mathrm{Re}(\Pi q^*)$, where * means complex conjugate, and the time-averaged energy flux of the incident wave per unit length of crest is $E_i=|A|^2\rho g\omega/(4mK)$. The efficiency of the power extraction is then given by $\varepsilon = W(E_i b)^{-1}\cos\beta$. Replacing Π and q by their expressions derived above, we find

$$\varepsilon = 4(S-\chi S^2-\chi R^2), \tag{20}$$

where

$$\{R,S\} = \frac{2m\gamma_j \sin^2(Ka\cos\beta)}{[(u+r)^2+(v+s)^2]\cos^3\beta}\{(u+r),(v+s)\}, \tag{21}$$

$$\chi = \frac{v\cos^3\beta}{2m\gamma_j^2\sin^2(Ka\cos\beta)} = 1+\frac{a_{-j}}{\gamma_j}+\frac{\cos^3\beta}{\gamma_j^2\sin^2(Ka\cos\beta)}\sum_{\substack{n=1\\ n\neq j}}^{N}(a_n+a_{-n})\gamma_n\frac{\sin^2(Ka\cos\tau_n)}{\cos^3\tau_n}. \tag{22}$$

It can easily be found that the efficiency ε, for a given χ, reaches its maximum value, equal to χ^{-1}, when R=0 and $S=(2\chi)^{-1}$. For given values of Ka, Kb, Kh and β, these conditions yield the following expressions for the optimum value of the turbine coefficient $C e^{i\theta}$

$$C = \frac{v}{\cos\beta}\left[\frac{\tanh Kh}{Ka}\right]^{1/2}, \qquad \tan\theta = u/v. \tag{23a,b}$$

It can be found from (22) that it is $\chi=1$ (and hence $\varepsilon_{max}=1$) if $(j/\ell)\sin\beta < 1+\sin\beta$, which is equivalent to (11), and that $\chi>1$ (i.e. $\varepsilon_{max}<1$) otherwise. Relation (11) was seen to be the condition for the radiated-wave far-field

($x \to -\infty$) to consist of a single regular wave whose crests make an angle $-\beta$ with the z-axis and so propagates in the same direction as the reflected wave. This allows the radiated and the reflected wave fields to cancel out each other, resulting in the total absorption of the incident wave flux, provided that the ratio Π/A, of the surface-pressure amplitude to the incident wave amplitude, has the right (complex) value. The latter condition if equivalent to (23a,b). If on the contrary (11) is not satisfied, the radiated-wave far-field will contain, in addition, other waves of nonzero amplitude propagating at different angles, therefore making total cancellation by the reflected wave impossible.

Equation (23a) gives the values of C that maximize the efficiency ε, for any given values of Kh, Ka, Kb, β and θ. It can be found that this is true even if θ does not satisfy condition (23b), and in particular if $\theta=0$ (turbine allowing no phase difference). Obviously, in the latter cases the maximum efficiency is in general less than unity, even if condition (11) is satisfied.

It can easily be found that the values of the pressure amplitude Π and of the efficiency ε, derived in this section for a row of chambers of length a with reflector, are exactly twice the corresponding values for a row of chambers of length 2a without reflecting submerged wall (i.e. the geometry considered first in section 2), the latter having the same values of Kh, Ka, Kb, β and θ, and a turbine constant C two times larger. This means that (17) and the following results can be applied to this case, provided that Π, ε and C are replaced by $\frac{1}{2}\Pi$, $\frac{1}{2}\varepsilon$ and 2C, respectively.

4. Numerical results

The following numerical results refer to a row of chambers of length a ($-a<z<0$) with a reflecting submerged vertical wall at $z=0$.

Figures 1 and 2, valid for deep water, show the resonance values of θ and C (as given by (23a,b)) respectively as functions of a/λ, for several values of the angle of incidence β. It was shown in [1], for normal incidence, that using a turbine capable of nonzero phase-difference (more precisely $\theta<0$) provides the means for reducing the optimum dimensionless value of the chamber length, a/λ, (i.e. which allows the whole incident energy to be extracted). A way that as been proposed for achieving $\theta\neq0$ consists in employing an air turbine of the Wells type, with controlable rotor-blade stagger angle [7]. A nonzero phase-difference implies that, during a fraction (equal to θ/π)

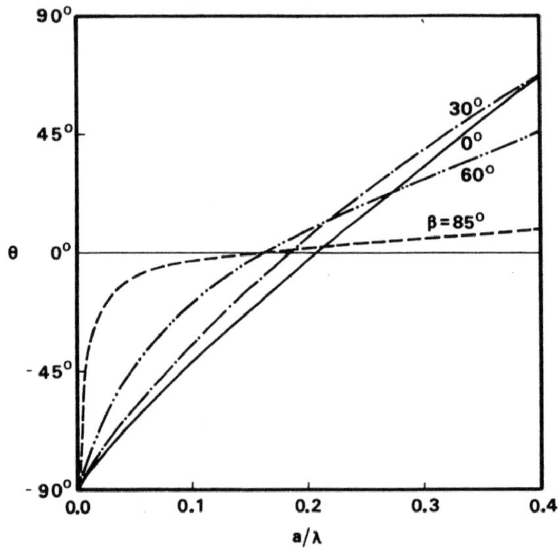

Fig.1. Resonance value of turbine phase difference θ as a function of dimensionless chamber length a/λ, for deep water and several values of the angle of incidence β (23b), and b=a.

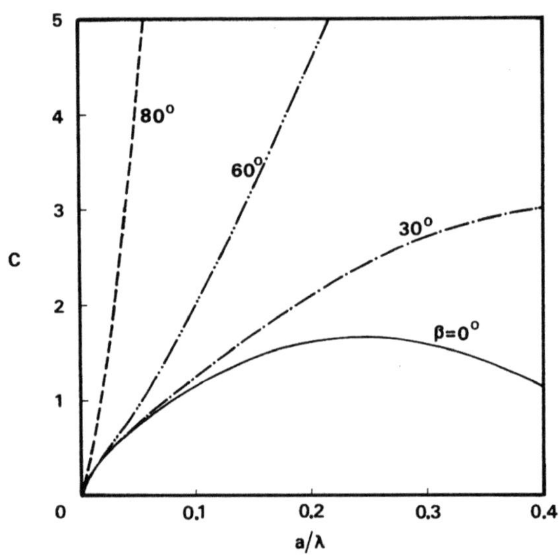

Fig.2. Resonance value of turbine coefficient C as a function of dimensionless chamber length a/λ, for deep water, b=a, and several values of the angle of incidence β.

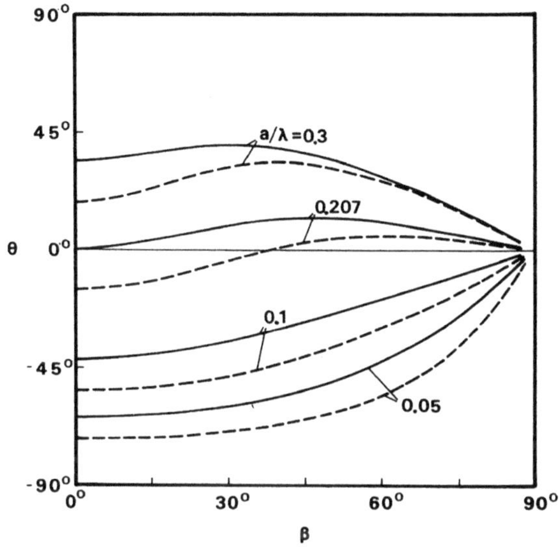

Fig.3. Resonance value of turbine phase-difference θ as a function of the angle of incidence β for deep (solid line) and shallow (broken line: h/λ=0.1) water, chamber width-to-length ratio b/a=1 and several values of dimensionless chamber length a/λ.

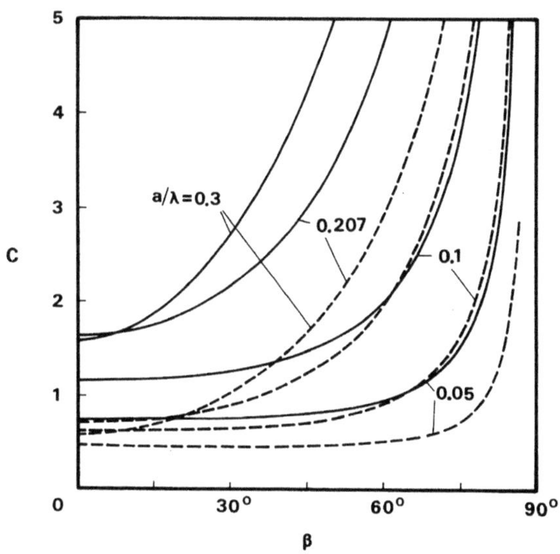

Fig.4. Resonance value of turbine coefficient C as a function of the incidence angle β for deep (solid line) and shallow (broken line: h/λ=0.1) water, chamber width-to-length ratio b/a=1, and several values of dimensionless chamber length a/λ.

of the period, the air pressure increases across the turbine, which in fact then works as a compressor supplying energy to the wave field. The ratio τ between the work of compression and the work of expansion done by the turbine during one cycle can easily be found to be given by $\tau=(\pi|\tan\theta-\theta|^{-1}+1)^{-1}$ for incompressible air. As noted in [1], the dissipation of energy in any real turbine will severely reduce the net power produced by the turbine if τ is not restricted to small values (say $\tau<0.15$, which corresponds to $|\theta|<57.2^{\circ}$). Figures 1 and 2 show that, for any fixed value of a/λ less then about 0.19 and for the chosen values of $b/a=1$ and $h/\lambda=\infty$, the required value of θ for $\varepsilon=1$ approaches zero with increasing β, whereas the required value of C increases with increasing β (more markedly so for larger values of a/λ). It is to be noted that a higher value of C will in general require a larger turbine (see [1]). If no phase-difference is allowed ($\theta=0$) then $\varepsilon=1$ is attainable only for a given value (in some cases a finite set of values) of a/λ depending of h/λ, b/a and β. Figure 1 shows, for $b=a$ and $h/\lambda=\infty$, that this value of a/λ (or the smalest one if there are several) decreases from 0.207 to about 0.16 when β increases from 0° to 85°; it can be found from figure 2 that the corresponding value of C increases with β. If a value of a/λ less then 0.207 is to be chosen, such that $\varepsilon=1$ is always attainable for any angle of incidence β without the phase-difference θ exceeding an acceptable value (say $|\theta|<57.2^{\circ}$), then figure 1 shows that such value of a/λ is the one given by the curve $\beta=0^{\circ}$. Figures 3 and 4 show the resonance values of θ and C (required if $\varepsilon=1$ is to be attained) as functions of β for deep and shallow ($h/\lambda=0.1$) water, for several values of a/λ and $b/a=1$. Note that the curves of θ for deep water are always above those for shallow water, that they converge to $\theta=0^{\circ}$ as β approaches 90°, and that for the smaller values of a/λ (0.1 and 0.05) θ is an increasing function of β. Figure 4 shows that, for any β, C is larger for deep then for shallow water. This extends a result found in [1] for normal incidence.

The effect of varying b/a and β upon the efficiency is illustrated in figure 5, for $h/\lambda=0.1$ and $a/\lambda=0.25$, and for a turbine with zero phase-difference ($\theta=0^{\circ}$). The turbine coefficient, $C=0.68$, is the value given by (23a) for $\beta=0$. It can be seen that the efficiency ε decreases with increasing β for given b/a, as should be expected. We further note that, for fixed $\beta>0$, it is obvious that ε will ultimately decrease to zero as $b/a\to\infty$; figure 5 shows that ε is not a monotonic function of b/a.

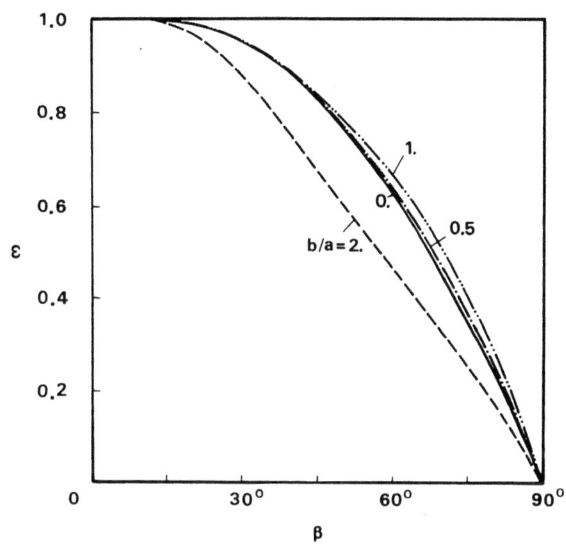

Fig. 5. The efficiency ε plotted against the angle of incidence β for several values of the chamber width-to-length ratio b/a, for h/λ=0.1 and a/λ=0.25, and for a turbine with zero phase difference (β=0) and with a coefficient C=0.68 (given by (23a) for β=0).

REFERENCES

1. Sarmento, A.J.N.A.; Falcão, A.F. de O.: Wave generation by an oscillating surface-pressure and its application in wave-energy extraction. J. Fluid Mech. 150 (1985) 467-485.
2. Evans, D.V.: Wave-power absorption by systems of oscillating surface-pressure distributions. J. Fluid Mech. 114 (1982) 481-499.
3. Wehausen, J.V.; Laitone, E.V.: Surface waves. In Handbuch der Physik, vol. 9, 446-778. Berlin: Springer-Verlag 1960.
4. Gradshteyn, I.S.; Ryzhik, I.W.: Tables of integrals, series and products, p. 736. New York, London: Academic Press 1965.
5. Apostol, T.M.: Mathematical analysis, 2nd ed. Reading: Addison-Wesley 1974.
6. Falnes, J.; Budal, K.: Wave-power conversion by point absorbers. Norwegian Marit. Res. 6 (1978) 2-11.
7. Gato, L.M.C.; Falcão, A.F. de O.: Aerodynamics of self-rectifying turbines for wave-energy extraction (in Portuguese). In Proc. 3rd Portuguese Congr. Theor. Appl. Mech., Lisbon 1983.

Hydrodynamic Analysis and Design Method of OWC Wave Energy Convertors

J. Q. CHEN, G. K. CHEN

Shipbuilding Department
South China Inst. of Technology
Guangzhou, China

Summary

In this paper, using potential flow theory and supposing that the pressure in the air cabin is proportional to the vertical velocity of the water column, we established a hydrodynamic model for OWC wave energy convertors, resulting in a collection of illustration plates. Hence, the behavior of an OWC and the relation between its parameters are discussed.
Optimization theory was used to design an optimum convertor which was working in the irregular waves.
It was found that the numerical results fit well the experimental data.

Forewords

For an Oscillating Water Column (OWC) convertor, the first stage, the most important stage, is that the wave energy is converted to the pneumatic energy in cabin. Generally, an OWC is of resonance. However, an OWC open on one side as shown in Fig. 1 has a non-resonance character. Thus, a simple spring-mass model will not be suited to be adopted. Some people [1] introduced a new conception zero-mass float (piston). This conception is useful to explain the behaviour of the OWC but is of no use for quantitative analysis, because the added mass which plays an important role is hard to determine. Therefore, the hydrodynamic method is required to make a correct calculation of an OWC with a side opening. This paper is devoted to this purpose.

Hydrodynamic Model

A chamber section of a typical OWC is shown in Fig. 1

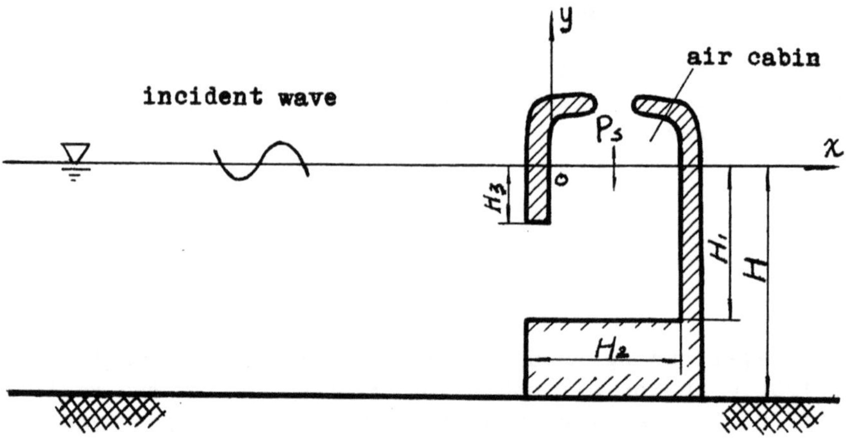

Fig. 1 Chamber Section and Coordinates

where H - water depth

H_1 - water depth in chamber

H_2 - chamber width

H_3 - draft

P_s - pressure in air cabin.

Taking into account that an OWC convertor of this type usually has a big broadness and is broadwisely located on the seabed, our problem can be regarded as two-dimensional. As we know, in linearized wave theory, the velocity potential $\Phi(x,y,t)$ must satisfy Laplace's equation and boundary conditions [2]:

$$\left.\begin{array}{ll} \Delta\Phi = 0 & \\ g\frac{\partial\Phi}{\partial y} + \frac{\partial^2\Phi}{\partial t^2} = \begin{cases} 0 & y=0,\ -\infty < x \le 0 \\ -\frac{1}{\rho}\frac{\partial P_s}{\partial t} & y=0,\ 0 < x \le H_2 \end{cases} & \\ \frac{\partial\Phi}{\partial n} = 0 & \text{on all rigid surface} \end{array}\right\} \quad (A)$$

The difficulty in solving the problem (A) above lies in dealing with the pressure P_s. Indeed this appears as an unknown quantity in the boundary condition. Therefore, it seems that adopting a repeated iteration will not be avoidable.

However, if we write the total potential Φ in the form

$$\Phi(x,y,t) = \Phi_I(x,y,t) + \check{\Phi}(x,y,t) + P_{so}\hat{\Phi}(x,y,t) \tag{1}$$

where

Φ_I - incident wave potential with amplitude A and circular frequency ω

$$\Phi_I = \frac{-iAg\, ch(m_1(y+H))}{\omega\, ch(m_1 H)} e^{i(m_1 x - \omega t)} \quad , \tag{2}$$

here $\omega^2/g = m_1 th(m_1 H)$;

$\check{\Phi}$ - disturbance wave potential,

$\hat{\Phi}$ - radiation wave potential due to the pressure in cabin P_s,

the difficulty mentioned above will be overcome. In fact, the Φ satisfies all conditions in problem (A), provided the $\check{\Phi}$ and $\hat{\Phi}$ respectively are the solution of following problems:

$$\left.\begin{array}{ll} \Delta\check{\Phi} = 0 & \\ g\dfrac{\partial\check{\Phi}}{\partial y} + \dfrac{\partial^2\check{\Phi}}{\partial t^2} = 0 & y=0 \\ \dfrac{\partial(\Phi_I + \check{\Phi})}{\partial n} = 0 & \text{on all rigid surface} \\ \lim\limits_{x\to-\infty}\left[\dfrac{\partial\check{\Phi}}{\partial x} - \dfrac{m_1}{\omega}\dfrac{\partial\check{\Phi}}{\partial t}\right] = 0 \;, & \end{array}\right\} \quad (B)$$

$$\left.\begin{array}{ll} \Delta\hat{\Phi} = 0 & \\ g\dfrac{\partial\hat{\Phi}}{\partial y} + \dfrac{\partial^2\hat{\Phi}}{\partial t^2} = \begin{cases} 0 & y=0, \quad -\infty < x \le 0 \\ \dfrac{i\omega}{\rho} e^{-i\omega t} & y=0, \quad 0 < x \le H_2 \end{cases} & \\ \dfrac{\partial\hat{\Phi}}{\partial n} = 0 & \text{on all rigid surface} \\ \lim\limits_{x\to-\infty}\left[\dfrac{\partial\hat{\Phi}}{\partial x} - \dfrac{m_1}{\omega}\dfrac{\partial\hat{\Phi}}{\partial t}\right] = 0 \;. & \end{array}\right\} \quad (C)$$

Now, there are no more unknown parameters in the boundary conditions of (B) and (C). As a result, the repeated permulating is unnecessary. The last condition in (B) and (C) is named the radiation condition. The P_{so} in (1) is the amplitude of P_s. Since it must have the same frequency as incident wave, we can write

$$P_s = P_{so}\, e^{-i\omega t} . \qquad (3)$$

On the other hand, it may not have the same phase as the wave, so the P_{so} must be complex.

The P_s can be expressed by $\check{\Phi}$ and $\hat{\Phi}$. In fact, according to our experiments, the P_s is proportional to $\bar{v}$, the average vertical velocity of the water surface in chamber, or

$$P_s = \mu \bar{v} , \qquad (4)$$

where

μ - the applied damping of the air cabin ,

$$\bar{v} = -\frac{1}{g}\frac{\partial^2 \bar{\Phi}}{\partial t^2}\bigg|_{y=0} - \frac{1}{\rho g}\frac{\partial P_s}{\partial t} . \qquad (5)$$

The sign "-" expresses the average within $0 \leq X \leq H_2$.

Substituting (1) (3) (5) in (4) and introducing

$$\Phi_I = \phi_I e^{-i\omega t} , \quad \check{\Phi} = \check{\phi} e^{-i\omega t} , \quad \hat{\Phi} = \hat{\phi} e^{-i\omega t} , \qquad (6)$$

we obtain

$$P_{so} = \frac{\overline{\phi_I} + \overline{\check{\phi}}}{\frac{g}{\mu\omega^2} - \overline{\hat{\phi}} - \frac{i}{\rho\omega}}\Bigg|_{y=0} . \qquad (7)$$

Now, we must focus our attention to the $\check{\Phi}$ and $\hat{\Phi}$. They are the solution of problems (B) and (C). After some deductions the theoretical solution of $\check{\Phi}$ and $\hat{\Phi}$ or $\check{\phi}$ and $\hat{\phi}$ are obtained. From (1) the final expression $\Phi(x,y,t)$ can be obtained too.

They are

$$\begin{aligned} \check{\phi} = {} & B_0 ch(m_2(y+H_1))(e^{im_2 x} + e^{im_2(2H_2-x)}) \\ & + \sum_{j=1}^{\infty} B_j \cos(n_j(y+H_1))(e^{n_j(x-2H_2)} + e^{-n_j x}) \\ & + \frac{Ag\, ch(m_1(y+H))}{\omega\, ch(m_1 H)}\Big[\frac{m_1}{im_2} e^{i(m_1 H_2 + m_2 H_2 - m_2 x)} \\ & - m_1 e^{im_1 H_2 - nH_2 + nx}\Big] , \qquad 0 < x \leq H_2 , \qquad (8) \end{aligned}$$

$$\hat{\Phi} = C_0 ch(m_2(y+H_1))(e^{im_2 x} + e^{im_2(2H_2-x)}) + \sum_{j=1}^{\infty} C_j Cos(n_j(y+H_1))(e^{n_j(x-2H_2)} + e^{-n_j x}) - \frac{i}{\omega\rho}, \quad 0<x\leq H_2 \quad (9)$$

and

$$\Phi(x,y,t) = \begin{cases} \Phi^{I}(x,y)\, e^{-i\omega t} & 0<x\leq H_2 \\ \Phi^{II}(x,y)\, e^{-i\omega t} & -\infty<x\leq 0 \end{cases} \quad (10)$$

$$\begin{aligned} \Phi^{I} = {} & B_0 ch(m_2(y+H_1))(e^{im_2 x} + e^{im_2(2H_2-x)}) \\ & + \sum_{j=1}^{\infty} B_j Cos(n_j(y+H_1))(e^{n_j(x-2H_2)} + e^{-n_j x}) \\ & + \frac{Ag\, ch(m_1(y+H))}{\omega\, ch(m_1 H)}\left(\frac{m_1}{im_2} e^{i(m_1 H_2 + m_2 H_2 - m_2 x)} - i e^{im_1 x} - m_1 e^{i(m_1 H_2 - nH_2 + nx)}\right) \\ & + P_{so}\Big[C_0 ch(m_2(y+H_1))(e^{im_2 x} + e^{im_2(2H_2-x)}) \\ & + \sum_{j=1}^{\infty} C_j Cos(n_j(y+H_1))(e^{n_j(x-2H_2)} + e^{-n_j x}) - \frac{i}{\omega\rho}\Big] \end{aligned} \quad (11)$$

$$\begin{aligned} \Phi^{II} = {} & \frac{-iAg\, ch(m_1(y+H))}{\omega\, ch(m_1 H)} e^{im_1 x} + A_0 ch(m_1(y+H)) e^{-im_1 x} \\ & + \sum_{j=1}^{\infty} A_j Cos(k_j(y+H)) e^{k_j x} + P_{so}\Big[D_0 ch(m_1(y+H)) e^{-im_1 x} \\ & + \sum_{j=1}^{\infty} D_j Cos(k_j(y+H)) e^{k_j x}\Big]. \end{aligned} \quad (12)$$

Parameters in these expressions have the following relations

$$\omega^2 = -g n_j Tg(n_j H_1) \quad ,$$
$$\omega^2 = g m_2 th(m_2 H_1) \quad ,$$
$$\omega^2 = -g k_j tg(k_j H) \quad .$$

The velocity potencial (10) satisfies problem (A) everywhere except at the chamber opening. So the unknown coefficents B_0, A_0, C_0, D_0, B_j, A_j, C_j, D_j, $(j=1,2, \ldots\ldots)$ in (8), (9), (11), (12) can be determined by mate - condition at this place, i.e.

$$\left.\begin{aligned} \frac{\partial \Phi^{I}}{\partial x} &= \frac{\partial \Phi^{II}}{\partial x} \\ \Phi^{I} &= \Phi^{II} \end{aligned}\right\} \qquad x=0, \; -H_1 \leq y \leq H_3 .$$

In South China Inst. of Technology different chambers varying in the dimensionless ratios were calculated, resulting in a series of illustration plates [3]. Hence, the interrelation between the energy conversion efficiency, geometry parameters, wave conditions and the pressure in the cabin are clearly demonstrated. Fig. 2 is one of these plates, where horizontal coordinate is dimensionless period $\Pi_1 = \frac{T^2 g}{2\pi H}$ and vertical coordinate is the efficiency of the chamber η. The parameter indicated at every curve is dimensionless applied damping

$$\Pi_2 = \frac{\mu}{\rho\sqrt{2\pi g H}} .$$

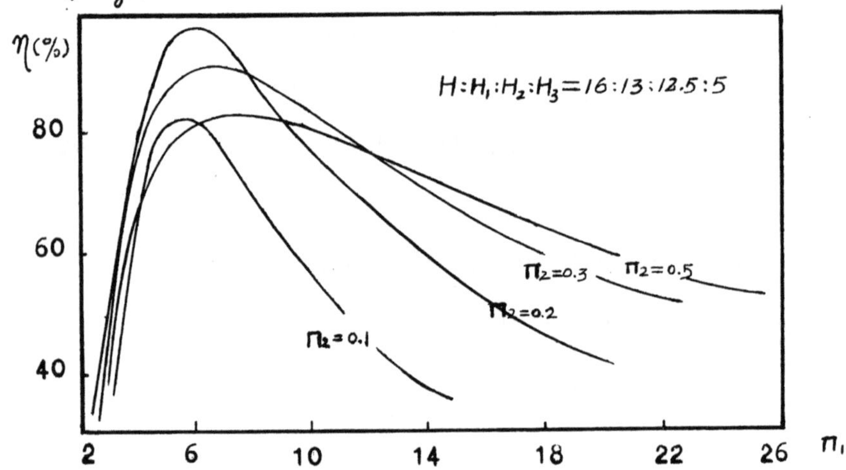

Fig. 2 dimensionless characters of a chamber

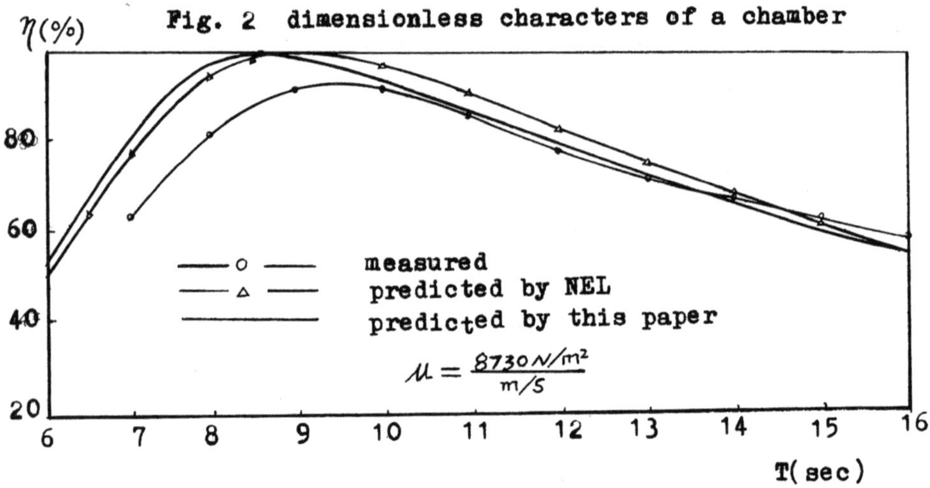

Fig. 3 comparison between theoritical and experimental data

As shown in Fig.3 , these theoretical solutions fit the experimental data well.

Fig.4 proved it to be reasonable that μ in (4) was regarded as a constant for a cabin with definite nozzle ratio.

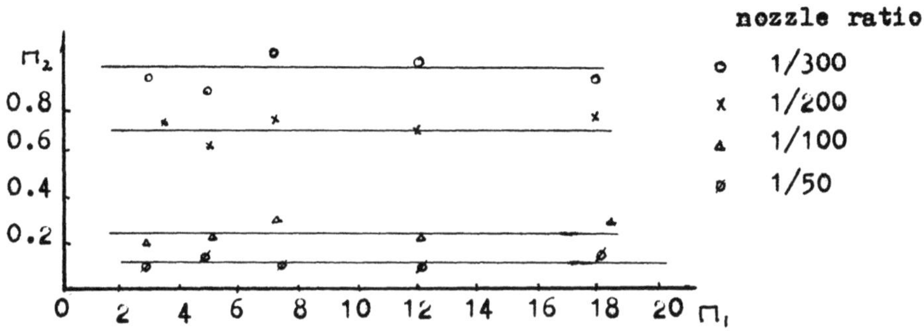

Fig.4 verifing Π_2 to be constant

With these plates designers will be in a position to design a chamber which has the optimum response to the "design wave", a regular wave.

Optimum design

Because of the irregularity of ocean waves, it is not a rational aim to design a chamber with the good response only to a regular wave. Obviously, the rational objective function is the total power taken out from the definite ocean waves depicted by its wave spactrum $S(\omega)$. The input of a wave energy convertor is energy flow of waves. So the right design view must be that a wave energy convertor has to match with the density spectrum of energy flow $\overline{S}(\omega)$ instead of energy content $S(\omega)$. Only in this way can we get the maximum power. The relation between $\overline{S}(\omega)$ and $S(\omega)$ is

$$\overline{S}(\omega) = \frac{1}{\omega} S(\omega) .$$

Therefore, the objective function for optimum design is

$$Q = \int_0^{\infty} \eta(\omega)\, \overline{S}(\omega)\, d\omega ,$$

By the application of this function, South China Inst. of Technology put forward an optimum design program.
From many numerical examples computed by this program it is indicated that for definite wave conditions, the optimum period corresponding to peak efficiency will be longer than the average period of ocean waves. This problem will be further discussed in another paper presented by the authors [4].

Conclusions

In the linearized theory a potential flow model was successfully used to study an OWC chamber with a side opening. The solution is relatively simple and requires a very short time to compute. From a great number of examples we found that:

1. For a cabin with definite nozzle ratio, it is reasonable to assume the applied damping μ to be constant.
2. A well designed OWC with a side opening is almost non-resonant.
3. Whatever size of this OWC chamber, theoretically, it has a maximum efficiency near 100%. The period corresponding to this efficiency depends mainly on the chamber's width. The longer the width the longer the period.
4. An optimum chamber should match the density spectrum of energy flow instead of energy content. In this point of view, by optimization theory, the optimum period can be calculated and demonstrated to be longer than average period.

Reference

1. Moody, G.W. : The NEL Breakwater - A Realistic Approach to Wave Energy Conversion. Symposium of Marintec offshore. Shanghai China (1983)

2. Newman, J.N. : Marine Hydrodynamics. US : The Massachusetts Institute of Technology Press 1977

3. Chen, G.K. : Hydrodynamic Analyses of OWC Wave Energy Convertors. Master Dissertation of South China Inst. of Technology. China (1984)

4. Chen, J.Q. ; He,M.K. : The Behavior of Wave Energy Convertors in Irregular waves. Symosium of the Institute's Day. South China Inst. of Technology. China. (1984)

Wave Interaction with Oscillating Bodies and Water Columns

J. FALNES
Division of Experimental Physics
University of Trondheim
N-7034 Trondheim - NTH
Norway.

and

P. McIVER
School of Mathematics
University of Bristol
Bristol BS8 1TW
United Kingdom.

Summary.

A linear hydrodynamical theory for a system of oscillating bodies and oscillating water columns (OWCs) is described and illustrated by considering a system of one OWC and one body oscillating in a single mode only. A theoretical relationship between added mass and the energy of the near field motion is generalised to include OWCs. Numerical results are given for a two-dimensional example where the structure oscillates in the surge mode.

Introduction

Many proposed devices for the conversion of ocean-wave power contain an OWC within an immobile structure which may be at the shore-line or on the sea bed. As early as 1910 useful power was produced by a turbine driven by air which was pumped by oscillations of sea-water in a vertical bore hole in a cliff on the coast of France [1]. Resonant OWC devices sited on shore were proposed in France in 1952 [2] and in Norway in 1974 [3]. A prototype device [4] presently under construction on a steep rocky shore near Bergen, Norway is expected to deliver power to the electricity grid by the end of 1985. It is a three-dimensional wave-power absorber acting more as a point than a line absorber. To the latter category belong the NEL breakwater device [5] and the submerged Vickers device [6].

Alternatively, an OWC may be contained in a floating structure which allows energy utilisation in locations not suitable for fixed structures. Masuda [7] has developed small wave-powered navigation buoys and an extension of this work has led to the ship-like Kaimei device with several air turbines [8]. Other floating OWC devices studied are line absorbers by groups at NEL [5] and at CEGB [9] and point absorbers by groups in Belfast [10] and in Trondheim [11].

In theoretical studies of an OWC it was usual to replace the internal water surface by an imaginary weightless piston and then to apply the known theory of oscillating rigid bodies. A few years ago, Evans [12] gave a more correct pressure-distribution theory which allows for the spatial variation of the internal water-surface elevation in an OWC. In this theory all structrues are assumed to be immobile. Obviously, there is a need to extend the theory to cater for a system consisting of both oscillating structures and pressure distributions. This extension is the subject of a recent study by Falnes and McIver [13] who consider an arbitrary number of oscillating bodies and pressure distributions. The main results will be presented here; for simplicity the case of one OWC and a single oscillating-body mode will be considered. Some new reciprocity relations given here have been derived independently by Fernandes [14] who considered several pressure distributions within a single oscillating body.

In the present work we assume that all fluid and body motions vary time-harmonically with a single frequency and that the linearised theory of an ideal, irrotational fluid may be used. It may be noted that several of the general relations presented here have well-known couterparts in the theory of reciprocal linear electric circuits.

Definition of excitation and radiation parameters for a two-oscillator system

Consider an OWC together with a body oscillating in a single mode. This body may either be the structure containing the OWC or, if this is immobile, it may be another body. Omitting the factor $e^{i\omega t}$ in the usual complex representation, where ω is the angular frequency, the state of the oscillating system will be described by the complex amplitudes p and u, where p represents the dynamic air pressure in the internal air chamber above the OWC and u represents the velocity of the body.

The total volume flux due to the oscillation of the internal water surface of the OWC may be expressed

$$Q_{tot} = Q - Yp - H_{pu}u \tag{1}$$

and the total force component pertaining to the mode of the body oscillation may be expressed by

$$F_{tot} = F - Zu - H_{up}p. \tag{2}$$

In the above, the excitation volume flux Q and the excitation force F represent the effect of the incident wave only, since they correspond to the situation when the system is not oscillating, that is when p = 0 and u = 0. We have

$$Q = qA \quad \text{and} \quad F = fA \tag{3}$$

where A is the complex elevation amplitude of the undisturbed incident wave at some conveniently chosen reference point. The frequency dependent complex excitation parameters q and f do not, in linear theory, depend upon A. Similarly, the frequency dependent complex "radiation" coefficients Y, Z, H_{pu} and H_{up} in equations (1) - (2) do not depend upon A, p or u. The terms $-Yp$ and $-H_{up}p$ are contributions to the volume flux and to the body force, respectively, due to radiation from the OWC corresponding to the oscillating air pressure p. Similarly, the terms $-H_{pu}u$ and $-Zu$ are the corresponding contributions caused by the body oscillation. It has been shown [13, 14] that

$$-H_{up} = H_{pu} \equiv H \ . \tag{4}$$

For real ω, we decompose the radiation admittance Y, the radiation impedance Z and the radiation coupling parameter H into real and imaginary parts

$$Y = G + iB, \quad Z = R + iX, \quad H = C + iJ \ . \tag{5}$$

By analogy with electric-circuit usage we may term G and R the radiation conductance and radiation resistance, respectively. Note that in hydrodynamical texts R is usually termed the radiation damping, a term which we prefer to reserve for the Hermitian matrix

$$\underset{\sim}{\Delta} = \begin{pmatrix} G & -iJ \\ iJ & R \end{pmatrix} . \tag{6}$$

Further, X/ω is the so-called added mass, a term which may be misleading since it is not associated only with kinetic energy in surface wave problems [15]. However, a corresponding term would be to call B/ω the "added compliance".

It should be noted that for a system of several OWCs and several oscillating body modes the excitation coefficients q and f are generalised to column vectors and the radiation coefficients Y, Z and H to matrices [13].

Power and energy relations for wave generation

Consider the case with p,u ≠ 0 but A = 0, that is the system is forced to oscillate in the absence of an incident wave. The only wave motion in this situation is that radiated due to the forced oscillations. The radiated power P_r can be shown to be [13]

$$P_r = \frac{1}{2} G \, |p|^2 + \frac{1}{2} R \, |u|^2 + \text{Im} \, \{Jpu^*\} \, , \tag{7}$$

where * indicates complex conjugate. The second term is well-known from the theory of oscillating bodies, while the first is known from the theory of oscillating pressure distributions [12]. The final term is new and represents the coupling when both types of oscillator act simultaneously.

In the far-field region of the radiated wave the surface density of the averaged potential energy equals that of the kinetic energy associated with the propagating wave. However, in the so-called near-field region around the oscillating system these energies may differ. It can be shown [13] that the difference between the mean kinetic energy T and the mean potential energy V stored in the near field region is

$$T - V = \frac{1}{2\omega} \left(\frac{1}{2} X \, |u|^2 - \frac{1}{2} B \, |p|^2 - \text{Re} \, \{Jpu^*\} \right) . \tag{8}$$

For the case of an oscillating body only (u ≠ 0, p = 0) this shows that the rare cases where the added mass X/ω is negative are associated with the potential energy exceeding the kinetic energy [15]. With an oscillating pressure distribution only (p ≠ 0, u = 0) the added compliance B/ω is positive when V exceeds T. Note that V includes a contribution from the hydrostatic buoyancy of the OWC. This contribution dominates at low frequencies, in fact

$$B/\omega \rightarrow S/\rho g \text{ as } \omega \rightarrow 0, \tag{9}$$

where g is the acceleration due to gravity, ρ the water density and S the area of the internal water surface of the OWC. In passing, we remark that the excitation volume flux coefficient

$$q = Q/A \rightarrow i\omega S \text{ as } \omega \rightarrow 0. \tag{10}$$

The last term in equation (8) represents the coupling when $p,u \neq 0$. Note that only five of the six real parameters defined in equations (5) are involved in the power and energy relations (7) - (8).

Absorbed power

When the system responds to an incident wave the absorbed power may be expressed as

$$P = \frac{1}{2} \text{Re} \{pQ^*_{tot} + F_{tot}u^*\} \tag{11}$$

where the two terms represent the power absorbed through the OWC and the oscillating body, respectively. Substituting for Q_{tot} and F_{tot} from equations (1) - (2) gives

$$P = P_e - P_r \tag{12}$$

where the radiated power P_r is given by equation (7) and the excitation power P_e by

$$P_e = \frac{1}{2} \text{Re} \{pQ^* + Fu^*\}. \tag{13}$$

The actual values of p and u depend upon the excitation parameters (Q,F) and on the applied loading mechanisms of the power conversion devices, the details of which will not be considered here. If the excitation coefficients (q,f) and the radiation coefficients (Y,Z,H) for the system are known, and if the values p, u and A are known, from measurements for instance, then the absorbed power is given by the above equations.

Introducing the column vectors

$$\underset{\sim}{\pi} = \text{col}(p,u), \quad \underset{\sim}{\kappa} = \text{col}(Q,F) \tag{14}$$

and their transposes $\tilde{\pi}$ and $\tilde{\kappa}$, we may write

$$P = P_e - P_r = \frac{1}{4} (\tilde{\kappa}\underset{\sim}{\pi}^* + \tilde{\kappa}^*\underset{\sim}{\pi}) - \frac{1}{2} \tilde{\pi}\underset{\approx}{\Delta}\underset{\sim}{\pi}^* \tag{15}$$

where the radiation damping matrix $\underset{\sim}{\Delta}$ is given by equation (6). From equation (15) it can be shown [13] that, if the incident wave, and hence

$\underset{\sim}{\kappa}$, are given then

$$P_{max} = P_{r,opt} = \frac{1}{2} P_{e,opt} = \frac{1}{2} \tilde{\pi}_o \underset{\sim}{\Delta} \underset{\sim}{\pi}_o^* = \frac{1}{4} \tilde{\kappa} \underset{\sim}{\pi}_o^* \qquad (16)$$

where the optimum oscillation amplitudes ($\underset{\sim}{\pi} = \underset{\sim}{\pi}_o$) correspond to a solution of

$$\underset{\sim}{\Delta}^* \underset{\sim}{\pi}_o = \frac{1}{2} \underset{\sim}{\kappa}. \qquad (17)$$

If $\underset{\sim}{\Delta}$ is not singular, $\underset{\sim}{\Delta}^{-1}$ exists, and there is a unique optimum value

$$\underset{\sim}{\pi}_o = \frac{1}{2} (\underset{\sim}{\Delta}^*)^{-1} \underset{\sim}{\kappa} \text{ giving } P_{max} = \frac{1}{8} \tilde{\kappa} \underset{\sim}{\Delta}^{-1} \underset{\sim}{\kappa}^*. \qquad (18)$$

This result specialises to well-known results for oscillating bodies and for oscillating pressure distributions [12]. In both of those cases the radiation damping matrix is real and symmetric, in contrast to the more general case considered here where it is complex and Hermitian.

Hydrodynamical relations

The interaction between the waves and the oscillating system may be analysed using potential theory. The required potential must satisfy Laplace's equation within the fluid domain and certain boundary conditions on the fluid boundaries, the free surface and the wetted body surfaces. It may be decomposed into a scattering potential ϕ_s and a radiation potential ϕ_r. The latter may be decomposed further into parts corresponding to each component of the oscillating system, $\phi_r = \phi_p p + \phi_u u$.

The excitation volume flux Q and the radiation coefficients Y and H may be expressed by integrals of the vertical gradient of the different potential components (ϕ_s, ϕ_p, ϕ_u) over the internal free surface of the OWC [13]. Similarly, the excitation force F and the radiation coefficients H_{up}(= − H) and Z may be expressed as integrals of the potential components over the wetted surface of the oscillating body.

Further, from Green's theorem, the following relationship between the radiation coefficients and the excitation parameters may be proved [13]:

$$\underset{\sim}{\Delta} = \frac{\omega k}{4\pi\rho g^2 D |A|^2} \int_{-\pi}^{\pi} \underset{\sim}{\kappa}(\beta) \tilde{\kappa}^*(\beta) d\beta. \qquad (19)$$

Here, D = D(kh) is a depth function equal to unity for deep water, h is the water depth and k is the angular repetency (wavenumber) related to ω by the dispersion relation

$$\omega^2 = gk \tanh (kh). \tag{20}$$

Note that the excitation vector $\underset{\sim}{\kappa}$ defined in equation (14), depends upon the angle of wave incidence β. The two-dimensional analogue of equation (19) is

$$\underset{\sim}{\Delta} = \frac{\omega}{2\rho g^2 D |A|^2} \{\underset{\sim}{\kappa}(0)\ \tilde{\kappa}^*(0) + \underset{\sim}{\kappa}(\pi)\ \tilde{\kappa}^*(\pi)\}. \tag{21}$$

Equations (19) and (21) specialise to well-known relationships for an oscillating body only and for an OWC only [12].

Two-dimensional example

An axisymmetric system consisting of an OWC within a ring-shaped buoy has been studied in some detail by Fernandes [14] and by Oltedal [16]. For this case the radiation damping matrix $\underset{\sim}{\Delta}$ is singular.

We shall consider the two-dimensional example illustrated in Fig. 1, where the barriers are constrained to move in the surge mode only. In this case $\underset{\sim}{\Delta}$ is non-singular. The barriers are chosen to have unequal submergences so that the radiation coupling coefficient is not identically zero.

The solution is by the method of matched eigenfunction expansions. For each of the three regions shown in Fig. 1 an expansion containing unknown coefficients is assumed in terms of a complete set of eigenfunctions [17]. The coefficients are determined by satisfying the boundary conditions and by matching the appropriate expansions on the boundaries between the regions. Details of the derivation and numerical computations are given elsewhere [18].

For the scattering problem results are given in terms of the excitation volume flux Q in Fig. 2a and the excitation force F in Fig. 2b. The curves are sharply peaked as a result of a resonance occuring at the fundamental frequency of oscillation for the column of water between the barriers, this resonance has been discussed for barriers of equal length by Newman [19]. A similar example of a pressure distribution between two f i x e d barriers of equal length has been analysed by Masuda, Maeda and

Kato [20]. In this case they found a zero of the exciting force while results for the internal water surface elevation qualitatively agree with the present Fig. 2a.

Results for the radiation properties are presented in Figs. 3(a-c). Again the effects of the resonance are clear. In Fig. 3b, the radiation reactance X (X/ω is the added mass) is just negative, in one case, for a small range of ka, see e.g. [21] for a similar observation. It may also be seen from Fig. 3c that the coupling, as described by H, is most significant in the region of the resonance.

Conclusion

In the present paper an outline has been given of a linear theory [13] which describes the hydrodynamic interaction between a wave and a system of OWCs and oscillating bodies. Such an analysis is an essential part of the complete description of such a system when used for conversion of wave power. It is also necessary to have a realistic mathematical model of the energy conversion machinery in order to make a reasonable choice at the design stage. In this design procedure hydrodynamic theories of the present type are indispensable.

The financial support of the UK Science and Engineering Reseach Council and the Norwegian Royal Ministry of Petroleum and Energy is acknowledged.

References

1. Palme, A.: Wave motion turbine. Power 52 (1920) 700-701.

2. French patent application filed 1952-04-11. See GB patent specification 741, 494 published 1955-12-07.

3. Budal, K.; Falnes, J.: Proposals for conversion of the energy in ocean waves. Internal Report, March 1974, Division of Experimental Physics, University of Trondheim.

4. Malmo, O.; Reitan, A.: Development of the Kvaerner multiresonant OWC. This symposium.

5. Moody, G.W.; Elliott, G.: The development of the NEL breakwater wave energy converter. 2nd Int. Symp. Wave Energy Utilisation, Trondheim (1982) 421-451.

6. Drew, S.D.: Recent developments with the Vickers OWC devices. 2nd Int. Symp. Wave Energy Utilisation, Trondheim (1982), 191-209.

7. Masuda, Y.: Study of wave activated generator and future view as an island power source. 2nd Int. Ocean Development Conf. (1972) Preprints 2, 2074-2090.

8. Ishii, S.; Miyazaki, T.; Masuda, Y.; Kai, G.: Reports and future plans for the Kaimei project. 2nd Int. Symp. Wave Energy Utilisation, Trondheim (1982) 305-321.

9. Count, B.M.; Miyazaki, T.: Study on floating attenuator wave energy devices. J. Soc. Naval Architects of Japan 155 (1984) 164-171.

10. Whittaker, T.J.T.; Robinson, R.W.; Murray, M.A.: Hydrodynamic study of an oscillating water column wave energy converter. Int. Conf. Future Energy Concepts (1981), IEE Conf. Pub. 192 143-146.

11. Budal, K.; Falnes, J.: Wave power conversion by point absorbers : A Norwegian Project. Int. J. Ambient Energy 3 (1982) 59-67.

12. Evans, D.V.: Wave power absorption by systems of oscillating surface pressure distributions. J. Fluid Mech. 114 (1982) 481-499.

13. Falnes, J.; McIver, P.: Surface wave interactions with systems of oscillating bodies and pressure distributions. App. Ocean Research, to appear.

14. Fernandes, A.C.: Reciprocity relations for the analysis of floating pneumatic bodies with application to wave power absorption. Proc. 4th Int. Offshore Mechanics and Arctic Eng. Symp., Dallas, Texas 1 (1985) 725-730.

15. McIver, P.; Evans, D.V.: The occurrence of negative added-mass in free surface problems involving submerged oscillating bodies. J. Eng. Maths 18 (1984) 7-22.

16. Oltedal, G.: Simulation of a pneumatic wave power buoy with phase control. This symposium.

17. Mei, C.C.: The Applied Dynamics of Ocean Surface Waves. New York : Wiley-Interscience 1983. Section 7.4.1.

18. Falnes, J.; McIver, P.: (Title as ref. 13). Report AM-84-03, School of Mathematics, University of Bristol (1984).

19. Newman, J.N.: Interaction of water waves with two closely spaced vertical obstacles. J. Fluid Mech. 66 (1974) 97-106.

20. Masuda, K.; Maeda, H.; Kato, W.: Fundamental study of an air-turbine type wave power absorber. Int. Conf. Management of Oceanic Resources - The Way Ahead, ECOR, London (1981).

21. Newman, J.: The motions of a floating slender torus. J. Fluid Mech. 83 (1977) 721-735.

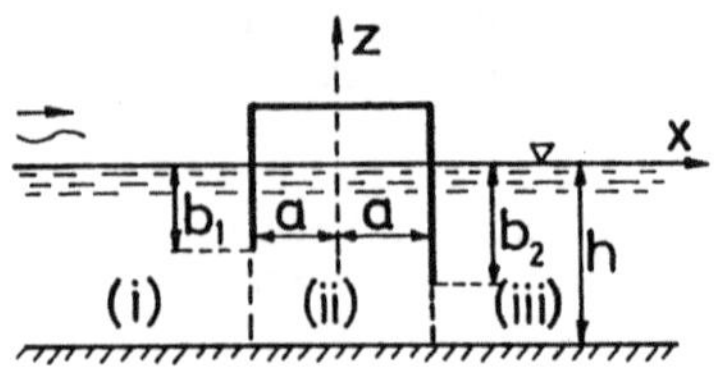

Fig. 1. Chamber for oscillating water column between two rigidly connected plane vertical barriers placed at $x = \pm a$. The barriers are submerged to depths b_1 and b_2 . The plane $z = -h$ is the sea bed and a plane wave is incident from $x = -\infty$. The solution for the potential is written as an eigenfunction expansion for each of the three indicated regions (i) , (ii) and (iii).

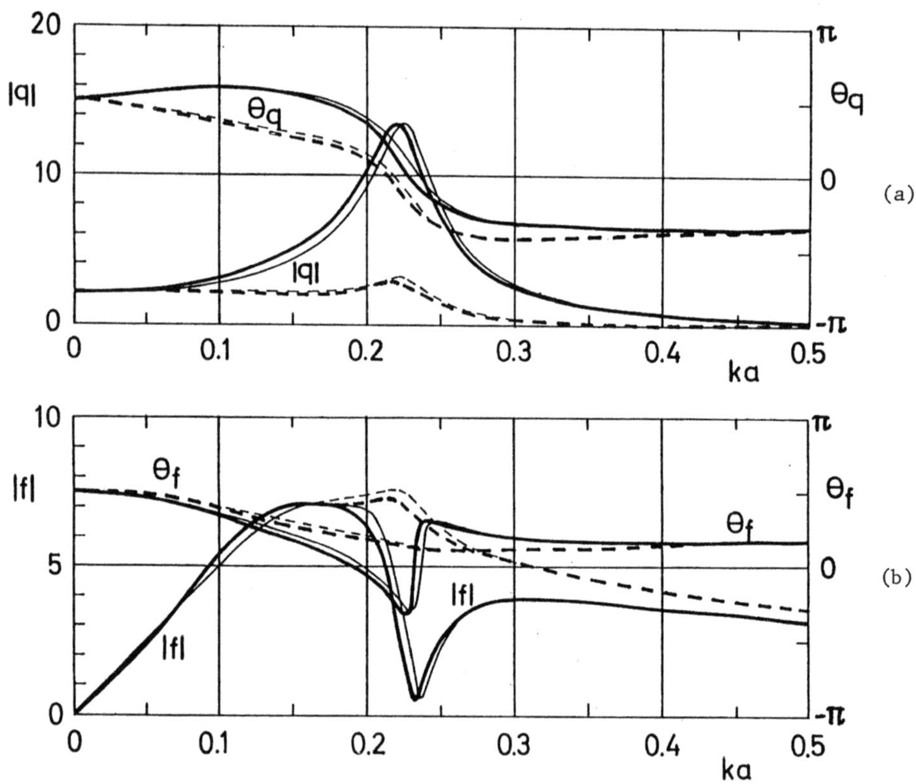

Fig. 2. Normalised scattering coefficients versus normalised barrier spacing ka for relative barrier submergences $b_1/a = 3$ and $b_2/a = 5$, and with relative water depth $h/a = 10$ (thinner lines) and $h/a = 20$ (thicker lines). For solid curves wave incidence is from the left in Fig. 1, and for the dashed curves from the right.

(a) Excitation volume flux: $Q/\omega aA = q = |q| \exp(i\Theta_q)$

(b) Excitation surge force: $F/\rho gaA = f = |f| \exp(i\Theta_f)$ measured in the direction of wave advance.

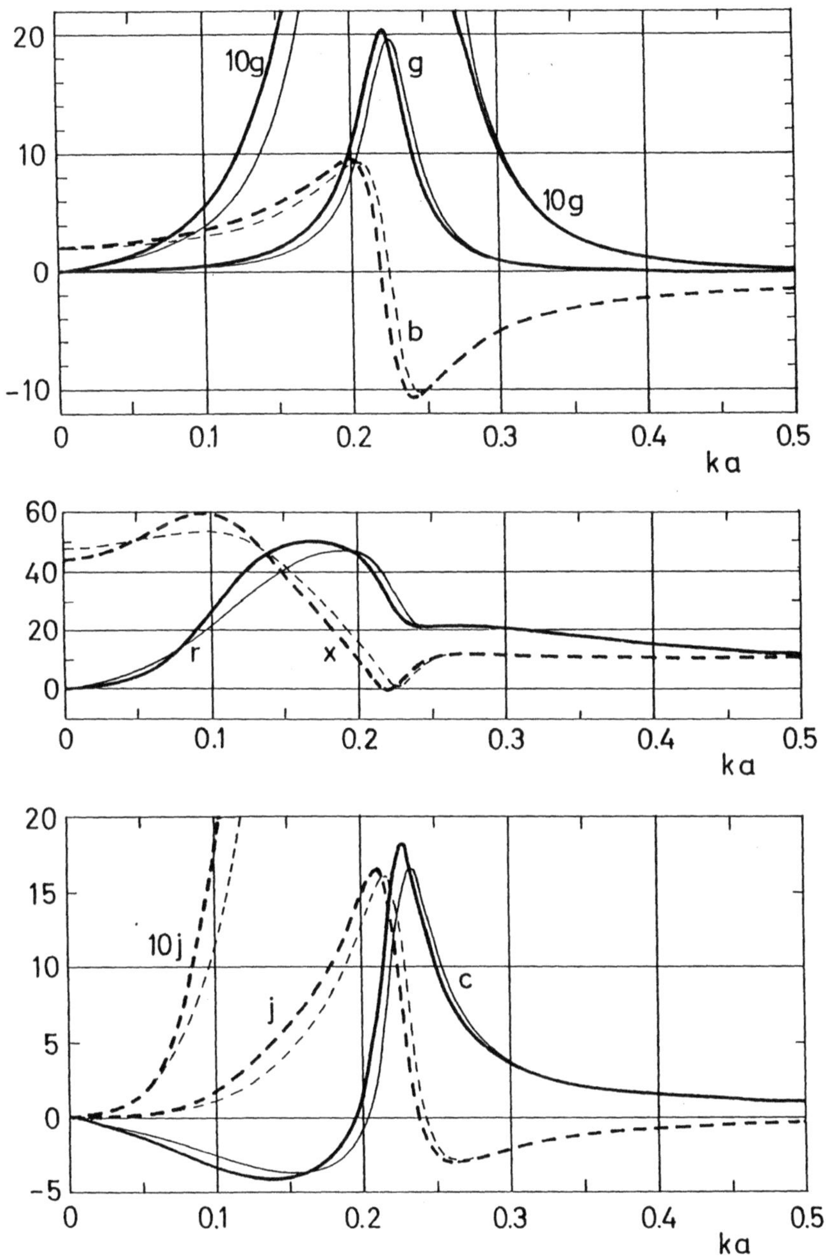

(a)

(b)

(c)

Fig.3. Normalised radiation coefficients versus normalised barrier spacing ka for relative barrier submergences $b_1/a = 3$ and $b_2/a = 5$ and for relative water depth $h/a = 10$ (thinner lines) and $h/a = 20$ (thicker lines)

(a) Radiation admittance: $Y/(\omega a/\rho g) = y = g + ib$

(b) Radiation impedance: $Z/\omega\rho a^2 = z = r + ix$. Note that x is the surge added-mass coefficient.

(c) Radiation coupling: $H/(\omega a\sqrt{a/g}) = h = c + ij$.

The Radiation and Scattering of Long Water Waves

M. J. SIMON and A. HULME

Department of Mathematics
University of Manchester, England

1. Introduction

One of the essential steps towards a theoretical understanding of wave-energy devices is an analysis of the hydrodynamic interaction between the structures, their modes of motion and the wave field. Classically, linear potential theory has been used to study this interaction, which allows its decomposition into monochromatic radiation and diffraction problems. There are well-established numerical procedures for the solution of such problems (see Yeung, 1985), methods which, although exact in principle, can involve large lengthy computations, which may be prohibitively expensive, when the bodies comprise difficult geometries without symmetry. The methods also have the feature that they give no insight into the effect of changes in the geometry on the device's hydrodynamic performance.

The motivation for the work of this paper was that many important ocean structures are small compared to the wavelength of typical incident waves (e.g. Budal et al 1982), and that a long-wave asymptotic analysis would be useful. This approach yields asymptotic expansions for the three important hydrodynamic quantities (the exciting force F, the damping B and the added mass A) for arbitrary bodies, and the geometric parameters enter explicitly. The numerical effort involved is very much smaller than in the finite-frequency numerical procedures. The technique can be used with two- or three-dimensional bodies in any mode of motion, and the bodies may be floating or submerged. In the case of finite depth the working is rather more involved, so the description of the method here will be limited to infinite-depth oceans. For simplicity, the technique will be described here as applied to the heaving motion of an axisymmetric three-dimensional body, and, as an example, the half-immersed sphere will be used for comparison with exact results.

In the asymptotic expansions the small parameter is the dimensionless wavenumber μ, and a typical term is $C_{mn}\mu^m(\ell n \cdot \mu)^n$; it is shown how the coefficients C_{mn} can be derived for any m and n, in principle, for the expansions of each of F, A & B. The method is in two parts; the first describes three relations between the three quantities of interest, and successive resubstitution reveals that all the coefficients can be obtained in terms of the geometrical parameters and an undetermined set of coefficients which multiply μ^m in the virtual mass A. In the second part it is shown how these coefficients can be easily determined via a hierarchy of integral equations, each of which has the same kernel involving only the simple zero-frequency Green's function.

2. The hydrodynamic quantities and their interrelations

Consider an incident plane wave of amplitude Ω, which is described by a potential

$$\mathrm{Re}[\frac{g\Omega}{\omega}\phi_I e^{-i\omega t}] \quad \text{with} \quad \phi_I = e^{-Ky-iKx} \tag{2.1}$$

where ω is the radian frequency, g is gravity and $K = \frac{\omega^2}{g}$ is the small wavenumber. Here x and z are in the horizontal and y is measured down from the undisturbed free surface.

The scattering of the incident wave by the body is described by a potential $\mathrm{Re}[\frac{g\Omega}{\omega}\phi_D e^{-i\omega t}]$ where ϕ_D is harmonic, satisfies the linearised free-surface condition

$$(\frac{\partial}{\partial y} + K)\phi = 0 \quad \text{on} \quad y = 0 \quad \text{outside the body,} \tag{2.2}$$

behaves like outgoing waves at large distance from the body, and satisfies

$$\frac{\partial\phi_D}{\partial n} = -\frac{\partial\phi_I}{\partial n} \quad \text{on the body} \tag{2.3}$$

Here $\underline{n}$ denotes the normal to the body directed into the fluid.

The heave force on the body is then

$$\mathrm{Re}[ig\rho\Omega F e^{-i\omega t}] \tag{2.4}$$

where
$$F = \int_{S_B} (\phi_I + \phi_D) n_y \, dS \qquad (2.5);$$

the integral is over the wetted surface S_B of the body.

The heave radiation potential ϕ_R is harmonic and satisfies the same conditions as ϕ_D on the free surface and at large distance, and the boundary condition

$$\frac{\partial \phi_R}{\partial n} = n_y \quad \text{on the body} \qquad (2.6).$$

The non-dimensional damping B and virtual mass A are defined by

$$V(A + iB) = - \int_{S_B} \phi_R \, n_y \, dS \qquad (2.7),$$

where V is the volume of the body.

As is well known, the reciprocity between radiation and diffraction now yields

$$B = \frac{K}{2V} |F|^2 \qquad (2.8).$$

This is the first of three key relations between F, A and B; in contrast to this exact result, the next important link is a formal asymptotic one, and no attempt is made to justify it rigorously. We shall use the result

$$\int_{S_B} H(x,y,z) n_i \, dS = \int_V \frac{\partial H}{\partial x_i} \, dV + \delta_{2i} \int_{S_A} H \, dS \quad (i = 1,2,3) \qquad (2.9)$$

with S_A = that part of $y = 0$ inside the body, and the Kronecker delta has been used.

The formal expansion of ϕ_I as $K \to 0$ is

$$\phi_I \sim 1 - K(y + ix) + \tfrac{1}{2}K^2(y + ix)^2 - \frac{1}{6} K^3(y + ix)^3 + O(K^4) \qquad (2.10)$$

The Froude-Krylov heave force is thus

$$F_I = \int_{S_B} \phi_I \, n_y \, dS = \int_V \frac{\partial \phi_I}{\partial y} \, dN + \int_{S_A} \phi_I \, dS$$

$$\sim S - KV + K^2(V_y - \tfrac{1}{2}S_{xx}) - \tfrac{1}{2}K^3(V_{yy} - V_{xx})$$

$$+ O(K^4) \qquad (2.11),$$

where use has been made of the notation

$$V_y = \int_V y dV, \quad V_{xx} = \int_V x^2 dV, \quad \text{etc.}$$

$$\text{and} \quad S = \int_{S_A} dS, \quad S_{xx} = \int_{S_A} x^2 dS, \quad \text{etc.} \qquad (2.12)$$

Note that $V_x = V_{xy} = S_x = 0$ by symmetry. It remains to evaluate $F_D = \int_{S_B} \phi_D \, n_y \, dS$ asymptotically, and we combine (2.3) and (2.10) to give, formally,

$$\frac{\partial \phi_D}{\partial n} \sim K\left(\frac{\partial y}{\partial n} + i \frac{\partial x}{\partial n}\right) - \tfrac{1}{2}K^2 \frac{\partial}{\partial n} (y + ix)^2 + \frac{1}{6} K^3 \frac{\partial}{\partial n} (y + ix)^3$$

$$+ O(K^4) \qquad (2.13).$$

We can pick out the axisymmetric component here, since only the evaluation of F_D matters; consequently we can write

$$\phi_{D,axi} \sim K\phi_R - \tfrac{1}{2}aK^2\phi_R^{(2)} + \frac{1}{6} a^2K^3\phi_R^{(3)} + O(K^4) \qquad (2.14),$$

where $\phi_R^{(2)}$ and $\phi_R^{(3)}$ are associated radiation potentials with boundary conditions

$$\frac{\partial \phi_R^{(2)}}{\partial n} = U^{(2)} = \frac{1}{a} \frac{\partial}{\partial n} (y^2 - \tfrac{1}{2}r^2), \quad \frac{\partial \phi_R^{(3)}}{\partial n} = U^{(3)} = \frac{1}{a^2} \frac{\partial}{\partial n} (y^3 - \frac{3}{2} r^2 y)$$

$$(2.15)$$

(The length-scale a of the body is introduced for dimensional reasons.)

We have thus

$$F_D \sim -KV(A + iB) + \tfrac{1}{2}aK^2V(A^{(2)} + iB^{(2)}) - \frac{1}{6}\, a^2K^3V(A^{(3)} + iB^{(3)})$$

$$+ O(K^4) \qquad (2.16),$$

where the associated damping and virtual masses are defined in the obvious fashion corresponding to (2.7)

$$- V(A^{(k)} + iB^{(k)}) = \int_{S_B} \phi_R^{(k)} n_y \, dS = \int_{S_B} \phi_R U^{(k)} dS \qquad (2.17).$$

Combining (2.11) and (2.16) gives

$$\frac{aF}{V} \sim g_1 - \mu(1 + A + iB) + \tfrac{1}{2}\mu^2(g_2 + A^{(2)} + iB^{(2)})$$

$$- \frac{1}{6}\, \mu^3(g \;\; + A^{(3)} + iB^{(3)}) + O(\mu^4) \qquad (2.18).$$

Here $\mu = Ka$ is the small expansion parameter, and

$$g_1 = \frac{aS}{V}, \quad g_2 = \frac{2V_y - S_{xx}}{aV}, \quad g_3 = \frac{3(V_{yy} - V_{xx})}{a^2V} \qquad (2.19)$$

are geometric parameters to be evaluated for the body in question. (The first two terms in (2.18) have been derived by Newman (1977) and others.).

The third important link used is between A & B, and is a correspondence between the coefficients in their asymptotic expansions. This set of equations comes from the integral-equation formulation of the radiation problem, as outlined briefly later.

The coefficient of $\mu^m(\ell n\ \mu)^n$ in the expansion of quantity X (which will stand for F, A, B, $A^{(k)}$ or $B^{(k)}$) will be denoted by X_{mn}. Then it can be shown that

$$B_{m0} = -\pi\, A_{m1} \ (m = 1,2,3,4),\ B_{m0} = -\pi(A_{m1} + 2\pi^2 A_{m3}) \ (m = 5,6,7,8)$$

$$B_{m1} = -2\pi A_{m2} \ (m = 3,4,5,6),\ B_{m1} = -2\pi(A_{m2} + 4\pi^2 A_{m4}) \ (m = 7,8)$$

$$B_{m2} = -3\pi A_{m3} \ (m = 5,6,7,8),\ B_{m3} = -4\pi\, A_{m4} \ (m = 7,8) \text{ etc.} \qquad (2.20).$$

These relations hold for arbitrary geometries and arbitrary modes of motion in the three-dimensional infinite-depth problem; axisymmetry is not necessary. Some of the relations may be trivially satisfied in some modes and for some bodies.

We can now use (2.8), (2.18) and (2.20) to successively determine coefficients in F, A and B.

3. An example - the heaving hemisphere

Since $F = O(1)$ as $\mu \to 0$, (2.8) shows that $B_{00} = 0$. Thus (2.18) gives

$$\frac{aF}{V} = g_1 - \mu(1 + A_{00}) + o(\mu) \quad \text{as} \quad \mu \to 0 \qquad (3.1);$$

and so (2.8), which can be rewritten as

$$\frac{1}{\pi} B = \left(\frac{V}{2\pi a^3}\right) \mu \left|\frac{aF}{V}\right|^2 \qquad (3.2),$$

yields

$$\frac{1}{\pi} B \sim \frac{V}{2\pi a^3} \{\mu g_1^2 - 2\mu^2 g_1 (1 + A_{00})\} \qquad (3.3)$$

Thus

$$A_{11} = -\frac{1}{\pi} B_{10} = -\frac{V}{2\pi a^3} g_1^2 = -\frac{S^2}{2\pi a V}$$

$$\text{and} \quad A_{21} = -\frac{1}{\pi} B_{20} = +\frac{V}{\pi a^3} g_1 (1 + A_{00}) = \frac{S}{\pi a^2} (1 + A_{00}) \qquad (3.4).$$

For the half-immersed sphere, with the length-scale a as the radius, we have $S = \pi a^2$, $V = \frac{2}{3} \pi a^3$ and so $g_1 = \frac{3}{2}$; also $g_2 = \frac{3}{8}$ and $g_3 = 0$. So

$$A_{11} = -\frac{1}{\pi} B_{10} = -\frac{3}{4}, \quad A_{21} = -\frac{1}{\pi} B_{20} = 1 + A_{00} \qquad (3.5).$$

We return to (2.18), which now gives

$$\frac{aF}{V} \sim \frac{3}{2} - \mu(1 + A_{00}) + \frac{3}{4} \mu^2 \ell n\, \mu + \mu^2 \left(\frac{3}{16} + \frac{1}{2} A_{00}^{(2)} - A_{10} - \frac{3i\pi}{4}\right)$$
$$+ o(\mu^2) \qquad (3.6).$$

Another use of (3.2) now yields

$$\frac{1}{\pi} B \sim \frac{3}{4} \mu - \mu^2(1 + A_{00}) + \frac{3}{4} \mu^3 \ln \mu$$
$$+ (\frac{3}{16} + \frac{1}{2} A_{00}^{(2)} - A_{10} + \frac{1}{3}[1+A_{00}]^2)\mu^3$$
$$+ o(\mu^3) \qquad (3.7),$$

and an application of (2.20) gives

$$A_{32} = -\frac{3}{8}, \quad A_{31} = -\{\frac{3}{16} + \frac{1}{2} A_{00}^{(2)} - A_{10} + \frac{1}{3}(1 + A_{00})^2\} \qquad (3.8).$$

It is clear that A_{00} and $A_{00}^{(2)}$ are required, as well as A_{10}. The integral-equation method of the next section gives these coefficients in succession, for general bodies, but it is worth noting that the first two can be evaluated explicitly for the hemisphere. They are

$$A_{00} = 3 \sum_{m=0}^{\infty} \frac{4m + 1}{2m + 1} \left(\int_0^1 xP_{2m}(x)dx\right)^2 = .830951$$

as given by Ursell (1963) and Hulme (1982), and

$$A_{00}^{(2)} = 2 \int_0^1 xP_2(x)dx = \frac{1}{4}.$$

We now write down the asymptotic expansions of A, F and B; each of these is given up to the order allowed by the determination, via integral equations, of A_{00}, A_{10}, A_{20}, A_{30} and the corresponding $A_{m0}^{(k)}$.

$$A \sim .830951 - \frac{3}{4} \mu \ln \mu - 1.112956\mu + 1.830951\mu^3 \ln \mu + .582908\mu^2$$
$$- \frac{3}{8} \mu^3 \ln^2 \mu - 2.542917\mu^3 \ln \mu + 4.006634\mu^3 \qquad (3.9)$$

$$-\frac{1}{\pi}\mathrm{Im}(\frac{aF}{v}) \sim \frac{3}{4}\mu^2 - 1.830951\mu^3 + \frac{3}{4}\mu^4 \ln \mu + 2.699167\mu^4 \qquad (3.10)$$

$$\mathrm{Re}(\frac{aF}{v}) \sim \frac{3}{2} - 1.830951\mu + \frac{3}{4}\mu^2 \ln \mu + 1.425456\mu^2 - 1.830951\mu^3 \ln \mu$$
$$- .634122\mu^3 + \frac{3}{8}\mu^4 \ln^2 \mu + 2.699167\mu^4 \ln \mu - 3.744528\mu^4 \qquad (3.11)$$

$$\frac{1}{\pi} B \sim \frac{3}{4}\mu - 1.830951\mu^2 + \frac{3}{4}\mu^3 \ln \mu + 2.542917\mu^3 - 2.746427\mu^4 \ln \mu$$

$$- 2.374082\mu^4 + \frac{9}{16}\mu^5 \ln^2\mu + 5.646816\mu^5 \ln\mu - .442638\mu^5 \quad (3.12)$$

Note that A_{42} and A_{41} are known, via (2.20) and (3.12), but A_{40} must be determined from another integral equation, before the expansion of A up to all terms in μ^4 can be written down.

4. The hierarchy of integral equations

For a smooth three-dimensional body the radiation potential ϕ_R, satisfying (2.6) and the free-surface condition, as well as Laplace's equation in the fluid and the radiation condition, can be obtained from the solution of the integral equation

$$2\pi\phi_R(\underset{\sim}{r}') - \iint_{S_B} \phi_R(\underset{\sim}{r})\frac{\partial}{\partial n}G(\underset{\sim}{r}', \underset{\sim}{r}; K)dS = - \iint_{S_B} G(\underset{\sim}{r}', \underset{\sim}{r}; K)n_y\, dS. \tag{4.1}$$

Here $G(r', r; k)$ is the Green function; that is, $\mathrm{Re}[Ge^{-i\omega t}]$ is the potential of an oscillatory point source in three dimensions with frequency $\omega = \sqrt{gK}$. Thorne (1953) gives an explicit expression for G, which can be written

$$G = \frac{1}{R} + \frac{1}{R'} + 2K \int_0^\infty J_0(kR'\sin\beta)e^{-kR'\cos\beta}\frac{dk}{k-K}. \tag{4.2}$$

(The path of integration passes under the pole at $k = K$). Here

$$R = |r-r'| = \{(x-x')^2 + (y-y')^2 + (z-z')^2\}^{\frac{1}{2}},$$

$$R' = \{(x-x')^2 + (y+y')^2 + (z-z')^2\}^{\frac{1}{2}}$$

$$\text{and} \quad R'\cos\beta = y + y' \quad (\beta \geqslant 0) \tag{4.3}$$

It is the difficulty involved in evaluating G and $\frac{\partial G}{\partial n}$ which makes the solution of (4.1) lengthy for general bodies.

Note that $G(r, r; 0) = \frac{1}{R} + \frac{1}{R'} = G_{00}$, say. Then

$$\frac{G - G_{00}}{2K} = \int_0^\infty J_0(kR'\sin\beta)e^{-kR'\cos\beta}\frac{dk}{k-K} \tag{4.4}$$

and the right side of this equation can be expanded as

$$-\sum_{n=0}^{\infty}(-1)^n\left(\frac{\partial}{\partial\nu}-i\pi\right)_{\nu=n}\left[\frac{(KR')^{\nu}P_{\nu}(\cos\beta)}{\Gamma(\nu+1)}\right] \tag{4.5}$$

(see Hulme (1982), appendix A). Using

$$\left(\frac{\partial P_{\nu}(z)}{\partial\nu}\right)_{\nu=n} = P_n(z)\ln\left(\frac{1+z}{2}\right) + (z-1)S_n(z) \tag{4.6}$$

where $S_n(z)$ is a polynomial of degree $n-1$ which can be found inductively, we can write

$$G \sim G_{00} + \sum_{m=1}^{\infty} K^m\{G_{m1}\ln K + G_{m0}\} \tag{4.7}$$

where

$$G_{m1} = -\frac{1}{\pi}\,\mathrm{Im}G_{m0} = 2(-1)^m\frac{(R')^{m-1}}{(m-1)}\,P_{m-1}(\cos\beta) \tag{4.8}$$

and

$$\mathrm{Re}\,G_{m0} = 2(-1)^m\frac{(R')^{m-1}}{(m-1)!}\,[\{\ln(R'+y+y') - \psi(m) - \ln 2\}P_{m-1}(\cos\beta)$$

$$+ (\cos\beta-1)S_{m-1}(\cos\beta)] \tag{4.9}$$

Specifically

$$G_{11} = -2,\; G_{10} = -2\{\ln(R'+y+y') + \gamma - \ln 2 - i\pi\}$$

$$G_{21} = 2(y+y'),\; G_{20} = -(y+y')G_{10} - 2R' \tag{4.10}$$

which uses $S_0 = 0$ and $S_1 = 1$.

Substitution of (4.7) into (4.1) suggests a corresponding expansion for ϕ, namely

$$\phi \sim \phi_{00} + \sum_{m=1}^{\infty} K^m\{\sum_{n=0}^{N(m)} \phi_{mn}(\ln K)^n\} \tag{4.11}$$

and separation of the powers of K and $(\ln K)$ gives

$$(2\pi I-K_0)\phi_{00} \equiv 2\pi\phi_{00}(r') - \iint_{S_B} \phi_{00}(r) \frac{\partial G_{00}}{\partial n} dS = - \iint_{S_B} G_{00}\, n_y\, dS \tag{4.12}$$

$$(2\pi I-K_0)\phi_{10} = - \iint_{S_B} G_{10}\, n_y\, dS + \iint_{S_B} \phi_{00} \frac{\partial G_{10}}{\partial n'} dS,$$

$$(2\pi I-K_0)\phi_{11} = - \iint_{S_B} G_{11}\, n_y\, dS = 2\iint_{S_B} n_y\, dS = 2S = \text{const}, \tag{4.13}$$

and $(2\pi I-K_0)\phi_{1n} = 0 \quad (n \geqslant 2)$.

Note that all the kernels are identical, involving only the (real) zero-frequency Green function G_{00}. Also it is possible to show that each integral equation has a unique solution; that means $\phi_{1n} \equiv 0 \quad (n \geqslant 2)$. Also, if the right side of the equation is a constant, C, say, then the solution will be identically constant $= C/4\pi$.

Further, the relation between G_{m1} and ImG_{m0} $(m \geqslant 1)$ gives the same relation between ϕ_{11} and $Im\phi_{10}$, since

$$(2\pi I-K_0)Im\phi_{10} = - \iint_{S_B} ImG_{10}\, n_y\, dS + \iint_{S_B} \phi_{00} \frac{\partial I_m G_{10}}{\partial n'} dS$$

$$= \pi \iint_{S_B} G_{11}\, n_y\, dS.$$

Hence $Im\phi_{10} = -\pi\phi_{11}$, and so $B_{10} = -\pi A_{11}$, as stated in (2.20). The remaining relations in (2.20) arise from subsequent members of the set of integral equations, for example

$$(2\pi I-K_0)\phi_{20} = - \iint G_{20}\, n_y\, dS + \iint \{\phi_{00} \frac{\partial G_{00}}{\partial n} + \phi_{10} \frac{\partial G_{10}}{\partial n}\} dS$$

$$(2\pi I-K_0)\phi_{21} = - \iint G_{21}\, n\, dS + \iint \{\phi_{00} \frac{\partial G_{21}}{\partial n} + \phi_{11} \frac{\partial G_{10}}{\partial n}\} dS \tag{4.14}$$

$$(2\pi I-K_0)\phi_{2n} = 0 \quad (n \geqslant 2).$$

Finally, it is easy to show by induction that $N(m)$ (in 4.11) $= \frac{1}{2}m$ or $\frac{1}{2}(m+1)$ whichever is an integer; this feature arises purely because $\frac{\partial G_{11}}{\partial n} = 0$.

All we are concerned with is evaluating A_{00}, A_{10} etc., so we only require $\mathrm{Re}\phi_{m0}$. It is clear that only $\mathrm{Re}G_{m0}$ are needed for this, and the four real integral equations used for (3.9)-(3.12) are very readily constructed and solved.

5. Results and discussion

Figures 1 and 2 show the comparisons between the series (3.9) and (3.12) and their exact values. It is clear that the asymptotic representations are poor for $\mu \gtrsim .3$, and so it might be expected that the method will only work when the diameter of the body in question is less than one tenth of the wavelength.

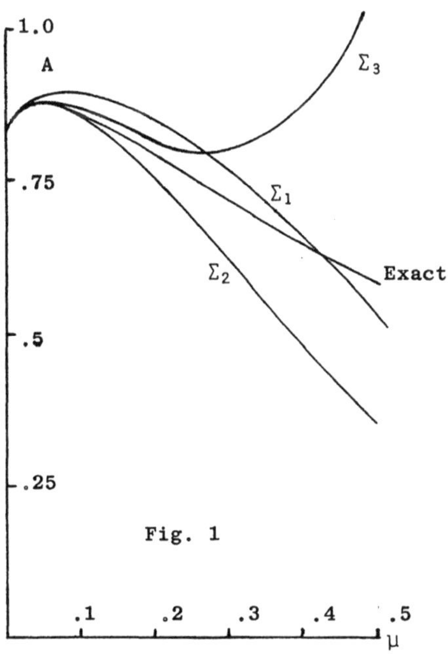

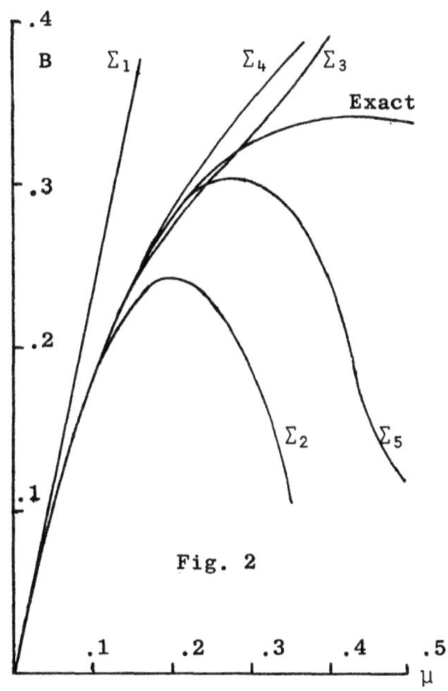

Figures 1 & 2: The asymptotic forms (3.9) and (3.12) compared with the exact values of virtual mass A and damping B. The labelling Σ_n denotes the sum up to all terms with μ^n .

However, it is known (Ursell, 1963) that $\Lambda = A + iB$ must be of the form

$$\Lambda(\mu) = \frac{E_1(\mu) + E_2(\mu)\mu L}{E_3(\mu) + E_4(\mu)\mu L} \qquad (L = \ell n\,\mu - i\pi) \qquad (5.1),$$

and this suggests the construction of a rational approximation

$$\Lambda(\mu) \sim \frac{(\alpha_1 + \beta_1\mu + \gamma_1\mu^2) + (\alpha_2 + \beta_2\mu)\mu L}{(1 + \beta_3\mu + \gamma_3\mu^2) + (\alpha_4 + \beta_4\mu)\mu L} \quad (5.2).$$

Using (2.20), (3.9) and (3.12), we can deduce that

$$\begin{aligned} &\alpha_1 = .830951, \ \alpha_2 = -\tfrac{3}{4}, \ \alpha_4 = 0 \\ &\beta_1 = -.098669, \ \beta_2 = \tfrac{1}{2}, \ \beta_3 = 1.220636, \ \beta_4 = -\tfrac{1}{2} \\ &\gamma_1 = -.500302, \ \gamma_3 = .331310 \end{aligned} \quad (5.3)$$

With these, (5.2) yields values for A and B as shown in Table 1. The agreement with the exact values is greatly improved, and the maximum of the damping is close to its correct position.

	from (5.2)&(5.3)		exact	
μ	A	B	A	B
.1	.8661	.1815	.8627	.1816
.2	.7893	.2782	.7938	.2793
.3	.7023	.3215	.7157	.3254
.4	.6172	.3320	.6452	.3410
.5	.5375	.3227	.5861	.3391
.6	.4631	.3017	.5381	.3271
.7	.3933	.2742	.4999	.3098
.8	.3271	.2437	.4698	.2899
.9	.2639	.2127	.4464	.2691
1	.2033	.1826	.4284	.2484

Table 1: A comparison of a rational approximation ((5.2) with (5.3)) and the exact values.

With higher-order rational approximation (involving some more terms in (3.12) and (3.9)), the agreement would be made even better. Thus the low-frequency asymptotic approximation method described in this paper provides rational approximation which gives reasonable agreement with exact results up to moderate wavenumbers.

References

1. Budal, K.; Falnes, J.; Iversen, L.C.; Liuebekken, P.M.; Oltedal, G.; Hals, T.; Onshus, T.; Hoy, A.S. "The Norwegian wave-power buoy project" in Proc. 2nd Intl. Symp. Wave Energy Utilization, Trondhein (1982).

2. Hulme, A.; "The wave forces acting on a floating hemisphere undergoing forced periodic oscillations". J. Fluid Mech., 121, 443-463 (1982).

3. Newman, J.N.; "Marine Hydrodynamics". M.I.T. Press (1977).

4. Thorne, R.C.; "Multipole expansions in the theory of surface waves". Proc. Camb. Phil. Soc., 49, 707-716 (1953).

5. Ursell, F.; "The periodic heaving motion of a half-immersed sphere: the analytic form of the velocity potential and the long-wave asymptotics of the virtual mass "Dept. of Mathematics Internal Rep., Univ. Manchester (1963).

6. Yeung, R.W.; "Comparative Efficiency of Numerical Methods" (in this Symposium).

Investigation of Wave Focusing over a Parabolic Step

R. Cengiz ERTEKIN*
Gerasime M. MONOPOLIS**

* Dept. of Naval Architecture and Offshore Engineering, Univ. of California,Berkeley,CA 94720, U.S.A. (Present address:Shell Development Company,PORD,P.O.Box 481 Houston ,Texas 77001, U.S.A.)

** Dept. of Civil Engineering/Hydraulic Engineering,Univ. of California,Berkeley,CA 94720, U.S.A.

Abstract

We report here upon an exploratory investigation of wave focusing over a bottom step which has a parabolic shape in the horizontal plane. Although building of a large-scale reflector can be a very costly project, the idea of using a parabolic reflector--either in the form of a wall or a step discontinuity at the bottom-- will be presented along with small-scale experimental results to test the feasibility of such systems. Some theoretical considerations for analytical description will be discussed.

Introduction

Considerable engineering effort has been devoted to the design of systems that convert wave energy to other usable forms of energy, and some promising concepts have been reported. At the same time, a similar effort has been launched on increasing the energy density of waves and, if possible, confining this energy to an area, in order to make the wave-energy converting systems more efficient and less costly and, therefore, commercially competitive.

The necessary tools to achieve wave energy focusing are available. Fortunately, for the Ocean Engineer, the similarity of analytic representation of water waves and light waves can instantly provide some interesting concepts derived directly from optics. A valuable report has been published on the use of the Fresnel Lens concept for wave energy focusing (Mehlum and Stamnes [1]). Another preliminary report has been given on the focusing of waves by means of a step change in depth (Monopolis et al. [2]).

A body of water surrounded by some reflective boundary may function as a wave reflector. Such a system, subjected to excitation from waves, gives out a response that is dependent on the characteristics of the excitation and the properties of the responding system. Excitation near the natural frequencies of a dynamic system can result in resonance, and in turn very large oscillations can be expected. This condition is very attractive in view of the wave-energy conversion problem. It is well known from optics that a light beam reflects and refracts after entering a medium with an index of refraction not equal to unity. Similarly, water waves reflect and refract when the bottom topography varies. This fact

Hydrodynamics of Ocean Wave-Energy Utilization
IUTAM Symposium Lisbon 1985
Editors: D. Evans, A.F. de O. Falcão

has been widely utilized by coastal engineers, mainly to divert wave energy. Our goal is just the opposite--we would like to focus as much of the energy as possible to a point. A parabolic reflector appears to be the best candidate to achieve this goal since a series of incoming waves parallel to the axis of the parabola will reflect and concentrate at the focus. In the following sections we shall investigate the response of this remarkable geometric shape to water waves.

Experiments

To determine the relative responses of various wave reflectors, four different geometries were investigated in laboratory tests; a semi-circular wall, a parabolic wall, a parabolic step and a parabolic step with a head-wall reflector. Experiments were performed at a ripple tank of the Hydraulic Engineering Department at the University of California at Berkeley. The ripple tank, whose side view is shown in Fig.1, is 6.1 m long, 1.13 m wide and 0.12 m deep. Detailed top view of the parabolic step is shown in Fig. 2 along with the incoming waves propagating from the shallower region with depth h_1 = 1.8 cm to the deeper region with depth h_2= 5.8 cm. The focal length of the parabola was 11 cm. Surface elevations were measured by resistance-type wave probes at various locations in the tank. Time histories of these elevations allowed us to determine the wave period (for a detailed discussion of the experimental setup and procedures see Monopolis et al.[2]).

All four designs had the same width and were tested in the same water depth for a range of wave frequencies 1.3 Hz $\leq f \leq$ 4.6 Hz. The corresponding wavelengths were in the range of 9 cm $\leq \lambda \leq$ 32 cm and the wave amplitudes were in the order of mm. A factor that greatly complicated the evaluation of the measurements stems from the small water depth of 1.8 cm. Because of the dissipation of wave energy through bottom friction wave amplitudes decay as the wave propagates along the tank. A typical example of this amplitude decay is shown in Fig. 3 for a range of frequencies measured at a fixed location 228.6 cm away from the wave maker. The normalized amplitude A_n is defined as the ratio of the measured wave amplitude at a given location to the incoming wave amplitude measured 30.5 cm away from the wavemaker. The wave decay along the tank (for a fixed wave frequency) was also measured at certain length intervals and was given by Monopolis et al. [2]. These wave-damping curves had been used when obtaining the relative amplification of waves at any desired point in the tank since any comparison regarding the efficiency of a reflector has to consider energy dissipation.

i) Parabolic-wall case:

A wall is placed along the parabolic step discontinuity shown in Fig.

2 for the experiments of the parabolic-wall case. The waves focused on the exact geometric focus point. Since the angle of incidence is large, Mach-stem patterns that travel along the parabolic wall were observed, as seen in Fig. 4. Surface elevations around the focus were measured at the nodes of a grid of 1 cm size. The contour plot of these measurements is shown in Fig. 5 for f=3.3 Hz. The normalized amplitude of 1.0 in this figure corresponds to a 7.6 amplification factor if wave dissipation is taken into account. It was observed that the behavior of the reflected system varied with the tuning factor, defined as the ratio of the focal length to the wavelength. Two predominant patterns were observed: one had concentric circles centered at the focus and the other had three revolving wave-intersection points that passed simultaneously over the focus. The experimental data of Fig. 5 correspond to the former one. The relative amplification at the focus is shown in Fig. 6. We note that the peak of this response curve around f = 3.5 Hz corresponds approximately to the wavelength being equal to the focal length.

ii) Parabolic-step case:

As in the parabolic-wall case, Mach stems were formed along the step discontinuity and they traveled towards the vertex of the parabola. Partial transmission was observed when the angle of incidence α_i (see Fig. 2) is less than the critical angle $\alpha_{ic}=34^0$ (this will be discussed in the next section). For $\alpha_i>\alpha_{ic}$ exponential decay of the waves was observed around the discontinuity (see the photographs of this case in Monopolis et al. [2]). In the deeper region wave patterns were quite confused owing to multiple reflection from the tank walls and back to the step. The amplified reflected wave converged in a region near the geometric focus but maxima occurred at two points symmetric about the axis of the parabola. As seen in Fig. 6, significant amplification occurs at low frequencies but there is no sign of wave and focal-length tuning as in the other cases. We note that the response curve of the parabolic-step case corresponds to measurements at the point of highest amplification near the focus.

iii) Parabolic step with short head wall case:

This configuration features a short wall--covering a strip of approximately 1/5 the width of the tank--which was placed at the apex and followed the shape of the step. The reason for testing this case was to reduce the energy leakage caused by the partial transmission observed in the parabolic-step case. The observed reflection patterns were closer to those observed in the parabolic-wall tests than the ones in parabolic step tests. But measurements showed no improvement in the long-wavelength range. As in the parabolic-wall case, wave length-focal length tuning is seen (Fig. 6) for this case also. The observed transmitted pattern was weaker than the case without wall, as expected, suggesting less energy leakage.

iv) Semi-circular wall case:

The parabolic bottom was replaced by a semi-circular bottom with a wall. A frequency-dependent amplification was observed but it did not exceed 3.0. As a result it was decided that further tests would be useless even though such a design is attractive because of its directional independence. The points of maximum amplification were along the middle of the tank, especially at a distance one half the radius from the center.

Some analytical considerations

It has been mentioned in the last section that in experiments of case ii) (parabolic step), at some values of the incidence angle α_i, partial transmission (in the form of progressive waves) was observed in the deeper region and at some other α_i's only an exponential decay of waves was observed. In order to understand better the physics of the problem we shall give some analytical expressions that explain various of the parabolic-step-case observations and attempt to obtain the amplification of the waves by a "strip-theory" approximation.

The complete solution to the problem of a series of incoming waves being diffracted by a step discontinuity such as the one shown in Fig. 2 is an involved theoretical and numerical task. The effects of viscosity (in the form of bottom friction) and nonlinearity (in the form of Mach-stems) are significant. Even though one assumes that the fluid is inviscid and the waves are infinitesimal, thereby allowing the use of a linear theory, the 3-D geometry of the step and the tank walls makes the problem very difficult to solve—even in the time-harmonic case. Nevertheless, many researchers have attacked the problem of a 2-D (in x-z plane) step discontinuity and obtained solutions under certain assumptions (see, for instance Lamb [3],§ 175, Bartholomeusz [4], Newman [5]). Monopolis et al. [2] extended the previous work and formulated the oblique step discontinuity (y' extends to $\pm\infty$) problem for any wave length. Here, we shall consider a strip, as shown in Fig. 2, which is parallel to the incoming wave orthogonal and assume that the solution of the problem for one strip is not affected by the presence of the adjacent strips. Furthermore, we shall assume that the problem for each strip is governed by the linear shallow-water wave equation. Even though it is expected that the shallow-water equation can produce accurate results for small-amplitude waves that are long, the assumption that there is no interaction between the strips is obviously a crude one.

Let us consider an incoming wave that approaches the step at an angle α_i. This wave will reflect with an angle $\alpha_r = \alpha_i$ and be transmited to the deeper region with an angle α_t. The analog of Snell's (or Descartes') law of refraction for the water waves is

$$\sin \alpha_t = k_1 \sin \alpha_i / k_2 \quad , \; k_1 = k_2 \sqrt{\mu} \; , \tag{1}$$

where $\mu = h_2/h_1$, and k_1 and k_2 are the wave numbers in the shallower and deeper regions, respectively. The relation $(1)_2$ between k_1 and k_2 can be obtained from the fact that the wave frequency ω is the same in both regions, along with the dispersion relation $\omega^2 = ghk^2$ in shallow water. With the help of $(1)_2$, $(1)_1$ can be written in another form:

$$\cos \alpha_t = \{1 - \mu \sin^2 \alpha_i\}^{1/2} \quad . \tag{2}$$

We note that if $\mu \sin^2 \alpha_i = 1$ then $\alpha_t = \pi/2$, i.e. the transmitted-wave orthogonal will be parallel to the step discontinuity, and when $\mu \sin^2 \alpha_i > 1$ $\cos \alpha_t$ will be pure imaginary, its argument being $\pi/2$ (for more discussion on the subject, see Mei [6], Chap. 4). The case $\mu \sin^2 \alpha_i = 1$ gives rise to the phenomenon of critical angle α_{ic}, as will be discussed shortly.

Now let us consider the well-known 2-D linear shallow water equation for a flat bottom:

$$\Phi_{tt} - c^2 (\Phi_{x'x'} + \Phi_{y'y'}) = 0 \quad , \; c^2 = gh \; , \tag{3}$$

where Φ is the velocity potential and c is the phase speed. The solution of (3) for a step change in depth is of the form

$$\begin{aligned} \Phi_1 &\propto (e^{i\alpha_1 x' - i\omega t} + \rho \, e^{-i\alpha_1 x' - i\omega t}) e^{i\beta y'} \quad , \; x' < 0 \; , \\ \Phi_2 &\propto \tau (e^{i\alpha_2 x' - i\omega t}) e^{i\beta y'} \quad , \; x' > 0 \; , \end{aligned} \tag{4}$$

where ρ is the reflection coefficient, τ is the transmission coefficient and

$$\alpha^2_{1,2} + \beta^2 = k^2_{1,2} \qquad , \qquad \omega^2 = gh_{1,2} k^2_{1,2} \quad . \tag{5}$$

To have a unique ρ and τ, continuity of wave elevation and mass at $x' = 0$ have to be employed (Lamb [3], § 175). Setting $\Phi_1 = \Phi_2$ and $h_1 \Phi_{1x'} = h_2 \Phi_{2x'}$ at $x' = 0$ one can obtain (see Tuck [7] on the discussion of Lamb continuity conditions)

$$1 + \rho = \tau \qquad \text{and} \qquad h_1 \alpha_1 (1 - \rho) = h_2 \alpha_2 \tau \quad . \tag{6}$$

We note that $\alpha_1 = k_1 \cos \alpha_i$, $\alpha_2 = k_2 \cos \alpha_t$ and $\beta = k_1 \sin \alpha_i = k_2 \sin \alpha_t$. From $(6)_{1,2}$

we obtain

$$\rho=\frac{\cos\alpha_i-\mu^{1/2}\cos\alpha_t}{\cos\alpha_i+\mu^{1/2}\cos\alpha_t}\quad,\ \tau=\frac{2\cos\alpha_i}{\cos\alpha_i+\mu^{1/2}\cos\alpha_t}\ . \tag{7}$$

Similar expressions were given by Christiansen [8]. We note that when $\alpha_t=\pi/2, \rho=1$, i.e. total reflection occurs, and from (2) the critical angle

$$\sin\alpha_{ic}=\mu^{-1/2}\ , \tag{8}$$

is obtained. The case of total transmission can be obtained by setting $\rho=0$ in (7) and gives rise to zero reflection angle:

$$\sin\alpha_{rc}=\{1-\mu/1-\mu^2\}^{1/2}\ , \tag{9}$$

where use has been made of (2). The variations of these two angles with μ are shown in Fig. 7. It is interesting to observe that in the region above the α_{ic} curve, total reflection takes place and under it both reflection and transmission take place except on the α_{rc} curve that corresponds to total transmission ($\tau=1$ from $(7)_2$).

If we set $\rho=|\rho|\exp(-i\psi)$, where ψ is the phase angle of the reflection coefficient, then one can obtain ρ and ψ from $(7)_1$ for the three angle-of-incidence regimes as follows:

$$\begin{aligned}
&|\rho|=|\ \text{Eq.}(7)_1|\ , && \psi=\pi && , \ 0\leq\alpha_i\leq\alpha_{rc}\ ,\\
&|\rho|=\ \text{Eq.}(7)_1\ , && \psi=0 && , \ \alpha_{rc}<\alpha_i\leq\alpha_{ic}\ ,\\
&|\rho|=1 && , \ \psi=2\tan^{-1}\left\{\left[\frac{\mu\sin^2\alpha_i-\mu}{1-\sin^2\alpha_i}\right]^{1/2}\right\}, && \alpha_i>\alpha_{ic}.
\end{aligned} \tag{10}$$

The variations of $|\rho|$ and Ψ are shown in Fig. 8 and Fig. 9 for three μ values. Similar curves can be obtained for τ and its phase angle but here we are concerned with reflection only.

The surface elevation η can be obtained through the linearized dynamic free-surface condition $\eta=-\Phi_t/g$. Using $(4)_1$, this condition gives (after taking the real part)

$$(\eta_1/A)=\cos(\alpha_1 x'+\beta y')+|\rho|\cos(\alpha_1 x'+\psi-\beta y')\ ,\ x'<0, \tag{11}$$

where A is the amplitude of the incoming wave. We note that harmonic time dependence ωt is omitted from (11) since the time a wave orthogonal takes to travel from the directrix to a point on the parabola and then back to the focus point is the same for any incidence angle (by definition of a parabola). The nondimensional wave amplitudes $(\eta_1)_i/A$ due to the incoming and reflected waves at the focus are shown for three μ values in Fig. 10 for each individual strip (represented by the incidence angle α_i) after proper coordinate transformation (from the local coordinates $O'x'y'$ to the global one Oxy) is made. It should be remarked that $(\eta_1)_i/A$ vanish at corresponding α_{ic} angles (this is not quite so in Fig. 10 due to the poor resolution of the plot) since when $\alpha_i=\alpha_{ic}$, $\rho=1$ and $\psi=0$ and thus the reflected angle cancels the incoming wave.

In an attempt to obtain the amplification of the waves at the focus we consider the energy density of the waves at the focus, given by

$$E_f=\frac{\rho g\ A^2}{L}\int_0^L((\eta_1)_i/A)^2\ dy\quad . \tag{12}$$

If now one assumes that the converged waves form a standing wave with an energy density of $\rho g\zeta^2/4$, where ζ is the amplitude of the standing wave, then from (12) one obtains

$$\zeta/A=[\frac{4}{L}\int_0^L((\eta_1)_i/A)^2\ dy]^{1/2}\quad , \tag{13}$$

where ζ/A is the amplification factor. This amplification is shown in Fig. 11 for μ=5.8cm/1.8cm and for a range of cyclic wave frequency f (Hz). This amplification is quite low compared with the parabolic-step experimental data shown in Fig. 6, especially in the low-frequency range. It also shows an oscillatory behavior in contrast with the continuous decay of experimental data.

Conclusions

The method of wave-energy focusing by means of parabolic reflectors is demonstrated in this study. Although the experimental results contain some uncertainty, emerging from the scale effects, presence of tank walls and, as explained in Monopolis et al. [2], inadequacy of the measuring equipment, they are very encouraging and provide valuable insight for the development of such designs. With regard to the accuracy of the wave gauges, it is recommended that sonic-type or hot-film wave probes be used in such small-scale experiments. A prototype design of a parabolic step can be realized as a ditch in shallow water whose width has to be determined; the requirement being that when the local disturbance reaches the deeper side and becomes a transmitted wave, the energy loss to this wave be minimum.

Various stages of the simplified "strip-theory" approach presented here can be a valuable tool in the preliminary experimental or prototype design of parabolic reflectors in order to adjust the critical angle and zero reflection angle, and thus maximize the amplification. Even though the upper bound of ζ/A is less than 4 from (13), it is expected that the solution of the complete boundary-value problem by a Green Function or a Hybrid Integral Method (see Yeung [9]) will provide more insight into the dependence of the reflector performance on the relative scales of waves and reflector geometry.

It is believed that the basic principles of wave focusing by this method have proven the need for a larger-scale investigation that will provide sufficient information for the evaluation of the economic feasibility of wave-energy conversion even though typical waves involve a large spectrum of frequencies and a mechanical system tuned to one frequency may not be particularly efficient when exposed to waves of much different frequencies.

The first author (RCE) would like to acknowledge partial support by ONR Contract N00014-84-K0026 for the preparation of this paper.

References

1. Mehlum, E. and Stamnes, J. (1980), "Power Production Based on Focusing of Ocean Swells," First Symposium on Wave Energy Utilization, Gothenburg: Chalmers Univ. of Tech., 1980, pp. 29-35.
2. Monopolis, G. M., Ansari, G. R. and Ertekin, R. C., (1980), "Wave Energy Focusing," Ocean Engineering: The Future, IDS Rep. No. 80-3, College of Eng., Univ. of Calif., Berkeley, pp. 183-238.
3. Lamb, H. (1932), "Hydrodynamics," 6th Edition, Cambridge University Press, 1932= Dover, New York, 1945, XV+738 pp.
4. Bartholomeusz, E. F. (1958),"The Reflexion of Long Waves at a Step," Proc. Camb. Phil. Soc., Vol. 54, pp. 106-118.
5. Newman, J. N. (1965), "Propagation of Water Waves over an Infinite Step," J. Fluid Mech., Vol. 23, part 2, pp. 399-415.
6. Mei, C. C. (1983), "The Applied Dynamics of Ocean Surface Waves," John Wiley & Sons, New York, xviii+740 pp.
7. Tuck, E. O. (1984), "Shallow-Water Waves at Discontinuities," Chap. 4 of Small-Gap Flows, Rep. No. NAOE 84-1, Dept. of Naval Arch. & Offshore Eng., Univ. of Calif., Berkeley, pp. 12-15.
8. Christiansen, P. L. (1976), "Diffraction of Gravity Waves by Ray Methods," Symposium on Waves on Water of Variable Depth, Proceedings (ed. Provis, D. G. and Radok, R.), Berlin: Springer Verlag, pp. 28-38.
9. Yeung, R. W. (1975), "A Hybrid Integral-Equation Method for Time-Harmonic Free-Surface Flow," Proc. First Int. Conf. on Num. Ship Hydro., Gaithersburg, Maryland, pp. 581-607.

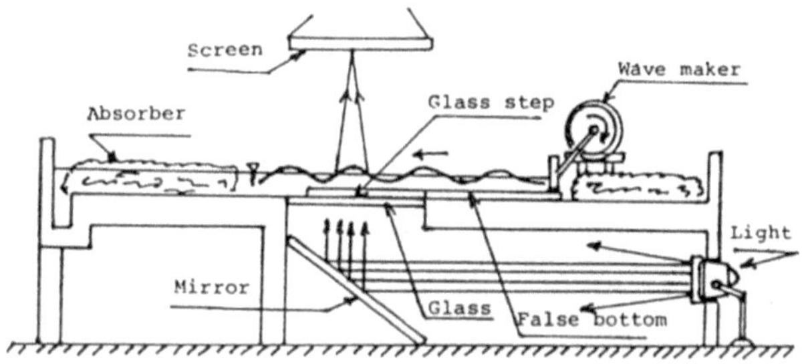

Fig.1 Side view of the tank.

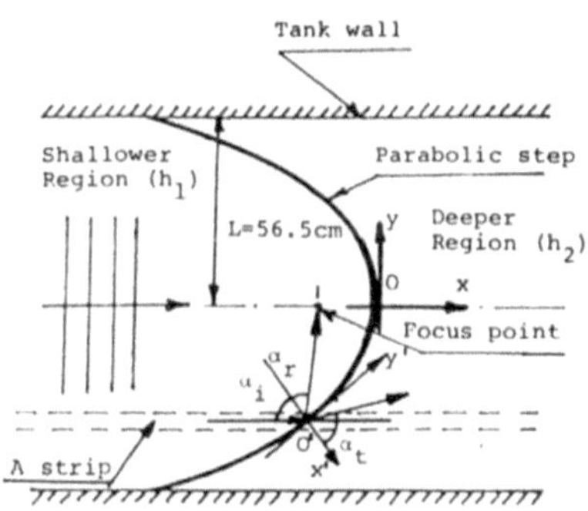

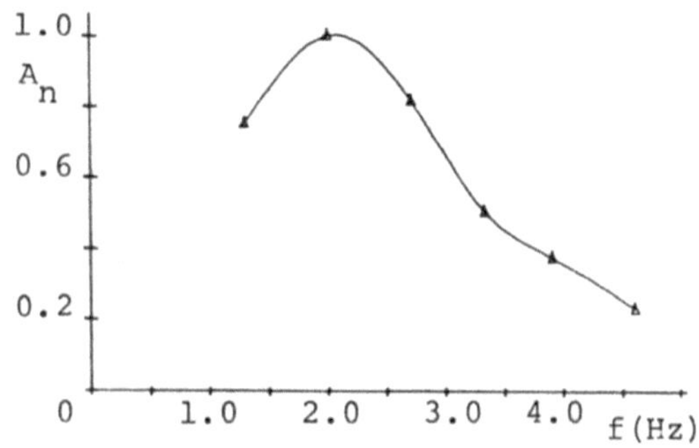

Fig.2 Top view of the tank. Fig.3 A typical wave-damping curve.

Fig.4 Focusing of waves (parabolic-wall case).

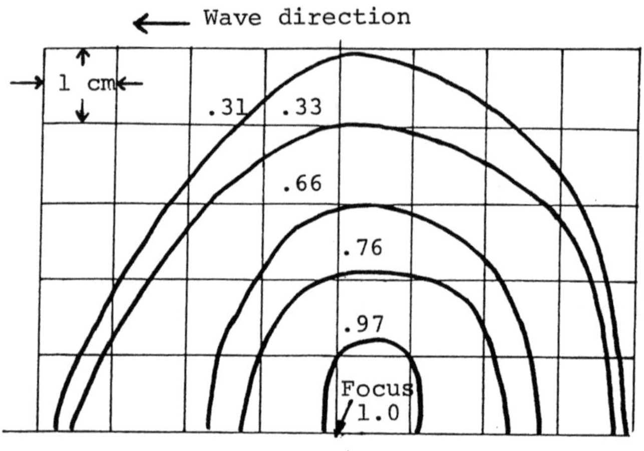

Fig.5 Surface-elevation contours, f=3.3 Hz.

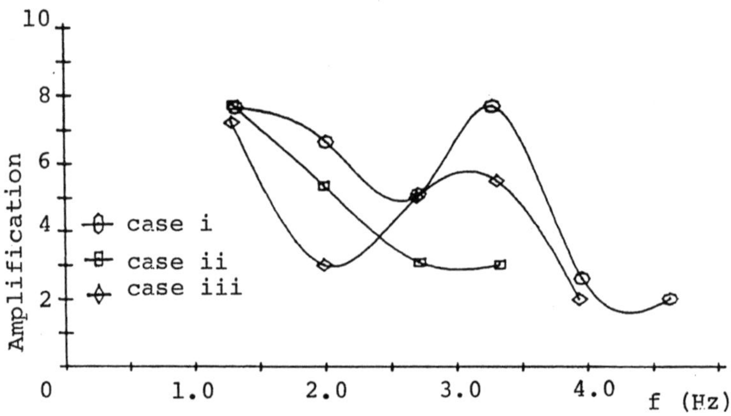

Fig.6 Amplification at the focus point.

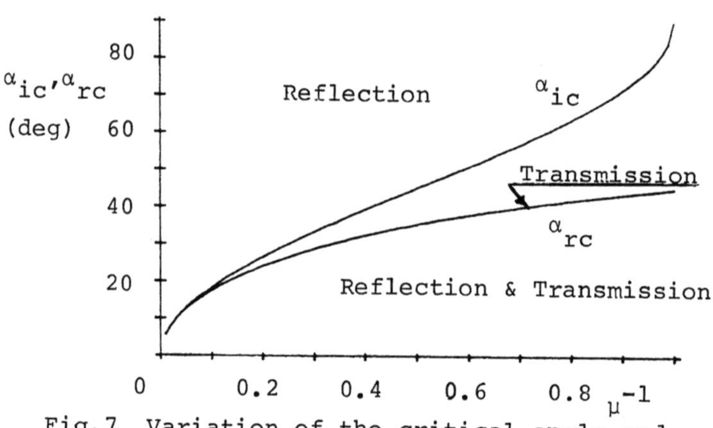

Fig.7 Variation of the critical angle and zero reflection angle with h_1/h_2.

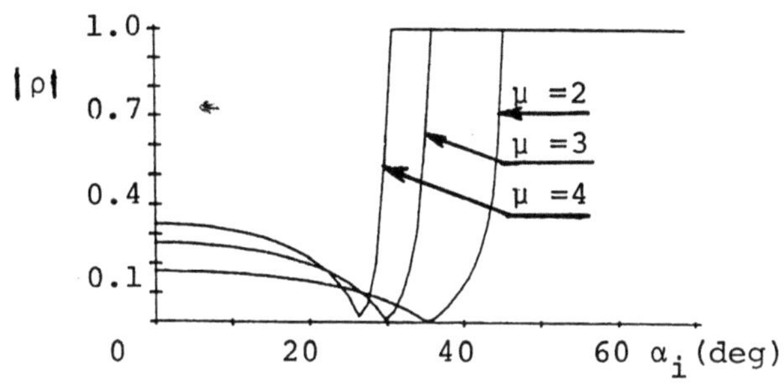

Fig.8 Reflection coefficient.

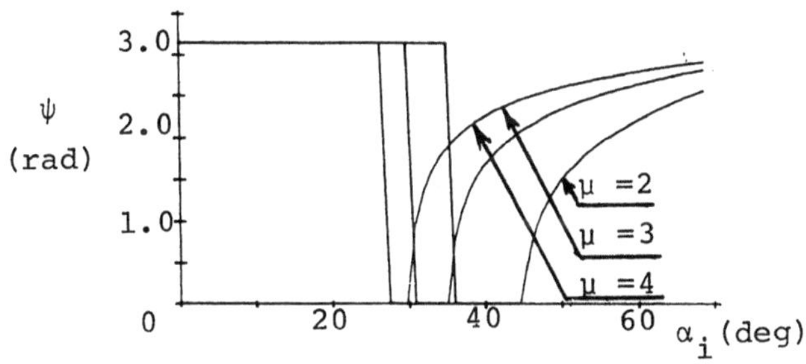

Fig.9 Phase angle of the reflection coefficient

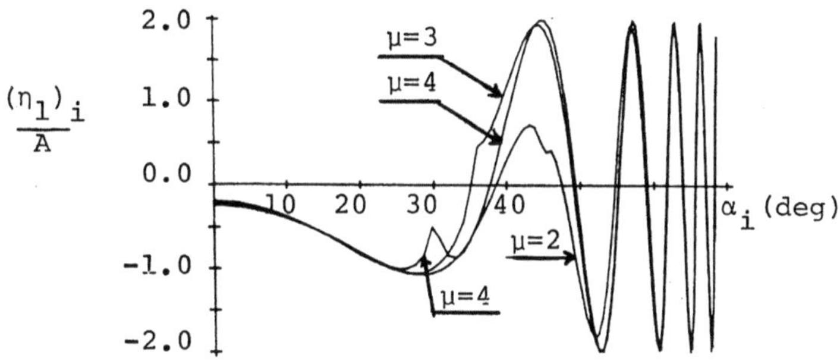

Fig.10 Amplitude contributions of each strip at the focus point.Wave frequency,f, is 2.0 Hz. h_1=1.0 cm.

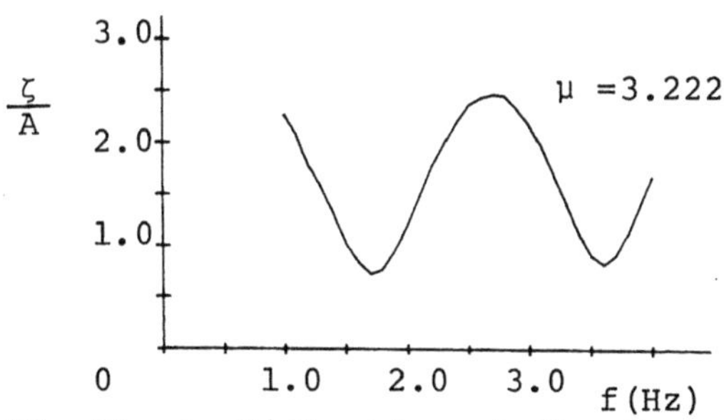

Fig.11 Amplification at the focus point.

An Experience of Wave Power Generator through Tests and Improvement

Yoshio Masuda

Chief Scientist. First Group. Marine Research and Development Department. Japan Marine Science and Technology Centre (JAMSTEC)

1. Introduction.

A wave power generator was a good target for an inventor. I began to invent a wave power generator in 1945, but has taken a long time to get a feasible device, and large scale application is still at the development stage.

I will introduce my experience through tests and mechanical improvements for wave power devices, and I wish to tell my vision of future development.

2. First stage studies.

(1) Study of three floats wave power generator.

I was able to conduct the first sea trial test of three floats wave power generator from April to June 1947 after some basic tests funded with support from my father Minoru Masuda. Fig. 1 shows its principle.

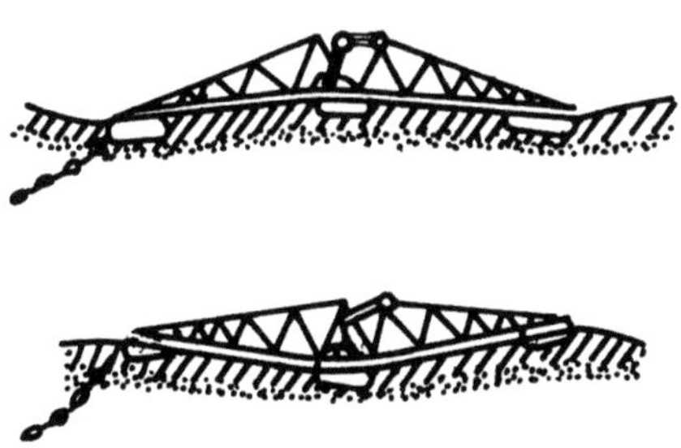

Fig. 1 Three Floats W. P. G.

As shown in Fig. 1, three floats are connected by two trusses with a hinge. The hogging and sagging forces produced by waves created an alternating rotation of the crank on the centre float which was geared up and changed to the same direction for an electric generator. The dimension of this device was 15 m in total length , and the generator output was 200 watts.

The reason for selection of a floating type was simplicity of construction and installation, but it was overturned by high wave at the first day in the test. Of course it was repaired, and it was tested over three months, with a maximum output of 200 watts in high sea condition. Throughout the test, troubles in the key of the crank shaft were caused by high torque and the alternating slow motion of the crank. I experienced my first design difficulty of the mechanical gear for a Wave Power Generator (W.P.G.). The test area was Genkai Nada off Fukuoka of Kiushu Japan. The basic principle of the floating type and attenuator type was followed by the Kaimei design.

(2) Front and Rear Moored Floating Wave Power Generator.

After the first failure of the three floats wave power generator, I stopped the wave power generator study, and I thought again, and I thought of the next device which is shown in Fig. 2.

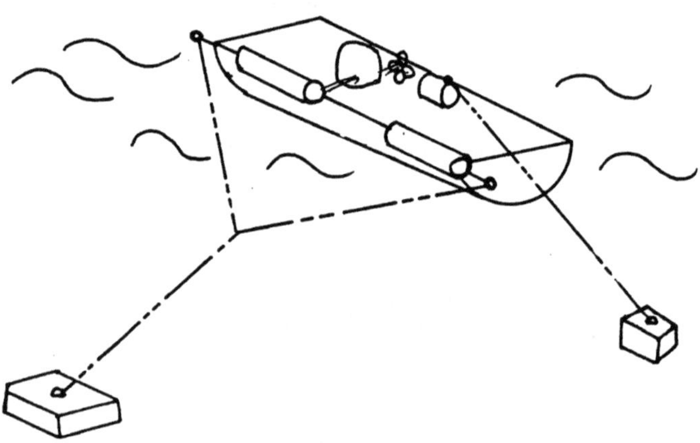

Fig. 2 Front and Rear Moored Floating type W.P.G.

The point of this idea is to use not only vertical wave motion but also horizontal wave motion. The buoy is moored both by front and rear lines. The lenght of the rear line is constant, and the front line is extendable by the buoy's oblique motion. This operates two high pressure pumps, and oil flows through a nozzle to rotate a Pelton wheel at high speed, producing electricity by a generator. A high pressure tank will smooth the oil flow.

I believe that an oil system is a good method of producing high speed rotation by slow motion at high torque.

In three years of development work I produced three models, two small models which operated successfully at sea, and a third model of one kW which did not succeed. The reason for this failure was the loss of a mooring line before installation of the model. This mooring line with its position buoy was lost in high waves during a typhoon.

Throughout these tests, I experienced a relatively high energy absorption ratio in both the front and rear moored buoys.

(3) Wave power generation for buoy use.

It is said that necessity is the mother of invention. There are a large number of buoys at sea for safety of navigation and other uses and all of them need electricity and if it is supplied by battery, maintenance is a hard job. This necessity gave the first use of wave power generation.

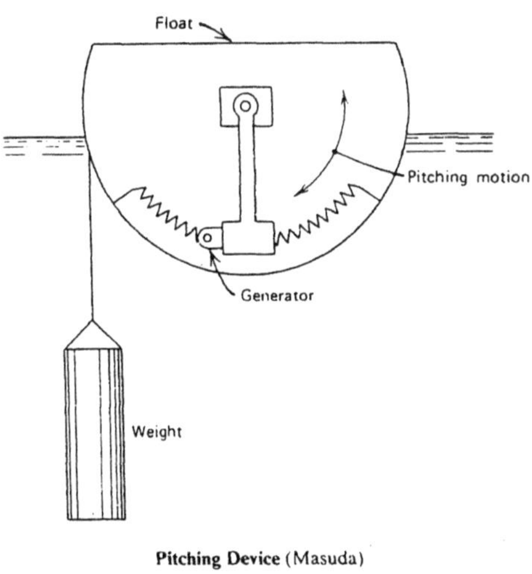

Fig. 3 Pendulum type W.A.G.

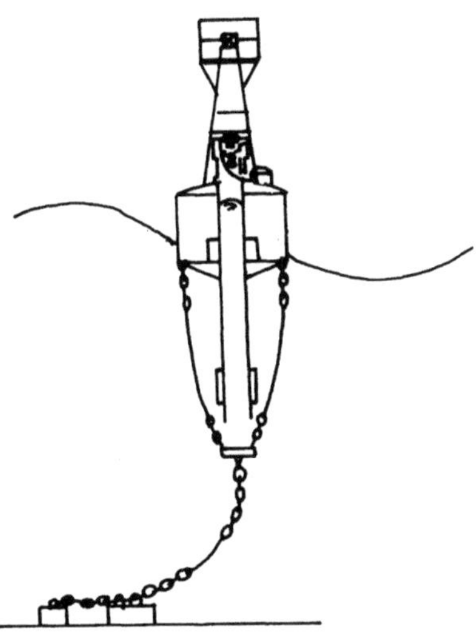

Fig. 4 Centre type W.A.G.

The Japan Research and Development Corporation started this study in 1964 and finished in 1965.
In this study, we tested two kinds of wave power generator, the first device being a pendulum type wave power generator as shown in Fig. 3.
The pendulum type uses the motion of the buoy by waves, and the pendulum weight produces gear rotation which is then geared up for generator. In order to produce a large motion, an underwater mass bucket is hanged on one side of buoy and the same weight without water mass is attached to the other side.

This device was thought to be promising, but the gear mechanism gave trouble after some years, and since it was a relatively high price this device was not used.

The centre type was used in practice (Fig. 4). It is an air turbine device, the buoy has a long vertical tube through the centre of the buoy, and the bottom of the tube opens to the sea in relatively deep water.

Since wave motion decreases with depth, and since the buoy moves up and down with the surface waves, a relative wave motion is produced in the centre pipe. The back and forth motion of the air in the pipe is rectified by valves, and rotates an impulse turbine.

Fortunately, this air turbine type had a very long life; the longest unit has operated more than 20 years at sea. The main reason is that sea water particles in the air are separated by the high speed rotation of turbine, and has no influence on the bearings, and such rotating machines have relatively high reliability.

The Japanese manufacturer is Ryokuseisha Company, who produced about 1,200. I gained confidence in air turbine systems through this practical experience.

3.Recent studies.

(1) The Wave Power Generator KAIMEI.

The oil crisis of the nineteen seventies created a steady demand for renewable energy. My first idea for a large scale W.P.G. was a circular ring buoy, as shown in Fig. 5. The diameter of the ring must be longer than a wave length, and so it is of the order of 100 m. This idea was adopted by JAMSTEC in 1975.

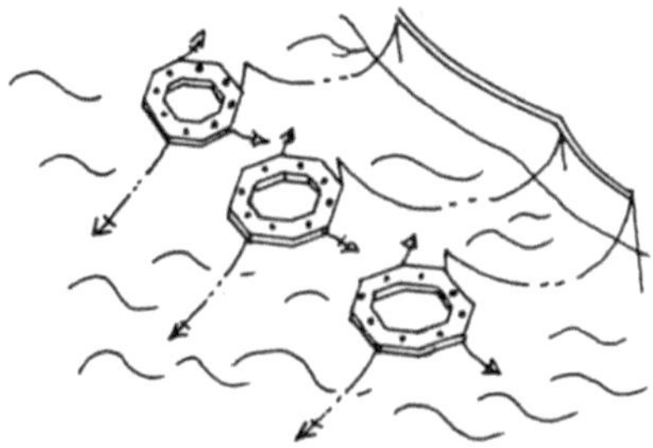

Fig. 5 Ring Buoy

Fig. 6 Wave power Generator KAIMEI (Attenuator type)

Kaimei is a ship-like floating-type wave power generator, and is an attenuator-type device.

In order to get safety in the first sea trial of this wave power device, I adapted a ship's long experiences on sea for wave power design. KAIMEI is 80 m in length, 12 m in width, 5-7 m in height, and its hull strength and stability etc. were designed by ship regulations.

It had 11 air chamber rooms and 4 buoyancy rooms and it was moored in 40 m of water.

It was tested two years at sea, and the last sea test will be held in

1985.
As a result of these tests, the following points were concluded. Safety mooring and safety operation on rough sea can be based on ship design. Safe and reliable operation of air turbine type wave power generator can be judged from a single small buoy unit. Power output was still smaller than expected, but there is considerable room for future development work.
I will describe improvements later.

(2) Shore fixed type Wave Power Generator.

The shore fixed type is different from floating type, but the next practical demand will be for island use, and so 40 kW test was planned by Japan Research and Development Corporation for 1984.
An air chamber was constructed and fixed to natural shore rock as shown in Fig. 7. Its dimensions were 8 m in width, 5 m in length, 40 m in area, and 5 m in height. The water depth in the water entrance was very shallow, being only 2 m.
Most of the waves broke before reaching the chamber, so that the maximum wave energy was limited to small amounts.
Mitui shipbuilding company carried out the design and construction.
Two Wells turbines were installed on both sides of the shaft of a 40 kW electric generator in order to cancel the thrust forces of the Wells turbine.

Fig. 7 Shore fixed 40 kW W.P.G.

A valveless turbine box constructed by Fuji Electric machinery company was installed in the chamber. I believe that this is the first sea trial of the Wells turbine at this scale. The tests took place over one year including winter at Sanje Turuoka city Yamagata Prefecture (very near to the Kaimei test site).

The test results were good. In particular the electric output was smoothed due to the flywhell effect of the high speed of the Wells turbine. I gained confidence to use this unit for island power supply.

4. Future view.

(1) Improvement of floating type.

A UK study concluded that a terminator is better than an attenuator from the view point of the energy absorption ratio. I agree with that conclusion, but an attenuator has several good points and I think that the Kaimei buoy can be improved by several methods.

Throughout the Kaimei tests, I noticed that the position of the buoyancy rooms influenced the energy absorption very much. The Kaimei used semi-catamaran buoyancy rooms, but if it is changed to separate buoyancy rooms, the energy absorption ratio will increase by about 2.5 times. Fig. 8 shows one proposal of the UK NEL attenuator having a bottom plate with the wave entering from front and side only. This idea will be one good method.

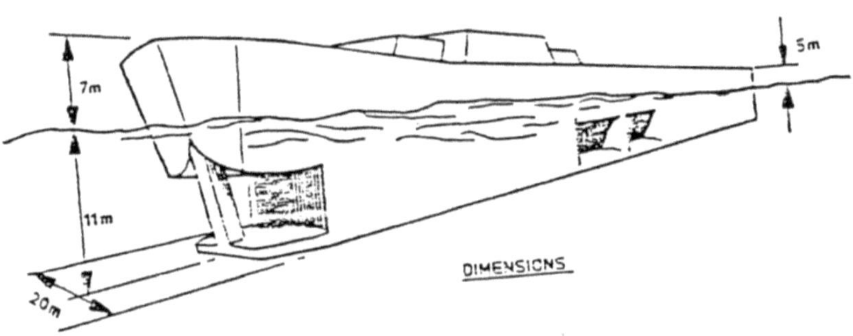

Fig. 8 Attenuator with front and side openning

I wish to add one new buoy shape which has an entrance at the back. I call it a backward bent duct buoy. It has a high efficiency and very small mooring force. This buoy will be used for navigational buoy at first, and it may be developed for medium scale buoy for island power supply.

(2) Shore fixed type.

The shore fixed type for island use will enjoy a good market for wave power device, since there is no example at this moment except tests in Japan and Norway. The first work will be to estimate demand from developing countries, and to develop engineering technology in order for it to be economical.

(3) Others.

The present situation for wave power study is not so good, but I believe that the wave power is valuable energy. If we continue the research work, we can find a feasible method for supplying on electricity.

MIX
Papier aus verantwortungsvollen Quellen
Paper from responsible sources
FSC® C105338

If you have any concerns about our products,
you can contact us on
ProductSafety@springernature.com

In case Publisher is established outside the EU,
the EU authorized representative is:
Springer Nature Customer Service Center GmbH
Europaplatz 3, 69115 Heidelberg, Germany

Printed by Libri Plureos GmbH
in Hamburg, Germany